Emerging Wireless Networks

Concepts, Techniques, and Applications

Emerging Wireless Networks

Concepts, Techniques, and Applications

Edited by
Christian Makaya and Samuel Pierre

CRC Press
Taylor & Francis Group
Boca Raton   London   New York

CRC Press is an imprint of the
Taylor & Francis Group, an **informa** business
AN AUERBACH BOOK

CRC Press
Taylor & Francis Group
6000 Broken Sound Parkway NW, Suite 300
Boca Raton, FL 33487-2742

© 2012 by Taylor & Francis Group, LLC
CRC Press is an imprint of Taylor & Francis Group, an Informa business

No claim to original U.S. Government works

Printed in the United States of America on acid-free paper
Version Date: 20111101

International Standard Book Number: 978-1-4398-2135-0 (Hardback)

Visit the Taylor & Francis Web site at
http://www.taylorandfrancis.com

and the CRC Press Web site at
http://www.crcpress.com

Preface

Mobile and wireless networks have gained much interest and their use has become a commodity nowadays. We have experienced an explosive increase in wireless data applications, such as real-time multimedia services (e.g., VoIP, video streaming, gaming). Furthermore, diverse emerging wireless technologies have been defined, and ongoing work and research activities are still very active for the design of architectures, protocols and standards. Among the popular emerging wireless technologies we will consider in this book are fourth generation (4G) networks, which include Long Term Evolution (LTE) and WiMAX. Legacy wireless technologies such as UMTS/HSPA, EV-DO, and WiFi will be part of 4G networks.

Although much work has been done on these different technologies, their coexistence leads to several challenges. In fact, the interworking of these diverse technologies, as well as the efficient delivery of value-added applications and services over these emerging wireless systems lead to challenging issues such as security, mobility management, architectures, quality of service (QoS) provisioning, and resources allocation. To solve these different issues related to heterogeneity of wireless access technologies and networks, several efforts have been made in order to propose innovative ideas in terms of novel protocols, architectures, and algorithms. The requirements of next-generation or emerging wireless networks open even more new opportunities for many interesting and comprehensive research topics targeting at concepts, methodologies, and techniques to support advanced mobile value-added services. Clearly, the development of new mechanisms, protocols, algorithms, applications, architectures, and systems will have a significant impact for the successful deployment of emerging wireless networks.

This book is a collection of surveys and research papers related to new concepts, techniques, and applications of emerging wireless networks. These networks are also called Next-Generation Wireless Networks (NGWN) and the fourth generation (4G) is an instance of such networks. The interworking of these networks should handle a set of specific requirements and challenges related to economy, scalability, seamless mobility, security, quality of service, service continuity, network selection, overall performance, and so on. This book aims at presenting the latest and most innovative concepts, techniques, and applications related to emerging wireless networks. It contains 12 chapters which can be outlined as follows.

Chapter 1 presents a survey of management challenges for emerging wireless networks. With the heterogeneity of emerging wireless networks their management is challenging in terms of mobility, resource, security, scalability, reliability, integrated and service management. Therefore, new management standards, architectures, theories, and technologies should be investigated to match the current requirements to manage emerging wireless networks. Furthermore, a set of enabling technologies is recognized as potential candidates for network management and can be based on policy-based management, probabilistic management, self-management, and context-aware and autonomic-management.

Chapter 2 provides the state of the art on mobility management for all-IP mobile networks, mainly at the network and link layers. It also discusses qualitative analysis on advantages and shortcomings among different approaches available in the literature. Various issues and challenges in mobility management in all-IP and ad-hoc mobile networks are studied, and solutions are proposed to either reduce partly, or remove completely these discovered problems. The case of next-generation mobile networks is also considered.

Advances in wireless networking and mobile computing enable users to benefit from diverse mobile communications systems. On the other hand, subscribers are intensifying demands for seamless roaming across different wireless systems while profiting by services anytime, anywhere, and from any device. Under the circumstances, disparate networks are expected to integrate with each other to provide ubiquitous and high data-rate services to roaming users. However, the integration of such systems brings about a variety of challenges because of the heterogeneities in security protocol, quality of service provisioning, mobility management, architecture, radio access technologies, etc. Therefore, designing new interworking and integrated architecture is a crucial task during the wireless system evolution. Chapter 3 proposes an integrated network architecture design for next-generation wireless systems, including wireless local, metropolitan, and wide area networks. This chapter provides telecom engineers, researchers, and scientists with a comprehensive technical guide and references for designing interworking architectures dedicated to next-generation wireless systems.

Chapter 4 analyzes the WLAN/3G networks convergence and advanced mobility features. More specifically, it proposes a service-level solution for 802.11/WiFi and 3G networks convergence based on the infrastructure proposed in the IP Multimedia Subsystem (IMS) for service delivery. The chapter also proposes the implementation of a mobility server that handles the session mobility/splitting between 3G networks and IEEE 802.11/WiFi. This mobility server will be deployed in an application layer and will have access to the admission policy of each domain, leading to better utilization of resources in these two networks technologies.

Chapter 5 examines challenges and issues related to mobile virtual private networks (VPN). A VPN is a concept used to build a secure and private communication path on top of public communication network such as Internet. Indeed, VPN is an overlay network that uses the public network to carry data traffic between corporate sites and users, maintaining privacy through the use of tunneling protocols and security procedures. VPNs were designed to support secure remote access for static connections and are based on the assumption that connection information (e.g., IP address) would not change. Mobile users, however, access information through wireless networks, and their underlying access networks may change over time. Due to their dependency on static, persistent connection information, traditional VPNs are not suitable for mobile users who wish to access secure information through wireless networks. Mobile VPN technology attempts to extend the VPN concept to support secure data access over public, unsecured wireless networks. Solutions and implementations for Mobile VPNs architectures are also discussed.

Chapter 6 analyzes the emerging broadband wireless standard IEEE 802.16, also widely known as WiMAX (Worldwide Interoperability for Microwave Access). It first presents an overview of this standard, followed by a survey of major related work proposed to enhance handover in 802.16, before describing a promising cross-layer design of WIMAX Handover, the FMIPv6 (Fast Mobile IPv6 Handover Protocol), and a newly proposed improved scheme. The proposed enhancement achieves considerably improvement over the existing schemes, especially under heavy network conditions and fast mobility speed.

Chapter 7 proposes a novel mechanism that provides service continuity to end-users in self-organizing IMS (IP Multimedia Subsystem) environments. A self-organizing IMS (SOIMS) enables IMS functional components and corresponding physical nodes to adapt dynamically and automatically, based on situations like network load, context awareness, and available system resources, while continuing IMS operation. The proposed mechanism is implemented in a real system (prototype) and its performance in terms of control signaling overhead and processing time during reconfiguration is analyzed. The proposed SOIMS approach guarantees service continuity to end-users transparently to any reconfiguration of IMS core network.

Chapter 8 presents challenges, solutions, and services of vehicular communication networks (VCN), a promising technology for improving traffic safety and traffic efficiency in different road systems, including Intelligent Transportation Systems (ITS). It introduces some important solutions, and presents an overview of a future ITS service called Start-to-Destination Driving Route Reservation (S2D-DRR). Finally, some open research issues are introduced with an objective to spark new research interests in the field of VCNs.

Chapter 9 presents a new transmission paradigm called network coding to improve the performance of the transmission control protocol (TCP). Compared to traditional routing protocols, network coding is bandwidth efficient and can achieve high throughput gains. By intelligently mixing (coding) packets together, fewer transmissions are required and bandwidth becomes available for new data. It is shown that when XOR network coding is used in conjunction with an opportunistic scheduling and the TCP window state, higher throughput can be achieved. A cross-layer approach is also proposed and simulation is carried out in order to evaluate theses approaches.

The integration of different radio access technologies becomes attractive with the rapid growth of wireless communication technologies and services. Chapter 10 presents a novel distributed scheme based on the stochastic optimization formulation of the network selection problem for heterogeneous wireless networks. It also proposes an optimal internetwork spectrum sharing scheme with spectrum pooling for heterogeneous wireless networks. Simulation results show significant performance improvement compared with the existing schemes.

The emerging 4G networks will be characterized by an heterogeneous environment where several access networks will be available. While one way to enhance the quality of service (QoS) offered to users in this context is through innovative protocols and new technologies, future trends should take into account as well the efficiency of resource allocation schemes. Chapter 11 describes issues and schemes for network selection and congestion avoidance in multi access networks. It analyzes techniques that may enable optimal distribution of resources in such networks. Fuzzy Logic-based schemes or Multi-Attribute Decision Making (MADM) methods are considered, and the guidelines for the exploitation of these tools are given. Finally, the chapter investigates the effectiveness of the proposed schemes in terms of pricing, apart from the provision of high levels of QoS.

The radio frequency spectrum is one of the most expensive and tightly regulated resources in the wireless communication domain. Recently, the spectrum scarcity problem has become more pronounced due to the crowding number of wireless devices and emerging wireless technologies. It is critical that the current wireless systems incorporate organic self-adaptive mechanisms to allow their coexistence in future wireless systems. Chapter 12 categorizes wireless systems as primary and secondary users. It also proposes a deterministic spatial spectrum reuse scheme by exploiting the fact that some well-established wireless systems have asymmetric characteristics in spectrum band utilization and wireless station deployment, suitable for the coexistence of secondary systems. More specifically, it examines the widely accepted and deployed GSM and WiFi systems as primary and secondary users, respectively, and the design of the coexistence system in such a way that they cause negligible interferences to each other. Experimental results show that the proposed scheme is a feasible approach to the spectrum scarcity problem.

Christian Makaya, Ph.D.
Samuel Pierre, Ph.D.
Montreal, September 2011

List of Figures

List of Tables

Contents

Bibliography   **97**

3 Integrated Network Architecture Design for Wireless Systems   **107**

Bibliography   **145**

4 WLAN/3G Convergence and Advanced Mobility Features   **151**

III Resource Management and Cognitive Networks 305

9 Network Coding Approach to Improving TCP Throughput in Wireless Networks 307

Tebatso Nage, Marc St-Hilaire, and F. Richard Yu

Part I

Mobility Management and Networks Convergence

1

Management Challenges for Emerging Wireless Networks

Jianguo Ding

Faculty of Sciences, Technology and Communication, University of Luxembourg, L-1359 Luxembourg, Luxembourg
Software Engineering Institute, East China Normal University, 200062 Shanghai, P.R. China
Email: jianguo.ding@ieee.org

Ilangko Balasingham

Interventional Center, Oslo University Hospital, Oslo, Norway
Department of Electronics and Telecommunications, Norwegian University of Science and Technology (NTNU), N-7491 Trondheim, Norway

Pascal Bouvry

Faculty of Sciences, Technology and Communication, University of Luxembourg, L-1359 Luxembourg, Luxembourg

CONTENTS

With the growth of emerging wireless networks in size, heterogeneity, pervasiveness, complexity of applications, network services, and the combination of rapidly evolving technologies and increased requirements from users, the emerging wireless networks are characterized by mobility, diffusion of heterogeneous nodes and devices, mass digitization, resource constraints, multifederated operations, scalability, dependability, context awareness, security, probability, new forms of user-centered content provisioning, new models of service, and the interaction with improved security and privacy. The wireless network management has to face the challenges in mobility management, resource management, security management, scalability management, reliability management, integrated management, and service management. Therefore, new management standard, architectures, theory, and technologies should be investigated to match the current requirements to manage emerging wireless networks. Furthermore, a set of enabling technologies is recognized as potential candidates for network management and can be based on policy-based management, probabilistic management, bio-inspired management, self-management, and context-aware and autonomic-management. This chapter presents a survey of management challenges for emerging wireless networks.

1.1 Introduction

Network management for emerging wireless networks (EWN) is facing new challenges with increased requirements from corporate customers and personal users. Wireless networking technology is ideal for many environments, including homes, airports, shopping malls, and personal telecommunications because it is inexpensive, easy to install (no wires), and supports mobile users. As a result, we have seen a sharp increase in the use of wireless in the form of WLANs, city wide meshes, sensor networks, and wireless home automation networks over the past few years. However, in using wireless technology effectively we encounter the emerging difficulties in networks management.

First, wireless links are susceptible to degradation (e.g., attenuation and fading) and to being interfered with, both of which can result in unpredictable barriers and poor performance. Second, since wireless deployments must share the relatively scarce spectrum resources that are available for public use, they often interfere with each other. These factors become especially challenging in deployments where wireless devices such as access points are placed in very close proximity [1]. Third, emerging dynamic wireless networks are more vulnerable than fixed networks because of the nature of the wireless medium and the lack of central coordination. Fourth, numerous wireless mobile devices raise plenty of concerns in energy efficiency for the sustainable QoS (Quality of Service).

In addition, in a multicasting scenario, traffic may pass through unprotected routers which can easily get unauthorized access to sensitive information. Last, dynamic topology and opportunistic network structures in an ad-hoc network create difficulties in security management, (for example, key management, distributed intrusion detection), service management, configuration management, and performance management. Moreover, the dynamics in wireless networks also results in uncertainty, such as deficient observability, uncertain event correlation (dependency), and misinterpretation of management information, thus making the network management more complex and challenging [12].

About 20 years ago, the classical agent manager centralized paradigm was the prevalent network management architecture, exemplified in the OSI reference model, the Simple Network Management Protocol (SNMP)[7, 20], and the Telecommunications Management Network (TMN) management framework [23]. The increasing trend toward extended network structure and applications, and enterprise application integration based on loosely coupled heterogeneous Information Technology (IT) infrastructures force a change in management paradigms from centralized and local to distributed management strategies and solutions that are able to cope with multiple autonomous management domains and possibly conflicting management policies [14].

In addition, based on emerging wireless networks and services, service-oriented business applications come with end-to-end application level QoS requirements and service level agreements (SLA) that depend on the qualities of the underlying IT infrastructure. The emerging environments with a large diversity of devices, wireless networks, providers, and service domains may be characterized as the networks and services of the future. The development of new models and architectures is particularly crucial to managing these networks and services as well enabling new management technologies that can cope with resource constraints, multifederated operations, scalability, dependability, context awareness, security, mobility, probability, etc. There is also a demand for novel business approaches to management of IT services in this scenario [14, 33].

This chapter will survey the evolution of wireless networks, the trends, the management challenges for emerging wireless networks and services, and the evolution of strategies and solutions in network management.

1.2   Advances in Wireless Networks

1.2.1   Classification of Wireless Networks

A wireless network refers to any type of computer network is commonly associated with a telecommunications network whose interconnections between nodes is implemented without the use of wires. Wireless telecommunications networks are generally implemented with some type of remote information transmission system that uses electromagnetic waves, such as radio waves, for the carrier, and this implementation usually takes place at the physical level or "layer" of the network.

With the evolution of technologies and applications in recently years, wireless networks have evolved into different type of networks. The marriage of the Internet and wireless technologies yields IP-based wireless networking and, further, Internet of Things (IoT). Figure 1.1 presents a classification of wireless networks and technologies [14]. The classification denotes the advances in wireless networks as well.

The varying wireless network technologies are depicted in Figure 1.2 together with the corresponding IEEE standards. Figure 1.3 [26] illustrates the current mosaic of wireless communication networks from the service coverage (range) standpoint. Two main network components are clearly distinguished, namely, wide area networks on one hand, and short-range networks on the other. Curiously, range wise, the development of wireless communication networks follows an ordered evolution from large to small networks, starting with very large distribution networks of up to hundreds of kilometers wide down to submeter short-range networks. Several reasons can be attributed to the development of increasingly smaller wireless networks, including the pressure to move toward unused (and typically higher) frequency bands of the spectrum and the need to support higher data throughput. In general, these two component networks were developed independently of each other but aiming, by design, to coexist.

Short-range wireless communications involve a very diverse array of air interface technologies, network architectures, and standards. The most well-known short-range wireless network technologies include wireless local area networks (WLAN), wireless personal area networks (WPAN), wireless body area networks (WBAN), wireless home networks (WHN), wireless sensor networks (WSN), car-to-car communications (C2C), Radio Frequency Identification (RFID), and Near Field Communications (NFC), ZigBee, Z-Wave, INSEON, Wavenis, and IPv6 over Low-Power Wireless PAN (6LoWPAN). As compared to long-range communications, short-range links require significantly lower energy per bits in order to establish a reliable link. They can thus achieve data throughput of several orders of magnitude higher than the typical values for cellular networks. Usually, short-range networks exploit distributed architectures using unlicensed spectrum.

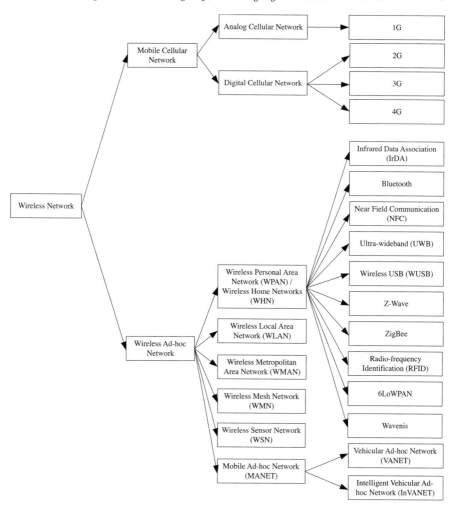

FIGURE 1.1
Classification of wireless networks and technologies.

Short-range communications encompass a large variety of different wireless systems with a great diversity of requirements. Figure 1.4 [28] illustrates the wide scope of short-range communications by classifying it according to the most common air interface technologies and network architectures, as well as supported mobility and data rates.

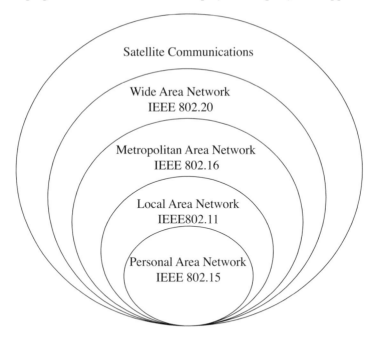

FIGURE 1.2
Wireless networks technologies based on IEEE standards.

1.2.2   Technical Advances for Wireless Networks

More recently, the evolution of wireless networks brings novel characteristics to today's networks and future networks [9], such as

- Use of flexible and efficient radio access, allowing ubiquitous access to broadband nomadic and mobile services;

- Managing new forms of ad-hoc communications with intermittent connectivity requirements and time-varying network topology;

- Integrating subnetworks at the edge, such as personal and sensor networks, and possibly integrated to the Internet for the benefit of humans;

- Eliminating the barriers to broadband access and enabling intelligent distribution of services across multiple access technologies with centralized or distributed control;

- Enabling seamless end-to-end network and service composition and operation across multiple operators and business domains;

- Supporting high-quality media services and critical infrastructure, (e.g., for energy and transport), with the existing networks significantly enhanced or even gradually replaced;

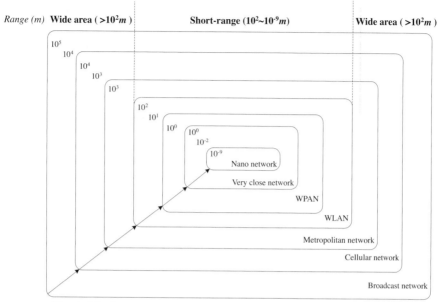

Range (m) **Wide area (>10²m)**   **Short-range (10²~10⁻⁹m)**   **Wide area (>10²m)**

FIGURE 1.3
Future wireless communications networks.

- Supporting reliable service across unreliable wireless networks.

The technical trends and computing models for wireless networks could be described as follows.

1.2.2.1 Nomadic Computing

Nomadic computing (mobile computing) [38] is the use of portable computing devices (such as laptop, handheld computers, mobile communication devices, and mobile sensors) in conjunction with mobile communications technologies to enable users to access the Internet and data on their home or work computers from anywhere in the world. People using such a system are sometimes referred to as tech-nomads, and their ability to use that system as nomadicity.

Nomadic computing provides a rich set of computing and communication capabilities and services to nomads as they move from place to place in a transparent, integrated, and convenient form. For wireless networks, nomadicity (mobility) in both the terminals and the services will have to be taken into consideration in future wireless network designs. The number of mobile networked devices as well as nomadic users will increase dramatically. Subsequently, more users and devices are connected and have direct dynamic communication link.

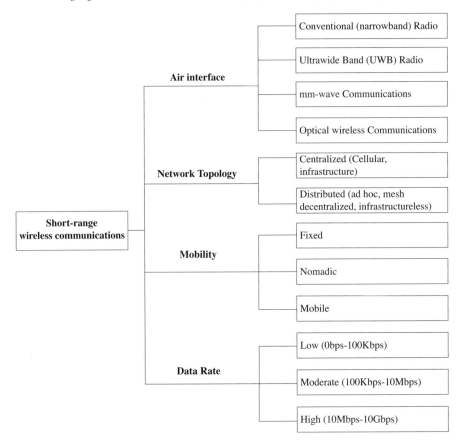

FIGURE 1.4
General classification of short-range communications.

1.2.2.2   Autonomic Computing

Autonomic Computing [51] is an initiative started by IBM in 2001. Its ultimate aim is to develop computer systems capable of self-management to overcome the rapidly growing complexity of computing systems management and to reduce the barrier that complexity poses to further growth. In other words, autonomic computing refers to the self-managing characteristics of distributed computing resources, adapting to unpredictable changes while hiding their intrinsic complexity from operators and users. An autonomic system makes decisions on its own, using high-level policies; it will constantly check and optimize its status and automatically adapt itself to changing conditions.

Besides enhanced user experience for human-to-human (H2H) or human-to-machine (H2M) interactions, autonomous machine-to-machine (M2M) communications has gained significant importance. In dealing with the complex wireless networks and extended applications, more and more business transactions and processes will be automated and will take place based on autonomous decisions without any human intervention. These will be often based on or influenced by context information obtained from the physical world, without the requirement of human input to describe the situation. The emergence of the Web2.0 and associated technologies are just a starting point in this development, and their apparent impact on economic development can be enormous.

The growing importance of context awareness, targeting-enriched experience, and intuitive communications services customized for mobile lifestyle and a mobilized workforce, will be incorporated to provide intelligent services in an invisible manner to users. Context awareness is used to design innovative user interfaces, and is often used as a part of ubiquitous and wearable computing [40].

1.2.2.3 Pervasive Computing

Pervasive computing [19] is the trend toward increasingly ubiquitous (another name for the movement is ubiquitous computing), connected computing devices in the environment, a trend being brought about by a convergence of advanced electronic and, particularly, wireless technologies and the Internet. Pervasive computing devices are not personal computers as we tend to think of them but very tiny, even invisible, devices, either mobile or embedded in almost any type of object imaginable, including cars, tools, appliances, clothing, and various consumer goods, all communicating through increasingly interconnected networks.

Dan Russell, director of the User Sciences and Experience Group at IBM's Almaden Research Center, and other researchers expect that in the future smart devices all around us will maintain current information about their locations, the contexts in which they are being used, and relevant data about the users. The goal of researchers is to create a system that is pervasively and unobtrusively embedded in the environment, completely connected, intuitive, effortlessly portable, and constantly available. Among the emerging technologies expected to prevail in the pervasive computing environment of the future are wearable computers, smart homes, and smart buildings. Among the myriad of tools expected to support these are application-specific integrated circuitry (ASIC); speech recognition; gesture recognition; system on a chip (SoC); perceptive interfaces; smart matter; flexible transistors; reconfigurable processors; field programmable logic gates (FPLG); and micro-electromechanical systems (MEMS).

Pervasive computing is a rapidly developing area of Information and Communications Technology (ICT). The term refers to the increasing integration of ICT into people's lives and environments, made possible by the growing availability of microprocessors with inbuilt communications facilities. Pervasive computing has many potential applications, from health and home care to environmental monitoring and intelligent transport systems. This briefing provides an overview of pervasive computing and discusses the growing debate over privacy, safety, and environmental implications.

A number of leading technological organizations are exploring pervasive computing. Xerox's Palo Alto Research Center (PARC), for example, has been working on pervasive computing applications since the 1980s. Although new technologies are emerging, the most crucial objective is not, necessarily, to develop new technologies. IBM's project Planet Blue, for example, is largely focused on finding ways to integrate existing technologies with a wireless infrastructure. Carnegie Mellon University's Human Computer Interaction Institute (HCII) is working on similar research in their Project Aura, whose stated goal is "to provide each user with an invisible halo of computing and information services that persists regardless of location." The Massachusetts Institute of Technology (MIT) has a project called Oxygen. MIT named their project after that substance because they envision a future of ubiquitous computing devices as freely available and easily accessible as oxygen is today.

Pervasive computing brings opportunities and challenges in interoperability on EWN. The emerging wireless network will become increasingly integrated with phones, television sets, home appliances, portable digital assistants, a range of other small hardware devices, and other heterogeneous networks, providing an unprecedented, nearly uniform level of integrated data communications. Users will be able to access and control this connected infrastructure from anywhere on the network. The consequent network interoperability applies at many different levels:

- Network interoperability is to provide the interoperation between heterogeneous networks (such as wireless networks and traditional networks) and devices (such as wireless terminals and sensors);

- Service interoperability to provide the ability to integrate largely standalone services with similar ones and with other services, for instance from the business domain;

- Semantic interoperability, so as to provide the (automated) understanding of the information exchanged and ensure quality of service;

- Interoperability of the service layer with network and application layers from different providers.

Expanded services (for both end users and network services) in EWN are likely to be comprised of a variety of components provided by a variety of players and running over a decentralized hosting (low-cost) infrastructure (including end-user devices, PC, servers, storage, computing, and networking/forwarding resources). This vision is expected to pave the way for a deep integration of service and network frameworks for network convergence among players (e.g., Network and Service Providers and Application Service

Providers) in new business models. Openness, broad federations of players, do-it-yourself innovative services, and knowledge management will allow people to be the true center of information society.

1.2.2.4 Cognitive Computing

Cognitive computing has its roots in the 1950s when computer companies first began to develop intelligent computer systems. Most of these systems were limited, however, because they could not learn from their experiences. Early artificial intelligence could be taught a set of parameters but was not capable of making decisions for itself or intelligently analyzing a situation and coming up with a solution. Enthusiasm for the technology began to wane, as scientists feared that an intelligent computer could never be developed.

However, with major advances in cognitive science, researchers interested in computer intelligence became enthused. Deeper biological understanding of how the brain worked allowed scientists to build computer systems modeled after the mind and, most importantly, to build a computer that could integrate past experiences into its system. Cognitive computing was reborn, with researchers at the turn of the 21st century developing computers that operated at a higher rate of speed than the human brain.

Cognitive computing is about engineering the mind by reverse engineering the brain. A cognitive network is an example of cognitive computing in wireless networks. A cognitive network is a network composed of elements that, through learning and reasoning, dynamically adapt to varying network conditions in order to optimize end-to-end performance. In a cognitive network, decisions are made to meet the requirements of the network as a whole rather than the individual network components. Cognitive networks can be characterized by their ability to perform their tasks in an autonomous fashion by using their self-attributes such as self-managing, self-optimizing, self-monitoring, self-repair, self-protection, self-adaptation, and self-healing to adapt dynamically to changing requirements or component failures while taking into account the end-to-end goals.

Cognitive networks use the self-configuration capability to respond and dynamically adapt to operational and context changes. The main function components of self-configuration are self-awareness and auto-learning, which are implemented by means of network-aware middleware and normally distributed across network components. Applications and devices adapt to exploit enhanced network performance and are agnostic of the underlying reconfigurations, in accordance with the seamless service provision paradigm.

Cognitive wireless access networks are those that can dynamically alter their topology and/or operational parameters to respond to the needs of a particular user while enforcing operating and regulatory policies and optimizing overall network performance. A cognitive infrastructure consists of reconfigurable elements and intelligent management functionality that will progressively evolve policies based on past actions [45].

1.2.2.5   Opportunistic Computing

Opportunistic computing can be described as distributed computing with the caveats of intermittent connectivity and delay tolerance. Indeed, mobile and pervasive computing paradigms are also considered natural evolutions of traditional distributed computing. However, in mobile and pervasive computing systems, the disconnection or sleep device situations are treated as aberrations, while in opportunistic computing, opportunistic connectivity leads to accessing essential resources and information [11].

Opportunistic computing exploits humans' mobility and their gregarious nature to enable a transmission only if two users are sufficiently close. Opportunistic computing can benefit from the ongoing and past research outcomes in pervasive and sensor systems, distributed and fault-tolerant computing, and mobile ad-hoc networking.

Technological advances are leading to a world replete with mobile and static sensors, user cell phones, and vehicles equipped with a variety of sensing and computing devices, thus paving the way for a multitude of opportunities for pairwise device contacts. Opportunistic computing exploits the opportunistic communication between pairs of devices (and applications executing on them) to share each other's content, resources, and services.

Opportunistic networks (Oppnets) are formed by small devices that communicate over a wireless link with each other. These devices are either mobile, that is, personal devices carried by a user, or fixed devices mounted at a dedicated location. In this sense, opportunistic networks are closely related to Mobile Ad-hoc Networks (MANETs) [21]. Oppnets have an outstanding potential for a truly beneficially "disruptive" effect on existing technologies. They can make applications more effective and efficient, in particular by providing these applications a wealth of communication modes, sensing devices, and other tools. By their very nature of relying on growth and expansion, Oppnets are highly adaptive, and can be exploited for achieving highly reliable and dependable operation in highly dynamic and unforeseeable situations [29].

1.2.2.6   Scalable Computing

Scalable computing involves using a computer system that can adapt to the need for more powerful computing capabilities. Scalability is also a desirable quality for a network, process, website, or business model [5]. The increasing number of wireless networks and heterogeneous networks brings new challenges to scalability. Examples include the extension of network physical topologies, modeling, validation and verification of business processes composed on SOA (Service Oriented Architecture); flexible evolution and execution of business processes; data, process and service mediation; reliable management of composed services; and brokering, aggregation, and data management.

1.2.2.7   Physical Computing

Physical (or embedded) computing, in the broadest sense, means building interactive physical systems by the use of software and hardware that can sense and respond to the analog world. While this definition is broad enough to encompass things such as smart automotive traffic control systems or factory automation processes, it is not commonly used to describe them. In the broad sense, physical computing is a creative framework for understanding human beings' relationship to the digital world. In practical use, the term most often describes handmade art, design or DIY hobby projects that use sensors and micro controllers to translate analog input to a software system, and/or control electro-mechanical devices such as motors, servos, lighting, or other hardware.

As a prospective example of physical computing, the "Internet of Things (IoT)" is the networked interconnection of objects from the sophisticated to the mundane through identifiers such as sensors, RFID (radio-frequency-identification) tags, and IP (Internet Protocol) addresses. Sensors form the edge of the electronics ecosystem in which the physical world interacts with computers, providing a richer array of data than is available from keyboards and mouse inputs. The IoT gets benefit from ubiquitous wireless communications and various wireless sensor networks.

The goals of the IoT (also known as the Internet of Objects) are to instrument and interconnect all things and to ensure that all those things are intelligent [10]. It is generally viewed as a self-configuring wireless network of sensors whose purpose would be to interconnect all things. The concept is attributed to the original Auto-ID Center, founded in 1999 and based at the time in MIT. The IoT is not restricted to fixed, formal network architectures. Ad-hoc wireless sensor networks can also form around personal mobile-communication devices.

The idea is as simple as its application is difficult. If all cans, books, shoes, or parts of cars are equipped with minuscule identifying devices, daily life on our planet will undergo a transformation. Things like running out of stock or wasted products will no longer exist as we will know exactly what is being consumed on the other side of the globe. Theft will be a thing of the past as we will know where a product is at all times. The same applies to parcels lost in the post.

If all objects of daily life, from yoghurt to an airplane, are equipped with radio tags, they can be identified and managed by computers in the same way humans can. The next generation of Internet applications (IPv6 protocol) would be able to identify more objects than IPv4, which is currently in use. This system would therefore be able to instantaneously identify any kind of object.

1.3 Management Challenges for Networks and Services

The network management tasks in EWN and services are investigated as follows [15].

1.3.1 Mobility Management

In EWN, such as wireless ad-hoc networks (WANET), mobility is one main characteristic of the networks. The nodes may not be static in space and time; this will result in a dynamic network topology. Nodes can move freely and independently. Also, some new nodes can join the network, and some nodes may leave the network. Individual random mobility, group mobility, and updating along preplanned routes can have major impact on the selection of a routing scheme and can thus influence performance. Multicasting becomes a difficult problem because mobility of nodes creates inefficient multicast trees and inaccurate configuration of network topology [31].

There are two distinct methods for mobility support in EWN networks [2]: Mobile IP and fast routing protocols. Mobile IP offers a pure network layer architectural solution to mobility support and isolates the higher layers form the impact of mobility. However, it does not work efficiently on a large-scale basis. Fast routing protocols are designed to cope with changes in the network topology. Routing in EWN is basically a compromise between the method of dealing with fast topology changes and keeping the routing overhead to a minimum.

- Routing Management

  Mobile wireless networks rely on a routing plane to fit with their dynamics and resource constraints. In particular, an ad-hoc routing protocol must be capable of maintaining an updated view of the dynamic network topology while minimizing the routing traffic. Dynamically changing network topology makes routing a complex issue and requires new approaches in the design of routing protocols for mobile wireless networks.

  Table-driven protocols and source-initiated protocols are two typical routing protocols for dynamic networks. Table-driven protocols include the Destination-Sequenced Distance Vector (DSDV) protocol [37], the Cluster-Head Gateway Switch Routing (CGSR) protocol [32], and the Wireless Routing Protocol (WRP) [34]. Source-initiated protocols include Dynamic Source Routing (DSR) protocol [24], Associative-Based Routing (ABR) protocol [46], Temporally Ordered Routing Algorithm (TORA) [50], and the Ad-Hoc On-Demand Distance Vector (AODV) protocol [36]. It is predictable that new optimized routing protocols will appear for particular application scenarios.

- Location Management

  Since there is no fixed infrastructure available for mobile wireless networks with nodes being mobile in three-dimensional space, location management becomes a very important issue. For example, route discovery and route maintenance are the challenges in designing routing protocols. In addition, finding the position of a node at a given time is an important problem. This leads to the development of location-aware routing, which means that a node will be able to know its current position.

  There are quite a few protocols that have been proposed in recent years specifically for location management. They can be classified into two major categories: location-assisted [41] and zone-based protocols [8]. Location-assisted protocols take advantage of the local information of hosts. In zone-based routing, the network is divided into several non-overlapping regions (zones), and each node belongs to a certain zone based on its physicals location.

1.3.2   Resource Management

Wireless networks use wireless medium to transmit and receive data. Nodes can share the same media easily. But wireless links have limited bandwidth and variable capacity. They are also error prone. In addition, most wireless nodes have limited power supply and no capability to generate their own power. Energy efficient protocol design (e.g., MAC, routing, resource discovery) is critical for the longevity of the mission.

Bandwidth and energy are scarce resources in EWN. The wireless channel offers a limited bandwidth, which must be shared among the network nodes. Mobile devices are strongly dependent on the lifetime of their battery.

- Spectrum Management [35]:

  Because of short communication links (multihop node-to-node communication instead of long-distance node to central base station communication), radio emission levels can be kept low. This reduces interference levels, increases spectrum reuse efficiency, and makes it possible to use unlicensed unregulated frequency bands. The goals of spectrum management include rationalizing and optimizing the use of the radio frequency (RF) spectrum; avoiding and solving interference; designing short- and long-range frequency allocations; advancing the introduction of new wireless technologies; and coordinating wireless communications with neighbors and other administrations.

- Power Management [4, 43]:

  Portable handset devices have limited battery power. These devices often participate as nodes in a mobile wireless network, and deliver and route packets. Whenever the power of a node is depleted, the wireless network may cease to operate or may not function efficiently. To minimize energy consumption and prolong the life of the nodes in mobile wireless networks, the design of the latter requires energy-awareness in each phase.

Power management for EWN is desired for (1) reducing overall energy consumption, (2) prolonging battery life for portable and embedded systems, (3) reducing operating costs for energy and cooling. Lower power consumption also means lower heat dissipation, which increases system stability, and promotes less energy use, which saves money and reduces the impact on the environment and is helpful in pursuing the goal of Green IT.

- Design of Energy-Efficient Protocols [25]

  Energy-efficient protocols are important for power management in EWN. In the design of energy-efficient protocols, the following aspects should be taken into account carefully to obtain the synthetical performance of EWN:

  - End-to-end delay: Shortest path routing reduces delay but may increase power consumption due to increased distance between two adjacent nodes in the path since transmit power is proportional to power of distance as mentioned in the previous section. Increased delay also means increased number of packets in the system, and this could become a problem in terms of network capacity.

  - Packet loss rate: Low transmitted power reduces power consumption but may increase the packet loss rate due to weaker signal strength. When the packet loss rate increases, it results in retransmissions of the packets, again increasing the total number of packets in the system.

  - Network capacity: Low transmitted power decreases the range of the transmitting node, which may result in less number of connections to neighboring nodes. Reduced number of links per node could also reduce the traffic-carrying capacity of the network; in worse cases, it can even break network connectivity and leave the network as disjoint subnetworks.

  - Relaying overhead: Routing in mobile wireless networks results in multi-hop transmissions because of the low transmitted power of the nodes, which means several intermediate nodes between source and destination. Multi-hop routing adds relaying overhead to each node in the routing path, as each node in the path needs to receive the packet from the physical layer, figures out the next hop, and retransmit through the network stack. But multi-hop routing in some cases consumes less power than single-hop routing.

- Interference: High power transmission also induces greater interference to neighboring nodes. Interference also results in unnecessary power consumption in the neighboring nodes, as they have to receive the signal even when the signal is not destined for them. Stronger interferences could cause increased number of collisions, thus increasing the number of retransmissions.

- Battery life: In minimum transmitted power routing, several other nodes for routing packets can use the same route. When this keeps happening over a period of time, the battery power on those nodes that are on the routing path will run out, thus reducing network lifetime.

- Connectivity: Sleep mode reduces power consumption by an order of magnitude. But when a number of nodes go into the sleep mode to save power, without coordination, it may disrupt network connectivity.

In applications, protocol design should consider the balance between power saving and communication efficiency.

1.3.3 Security Management

Emerging mobile wireless networks are more vulnerable than fixed networks because of the nature of the wireless medium and the lack of central coordination. Wireless transmissions can be easily captured by an intruding node. A misbehaving node can perform a denial-of-service attack by consuming the bandwidth resources and making them unavailable to the other network nodes. It is also not easy to detect a malicious node in a multi-hop ad-hoc network and implement denial-of-service properly. In addition, in a multi-casting scenario, traffic may pass through unprotected routers, which can easily get unauthorized access to sensitive information (as in the case with military applications). Dynamic topology and movement of nodes in EWN make key management difficult if cryptography is used in the routing protocol. Typical challenges for security management are

- Malicious Attacks [30, 47]

  The emerging wireless networks are even more vulnerable to attacks than their infrastructure counterparts. Both active and passive attacks are possible.

  In a passive attack, the normal operation of a routing protocol is not interrupted. Instead, an intruder tries to gather information by listening. Active attacks can sometimes be detectable and thus are less important. In an active attack, an attacker can insert some arbitrary packets of information into the network to disable it or to attract packets designated to other nodes. Typical attacks in EWN are as follows:

– Pin Attack

With the pin, or black-hole, a malicious node can pretend to have the shortest path to the destination of a packet. Normally, the intruder listens to a path setup phase and, when it learns of a request for a route, sends a reply advertising the shortest route. Then, the intruder can be an official part of the network if the requesting node receives its malicious reply before the reply from a good node, and a forged route is set up. Once it becomes part of the network, the intruder can do anything within the network, such as undertaking a denial-of-service attack.

The controlled node by the attacker on the path of communication might drop some or all the packets between the right role in need and the controller. In the worst case, it might forward packets containing insignificant information and drop packets containing critical information.

– Location-Disclosure Attack

By learning the locations of intermediate nodes in EWN, an intruder can find out the location of a target node. The location-disclosure attack is made by an intruder to obtain information about the physical location of nodes in the network or the topology of the network.

– Routing Table Overflow

Sometimes an intruder can create routes whose destinations do not exist. This type of attack, known as the routing table overflow, overwhelms the usual flow of traffic as it creates too many dummy active routes. This attack has a profound impact on proactive routing protocols, which discover routing information before it is needed but has a minimum impact on reactive routing protocols, which create a route only when needed. In EWN, the size of the routing table is limited. Thus the routing table overflow attack is vital to EWN.

– Energy-Exhaustion Attack

Battery-powered nodes can conserve their power by transmitting only when needed. Some attacks may target only critical weak devices. Such surgical attacks are capable of defeating the goal of networks, which is to maintain connectivity in crisis. For example, an intruder may try to forward unwanted packets or request repeatedly fake or unwanted destinations to use up the energy of the nodes' batteries.

– Denial-of-Service (DoS) Attack

False requests for help can be generated by malicious devices on weak links. They will keep the rescue team busy and unavailable for real emergencies. DoS attacks may target a "weak" device, such as a cellphone, that is critical to opportunistic network operation (e.g., if it is the only device that connects two parts of a city).

– ID Spoofing

Mapping some node properties (such as location) into node ID by a controller can be dangerous. A malicious device capable of masquerading can generate requests with multiple IDs, resulting in many false alarms for the rescue team. Services that need authentication can be misused if their IDs can be spoofed. A device capable of spoofing the ID of a trusted node or a node with critical functions can pose many kinds of attacks.

• Design of Security Routing Protocol [22]

In order to prevent EWN from attacks and vulnerability, the design of security-routing protocols should consider the following properties:

– Authenticity. When a routing table is updated, it must check whether the updates were provided by authenticated nodes and users. The most challenging issue in EWN is the lack of a centralized authority to issue and validate certificates of authenticity.

– Integrity of information. When a routing table is updated, the information carried to the routing updates must be checked for eligibility. A misleading update may alter the flow of packets in the network.

– In-order updates. Ad-hoc routing protocols must contain unique sequence numbers to maintain updates in order. Out-of-order updates may result in the propagation of wrong information.

– Minimum update time. Updates in routing tables must be done in the shortest possible time to ensure the credibility of the update information. A timestamp or time-out mechanism can normally be a solution.

– Authorization. An unforgeable credential along with the certificate authority issued to a node can determine all the privileges that the node can have.

– Routing encryption. Encrypting packets can prevent unauthorized nodes from reading them, and only those routers having the decryption key can access messages.

– Protocol immunization. A routing protocol should be immune to intruding nodes and be able identify them.

– Node-privacy location. The routing protocol must protect the network from spreading the location or other unpublic information of individual nodes.

- Self-stabilization. If the self-stabilization property of wireless network performs efficiently, it must stabilize the network in the presence of damages continually received from malicious nodes.

- Low computational load. Typically, a wireless node is limited in powers as it uses a battery. As a result, a node should be given the minimal computational load to maintain enough power to avoid any DoS attacks.

- Key Management [42]

Effective management of keys, or digital certificates holding the keys, is one of the key factors for the successful widespread deployment of public key cryptography. PKI (Public Key Infrastructure), an infrastructure for managing digital certificates, was introduced for this purpose. The most important component of PKI is the CA (Certificate Authority), the trusted entity in the system that vouches for the validity of digital certificates. PKI has been widely deployed for wired networks and some infrastructure-based wireless networks.

However, for wireless networks, connectivity, which was assumed to be good in previous PKI solutions, is no longer stable in ad-hoc networks. Unfortunately, maintaining connectivity is one of the main challenges in wireless networks since the inherent infrastructurelessness of EWN makes it hard to guarantee any kind of connectivity. Another serious problem present in wireless networks is the physical vulnerability of the nodes themselves. Considering that most wireless networks will be deployed with mobile nodes, the possibility of the nodes being captured or compromised is higher than that in wired networks with stationary hosts. Mobile nodes in an infrastructure-based wireless networks have the same vulnerability, but they can rely on the infrastructure for detection of compromised nodes and potentially some help with recovery. With an infrastructure-based solution, mobile nodes may store all sensitive information in the infrastructure and maintain minimal information in the device. Since there is no stable entity in an EWN, the vulnerability of wireless nodes is even higher.

Currently proposed solutions for providing PKI for EWN address the physical vulnerability of the nodes by employing the distribution of CA functionality across multiple nodes and using threshold cryptography. These approaches also increase the availability of the CA. The first challenge in such approaches is picking which nodes should be a part of the distributed CA. A random set of nodes from a network can be chosen to play the role of CA. Such a random choice may not be best for the security of the whole network. The second challenge is how to provide efficient, yet effective, communication between the mobile nodes and the CA nodes to create the illusion of an available CA, even in dynamic networks with possible compromises or partitions.

- Identity Management [18]

  Identity management broadly refers to the management of digital identities or digital identity data. Approaches to identity management differ in terms of management procedures (who is doing what and what are the possible operations on the data) and the types of data being stored and managed (e.g., comprehensive profiles of individuals or groups or a selection of roles or partial identity - the kind of personal information known to the system).

  In EWN, it is required to support cross-network identity and authentication: providing a trusted and efficient means of establishing identity is one of the key issues in cross-network connectivity.

1.3.4 Integration Management

A typical EWN is composed of various devices such as laptops, personal digital assistants, phones, and intelligent sensors. These devices do not provide the same hardware and software capabilities but have to interoperate in order to establish a common network and implement a common task. Seamless roaming and handoff in heterogeneous networks, multiple radio integration and coordination, communication between wireless and wired networks, and different administrative domains invoke EWN management systems to implement higher-level integrated and cooperative policies. The network maintenance tasks are shared among the mobile nodes in a distributed manner based on synchronization and cooperation mechanisms.

Middleware is an important architectural system to support distributed applications. The role of middleware is to present a unified programming model to applications and to hide problems of heterogeneity and distribution. It provides a way to accommodate diverse strategies, offering access to a variety of systems and protocols at different levels of a protocol stack. Recent developments in the areas of wireless multimedia and mobility require more openness and adaptivity within middleware platforms. Programmable techniques offer a feasible approach to avoiding time-varying QoS impairments in wireless and mobile networking environments. Multimedia applications require open interfaces to extend systems to accommodate new protocols. They also need adaptivity to deal with varying levels of QoS from the underlying network. Mobility aggravates these problems by changing the level of connectivity drastically over time.

1.3.5 Scalability Management

In EWN, overlay networks and extendable networks, the management plane must be easy to maintain and must remain coherent, even when the ad-hoc network size is growing, when the network is merging with another one or when it is splitting into different ones. Most of the routing algorithms are designed for relative small wireless ad-hoc networks. However, in some applications (e.g., large environmental sensor fabrics, battlefield deployments, urban vehicle grids, etc.), the ad-hoc network can grow to several thousand nodes. For example, there are some applications of sensor networks and tactical networks that require deployment of a large number of nodes. For wireless "infrastructure," networks scalability is simply handled by a hierarchical construction. The limited mobility of infrastructure networks can also be easily handled using Mobile IP or local handoff techniques. In contrast, because of the more extensive mobility and the lack of fixed references, pure ad-hoc networks do not tolerate mobile IP or a fixed hierarchy structure. Thus, mobility and scalability jointly with large-scale EWN becomes one of the most critical challenges in ad-hoc design [6, 3].

1.3.6   Reliability Management

Reliable data communications to a group of mobile nodes that continuously change their locations is extremely important, particularly in emergency situations [17]. Contrary to wired lines, the wireless medium is highly unreliable. This is due to the variable capacity, limited bandwidth, limited battery power, and attenuation and distortion in wireless signals, nodes motion and propensity for error. Thus, wireless systems must be designed with this fact in mind. Procedures for hiding the impairments of the wireless links from high-layer protocols and applications as well as development of models for predicting wireless channel behavior would be highly beneficial. Given the nature of mobile wireless networks, it is necessary to make the network management system survivable. This is due to the dynamic nature of such networks, as nodes can move in and out of the network and may be destroyed for various reasons such as battlefield casualties, low battery power, etc. Thus, there should be no single point of failure for the network management system.

1.3.7   Service Management

Incorporating QoS in EWN is a complex problem because of limited bandwidth and energy constraints. Designing protocols that support multi-class traffic and allows preemption, mobile nodes' position identification, and packet prioritization are some of the open areas of research. In order to provide end-to-end QoS guarantee, a coordinated effort is required for multi-layer integration of QoS provisioning. The success and future application of EWN will depend on how QoS will be guaranteed in the future.

For example, emerging multimedia services demand high bandwidth. There are some challenges and problems that need to be solved before real-time multimedia can be delivered over wireless links in EWN. The need for more bandwidth, less delay, and minimum packet loss are some of the criteria for high-quality transmission. However, the current best-effort network architecture does not offer any QoS. Hence, in order to support multimedia traffic, efforts must be made to improve QoS parameters such as end-to-end delay, packet loss rate, and jitter.

1.4   Management Strategies for Networks and Services

To conquer the management challenges, a set of enabling technologies are recognized to be potential candidates for EWN management.

1.4.1   Policy-Based Management

The task of managing information technology resources becomes increasingly complex as managers have to live with heterogeneous systems, different networking technologies, and distributed applications. As the number of resources to be managed grows, the task of managing these devices and applications depends on numerous system- and vendor-specific issues.

Policy-Based Network Management (PBNM) provides a means by which the administration process can be simplified and largely automated [48]. Strassner defined PBNM as a way to define business needs and ensure that the network provides the required services [44]. In traditional network management approaches, such as SNMP, the use of the network management system has been limited primarily to monitoring the status of networks. In PBNM, the information model and policy expressions can be made independent of network management protocols by which they are carried.

In PBNM, policies are defined as the rules that govern the states and behaviors of a network. The management system is tasked with

- the transformation of human-readable specifications of management goals to machine-readable and verifiable rules governing the function and status of a network;

- the translation of such rules to mechanical and device-dependent configurations; and

- the distribution and enforcement of these configurations by management entities.

For large networks with frequent changes in operational directives, policy-based network management offers an attractive solution that can dynamically translate and update high-level business objectives into executable network configurations. However, one of the key weaknesses in a policy-based network management lies in its functional rigidity. After the development and deployment of a policy-based network management, the service primitives are defined. By altering management policies and modifying constraints, we have a certain degree of flexibility in coping with changing management directives.

1.4.2 Bio-Inspired Management

Following the incredibly high similarity between emerging complex networks and biological systems, an increasing number of researchers try to draw inspiration from the living world and to apply biological models. Through time, biological systems are able to evolve and learn and adapt to changes that environment brings upon living organisms. Researchers try to investigate the various biological processes that exhibit self-governance and combine these different processes into a biological framework that can be applied to various communication systems to realize true autonomic behavior.

The properties of self-organization, evolution, robustness, and resilience are already present in biological systems. This indicates that similar approaches may be taken to manage different complex networks, which allows expertise from biological systems to be used to define solutions for governing future communication networks. Future wireless network applications are expected to be autonomous, scalable, adaptive to dynamic network environments, and simple to develop and deploy. In order to realize the potential network applications with such desirable characteristics, various biological systems have to develop the mechanisms necessary to achieve the key requirements of future network applications such as autonomy, scalability, adaptability, and simplicity [14]. The eternal goal of bio-inspired management is to establish robust and self-adaptive autonomic network systems with stable and optimal performance.

1.4.3 Probabilistic Management

For evolving a complex wireless network system, uncertainty is an unavoidable characteristics which comes from unexpected hardware defects, unavoidable software errors, incomplete management information, and a dependency relationship between the managed entities. An effective management system in networks should deal with uncertainty and suggest probabilistic reasoning for daily management tasks [13]. Due to the complexity of managed networks, it is not always possible to build precise models in which it is evident that the occurrence of a given set of alarms indicates a fault on a given element (object) [12].

The knowledge of the cause effect relations among faults and alarms is generally incomplete. In addition, some of the alarms generated by a fault are frequently not made available to the correlation system in due time because of losses or delays in the route from the system element that originates them. Finally, due to the fact that the configuration frequently changes, the more detailed a model is, the faster it will become outdated. The imprecision of the information supplied by specialists very often causes great difficulties. The expressions "very high," "normal," and "sometimes" are inherently imprecise and may not be directly incorporated into the knowledge basis of a conventional rule-based system. Fuzzy logic, Bayesian networks, and probability theory are popular models for probabilistic management, particularly in fault management.

1.4.4   Self-Management

Emerging wireless networks and related services normally have no centralized control, which implies that network management will have to be distributed across various nodes. The generated dynamic nature requires that the management plane is able to adapt itself to the heterogeneous capacity of devices and environment modifications. EWN is deployed in a spontaneous manner, and management architecture should be deployed in the same manner in order to minimize any human intervention. The complexity of EWN makes it imperative that any management system for such networks minimize the amount of human intervention required to achieve the desired network performance. This includes automatic fault detection and remediation as well as automation of component configuration. Self-management is appropriate in managing such dynamic and complex networks [16].

In a distributed and an infrastructure-less environment, self-management enables the EWN to autonomously determine its own configuration parameters including addressing, routing, clustering, position identification, neighborhood awareness, power control, etc. In some cases, special nodes (e.g., mobile backbone nodes) can coordinate their motion and dynamically distribute in the geographic area to provide coverage of disconnected islands.

Self-management includes the following four functional areas:

- Self-Configuration

  Automated configuration of components and systems follows high-level policies. The configuration process can be more specifically defined as follows:

  1. Installation: new installation of necessary components (OS, software, etc.).

  2. Reconfiguration: reconfiguration of installed components to fit unique situations.

  3. Update: version management of applications or modification of components to correct defects. This also includes reinstallation when parts of the configuration files have been corrupted due to virus attack or system error.

- Self-Healing

  The system automatically detects, diagnoses, and repairs localized software and hardware problems. A system is self-healing to the extent that it monitors its own platform, detects errors or situations that may later manifest themselves as errors, and automatically initiates remediation. Fault tolerance is one aspect of self-healing behavior, although the cost constraints of personal computing often preclude the redundancy required by many fault-tolerant solutions.

- Self-Optimization

  Components and systems continually seek opportunities to improve their own performance and efficiency. A system is self-optimizing to the extent that it automatically optimizes its use of its own resources to ensure optimal functioning with respect to the defined requirements. In a self-optimizing network, all these tasks (information collection, data analysis, configuration changes, and verification) should be accomplished automatically, and the automation makes it easier and faster to respond to the network dynamics.

- Self-Protection

  The system automatically defends itself against malicious attacks or cascading failures. It uses early warning to anticipate and prevent system-wide failures. A system is self-protecting to the extent that it automatically configures and tunes itself to achieve security, privacy, function, and data protection goals. This behavior is of very high value to personal computing, which is exposed to insecure networks, an insecure physical environment, frequent hardware and software configuration changes, and often inadequately trained end users who may be operating under conditions of high stress. Security is one aspect of self-protecting behavior.

1.4.5   Context-Aware and Autonomic-Management

A context-aware network is a form of computer network that is a synthesis of the properties of dumb network and intelligent computer network architectures. Dumb networks feature the use of intelligent peripheral devices and a core network that does not control or monitor application creation or operation. Such a network is to follow the end-to-end principle in those applications, which are set up between end peripheral devices with no control being exercised by the network. Such a network assumes that all users and all applications are of equal priority.

In general, context information can be static or dynamic and can come from different network locations, protocol layers, and device entities. Any conflict or undesired interaction must be handled by the independent applications. As such, the network is most suited to uses in which customization to individual user needs and the addition of new applications are most important. The pure Internet ideal is an example of a dumb network.

An intelligent network, in contrast to a dumb, network is most suited to applications in which reliability and stability are of great importance. The network will supply, monitor, and control application creation and operation. A context-aware network is a network that tries to overcome the limitations of the dumb and intelligent network models and create a synthesis that combines the best of both network models. It is designed to allow for customization and application creation, while at the same time ensuring that application operation is compatible not just with the preferences of the individual user but with the expressed preferences of the enterprise or other collectivity that owns the network. The Semantic Web is an example of a context-aware network. Grid networks, pervasive networks, autonomic networks, application-aware networks, and service-oriented networks all contain elements of the context-aware model.

In a context-aware network, new applications may be composed from existing network applications. Techniques for modeling applications allow for the identification of applications that satisfy specific functional requirements as well as necessary nonfunctional requirements. This method also allows applications to be described in terms of their overall purposes. For example, an application may describe a business process. The process can be linked to its larger objectives in the organization, including its priority and consequences of failure. The context-aware network can use these descriptions in its function to handle conflict between incompatible applications in the accessing of resources or in the violation of higher-level constraints. The context-aware network monitors application operation to ensure that they are compatible with higher-level requirements and constraints and that conflicts are resolved in their light as well.

A context-aware network is suited to applications in which both reliability and the need for system evolution and customization are required. It is finding increasing application in the development of the enterprise system for business processes, customer relations management, etc. Service-oriented architectures, which are a specialization of the context-aware model, are the current trend in enterprise computing [39].

So far, context-aware computing and self-managing systems are emerging independently. The application domains for which they are studied are different as well. The two approaches have several features in common. For example, both approaches aim at reducing human involvement: while context-aware computing aims at reducing the amount of explicit input a user should provide to computing systems, autonomous computing (self-management) aims at reducing the operational and maintenance cost of a system.

If one takes the conceptual framework of Kephart and Chess [27] as a reference framework of self-managing system, the sensing, actuating, and analysis component are typical components of context-aware computing. Much work has been done by the research community of context-aware computing to support context acquisition, context modeling, context representation, context reasoning, and context management. Researchers of autonomous computing can benefit a great deal by considering the usefulness of this work to develop self-managing systems [49].

1.5 Conclusion

The emerging wireless networks and services have some of the following features: mobility, diffusion of heterogeneous nodes and devices, mass digitization, resource constraints, multi-federated operations, scalability, dependability, context awareness, security, probability, new forms of user-centered content provisioning, new models of service, and the interaction with improved security and privacy. These features produce new technologies and networking architectures and exhibit huge challenges to render robust services, security, and management for EWN.

New management standard, architectures, theory, and technologies should be investigated to match the current and future requirements to manage emerging networks and services. Furthermore, a set of enabling technologies is recognized as potential candidates for network management and can be based on policy-based management strategies, artificial intelligence techniques, probabilistic approaches, bio-inspired approaches, etc. In desire to keep complex network systems under control, it is necessary for the IT industry to move to autonomic management, context-aware management, and self-management systems in which technology itself is used to manage technology.

Acknowledgment

This work is supported by projects STCSM 08ZR1407200, 08PJ1404600, NSFC 60803077, 973 2005CB321904, the Research Council of Norway project MELODY 187857/S10, and the National Research Fund (FNR) of Luxembourg.

Bibliography

[1] A. Aditya, J. Glenn, S. Srinivasan, and P. Steenkiste. Self-management in chaotic wireless deployments. *Wireless Networks*, 13(6):737–755, 2007.

[2] I.F. Akyildiz, J. McNair, J.S.M. Ho, H. Uzunalioglu, and W. Wang. Mobility management in next-generation wireless systems. *Proceedings of the IEEE*, 87(8):1347–1384, Aug. 1999.

[3] M. H. Ali, W. G. Aref, and I. Kamel. Scalability management in sensor-network phenomena bases. In *Proc. of 18th International Conference*

on Scientific and Statistical Database Management, pages 91–100, Los Alamitos, CA, 2006.

[4] M. Anand, E.B. Nightingale, and J. Flinn. Self-tuning wireless network power management. *Wireless Networks*, 11(4):451–469, 2005.

[5] A. B. Bondi. Characteristics of scalability and their impact on performance. In *Proc. of the 2nd International Workshop on Software and Performance*, pages 195–203, 2000.

[6] M. Burgess and G. Canright. Scalability of peer configuration management in partially reliable and ad hoc networks. In *Proc. of IFIP/IEEE Eighth International Symposium on Integrated Network Management*, pages 293–305, 2003.

[7] J. Case, M. Fedor, M. Schoffstall, and J. Davin. A Simple Network Management Protocol (SNMP). IETF RFC 1157, May 1990.

[8] C.-Y. Chang, K.-P. Shih, and S.-C. Lee. ZBP: A zone-based broadcasting protocol for wireless sensor networks. *Wireless Personal Communications*, 33(1):53–68, 2005.

[9] European Commission. *The Future of the Internet: A Compendium of European Projects on ICT Research, http://cordis.europa.eu/ict/ch1/*. European Commission Information Society and Information, 2008.

[10] M. Conner. Sensors empower the "Internet of Things", 2010.

[11] M. Conti and M. Kumar. Opportunities in opportunistic computing. *Computer*, 43(1):42–50, 2010.

[12] J. Ding. Probabilistic Management of Distributed Systems. In B.J. Kramer and A.H. Wolfgang, editors, *Contributions to Ubiquitous Computing*, volume 42, pages 221–248, Berlin, 2007. Springer Verlag.

[13] J. Ding. *Probabilistic Fault Management in Distributed Systems*. VDI Verlag, Dusseldorf, Germany, 2008.

[14] J. Ding. *Advances in Network Management*. CRC Press, Boca Raton, FL, 2010.

[15] J. Ding, I. Balasingham, and P. Bouvry. Management challenges for emerging networks and services. In *Proc. of the International Conference on Ultra Modern Telecommunications & Workshops, (ICUMT'09)*, pages 1–8, 2009.

[16] A. M. Hadjiantonis. *Policy-based self-management of wireless ad-hoc networks*. PhD thesis, University of Surrey, 2008.

[17] H. Hallani and S.A. Shahrestani. Improving the reliability of ad-hoc on demand distance vector protocol. *WTOC*, 7(7):695–704, 2008.

[18] R. Halperin and J. Backhouse. A roadmap for research on identity in the information society. *Identity in the Information Society*, 1(1):71–87, 2008.

[19] Uwe Hansmann. *Pervasive Computing: The Mobile World*. Springer, Berlin, 2003.

[20] D. Harrington, R. Presuhn, and B. Wijnen. An Architecture for Describing Simple Network Management Protocol (SNMP) Management Frameworks. IETF RFC 3411, Dec. 2002.

[21] A. Heinemann. *Collaboration in Opportunistic Networks*. VDM Verlag, Saarbruecken, Germany, 2007.

[22] Y.-C. Hu. Efficient security mechanisms for routing protocols. In *Proc. of NDSS'03*, pages 57–73, 2003.

[23] ITU-T. Principles for a Telecommunications Management Network. Technical Report ITU-T Recommendation M.3010, 1996.

[24] D. Johnson, Y. Hu, and D. Maltz. The Dynamic Source Routing Protocol (DSR) for Mobile Ad Hoc Networks for IPv4. IETF RFC 4728, Feb. 2007.

[25] C.E. Jones, K.M. Sivalingam, Prathima A., and J.-C. Chen. A Survey of energy efficient network protocols for wireless networks. *Wireless Networks*, 7:343–358, 2001.

[26] M.D. Katz, H. Frank, and P. Fitzek, editors. *Cooperative and Cognitive Networks: A Motivating Introduction*. Springer, Berlin, 2007.

[27] J.O. Kephart and D.M. Chess. The Vision of autonomic computing. *Computer*, 36(1):41–50, 2003.

[28] D. Klaus. *Technologies for the Wireless Future*. John Wiley & Sons, New York, 3rd edition, 2008.

[29] L. Leszek, A. Gupta, Z. Kamal, and Z. Yang. Opportunistic resource utilization networks: A new paradigm for specialized ad hoc networks. *Computers & Electrical Engineering*, 36(2):328–340, 2010.

[30] L. Leszek, Z. Kamal, V. Bhuse, and A. Gupta. *The Concept of Opportunistic Networks and their Research Challenges in Privacy and Security*, pages 85–117. Springer, Berlin, 2007.

[31] M. Li and K. Sandrasegaran. Network management challenges for next-generation networks. In *Proc. of the IEEE Conference on Local Computer Networks (LCN)*, pages 593–598, Washington, DC, 2005.

[32] W. Liu, C. Chiang, H. Wu, and C. Gerla. Routing in clustered multi-hop, mobile wireless networks with fading channel. In *Proc. of IEEE SICON'97*, pages 197–211, 1997.

[33] E. Madeira and B. Schulze. Managing networks and services of the future. *Journal of Network and Systems Management*, 17(1):1–4, 2009.

[34] S. Murthy and J. Garcia-Luna-Aceves. An efficient routing protocol for wireless networks. *Mobile Networks and Applications*, 1(2):183–197, 1996.

[35] J.M. Peha. Spectrum management policy options. *IEEE Communications Surveys & Tutorials*, 1(1):2–8, 1998.

[36] C. Perkins, E. Belding-Royer, and S. Das. Ad hoc On-Demand Distance Vector (AODV) Routing. IETF RFC 3561, Jul. 2003.

[37] C.E. Perkins and P. Bhagwat. Highly dynamic destination-sequenced distance vector routing (DSDV) for mobile computers. *ACM SIGCOMM Computer Commununications Review*, 24(4):234–244, 1994.

[38] T. Porta, K. Sabnani, and R. Gitlin. Challenges for nomadic computing: Mobility management and wireless communications. *Mobile Networks and Applications*, 1(1):3–16, 1996.

[39] D. Raz, A. Juhola, J. Serrat, and A. Galis, editors. *Fast and Efficient Context-Aware Services*. John Wiley & Sons, New York, 2006.

[40] A. Schmidt. *Ubiquitous Computing - Computing in Context*. PhD thesis, Lancaster University, 2003.

[41] M.H. Seung, M. Shiwen, Y.T. Hou, N. Kwanghee, and J.H. Reed. A location-assisted MAC protocol for multi-hop wireless networks. In *Proc. of IEEE Wireless Communications and Networking Conference (WCNC)*, pages 322–327, March 11-15, 2007.

[42] Y. Seung and R. Kravets. Key management for heterogeneous ad hoc wireless networks. In *Proc. of 10th IEEE International Conference on Network Protocols*, pages 202–203, 2002.

[43] A. Sinha and A. Chandrakasan. Dynamic power management in wireless sensor networks. *Design Test of Computers, IEEE*, 18(2):62–74, Mar/Apr. 2001.

[44] J.S. Strassner. *Policy-based Network Management: Solutions for the Next Generation*. Morgan Kaufmann, San Francisco, 2003.

[45] R.W. Thomas, L.A. DaSilva, and A.B. MacKenzie. Cognitive networks. In *Proc. of 1st IEEE International Symposium on New Frontiers in Dynamic Spectrum Access Networks (DySPAN)*, pages 352–360, 2005.

[46] C.-K. Toh. Associativity-based routing for ad hoc mobile networks. *Wireless Personal Communications*, 4(2):103–139, 1997.

[47] A. Traian, O. Seungchan, and H. Salim. Analyzing attacks in wireless ad hoc network with self-organizing maps. In *Proc. of 5th Annual Conference on Communication Networks and Services Research*, pages 166–175, 2007.

[48] D.C. Verma. Simplifying network administration using policy-based management. *IEEE Network*, 16(2):20–26, Mar/Apr. 2002.

[49] D. Waltenegus. *Context and Self-Management*, pages 1–14. Chapman & Hall/CRC, London, 2009.

[50] E. Weiss, G.R. Hiertz, and X. Bangnan. Performance analysis of temporally ordered routing algorithm based on IEEE 802.11a. In *Proc. of 61st IEEE Vehicular Technology Conference (VTC-Spring)*, volume 4, pages 2565–2569, 2005.

[51] J. Xiaolong and J. Liu. *Agents and Computational Autonomy: Potential, Risks, and Solutions*, volume 2969 of *Lecture Notes in Computer Science*, chapter From Individual Based Modeling to Autonomy Oriented Computation, pages 151–169. Springer, Berlin, 2004.

2

Mobility Management for All-IP Mobile Networks

Quan Le-Trung

Department of Informatics, University of Oslo, Norway
Email: quanle@ieee.org

Paal E. Engelstad

Department of Informatics, University of Oslo, Norway
Simula Research Laboratory, 1325 Lysaker, Norway
Telenor Research and Innovation (R&I), Fornebu, Norway
Email: paal.engelstad@telenor.com

Tor Skeie

Department of Informatics, University of Oslo, Norway
Simula Research Laboratory, 1325 Lysaker, Norway

Frank Eliassen

Department of Informatics, University of Oslo, Norway
Simula Research Laboratory, 1325 Lysaker, Norway

Amirhosein Taherkordi

Department of Informatics, University of Oslo, Norway

CONTENTS

One of the recent trends in networking has been concentrated on realizing all-IP mobile networks, together with new Internet applications and services. For the next generation of all-IP mobile networks, one of the challenging issues is mobility management. The contribution of this chapter is multifold. First, the state of the art in mobility management for all-IP mobile networks is presented, mainly at the network and link layers. Qualitative analysis of positive and negative points among alternative approaches are also discussed and shown. However, the cross-network or link-layer relations to control mobility management in all-IP mobile networks are normally specified at a high-level of abstraction. Thus, the second contribution of this chapter is to consider mobility management for all-IP mobile networks spanning a specific one, the wireless LAN (WLAN) IEEE 802.11 networks, in both network and link layers. Third, different issues in mobility management in all-IP mobile networks spanning into the mobile ad-hoc networks (MANETs) are considered. The reason is that WLANs are 1-hop access networks, while MANETs are multihop access networks. Due to multihop connectivity in MANETs, different problems appear consequent on the mobility of MANET nodes among different domains of all-IP mobile networks. The discovered problems and the proposed solutions to either reduce partly, or remove completely, these discovered problems, and constitute the forth contribution of this chapter. Finally, this chapter ends with conclusions and lays out some research directions in the next generation of all-IP mobile networks.

2.1 Introduction

All-IP mobile networks are defined as networks in which IP is employed from a mobile subscriber to the access points (APs) that connect the wireless networks to the Internet [40]. In the fourth-generation (4G) of wireless networks, the goal is to allow users to communicate reliably and cost effectively via any media, at any time, and anywhere [18, 4, 45]. Thus, 4G [1] systems will integrate existing and new wireless networks seamlessly, allowing mobile users to roam globally with no limit to underlying access technologies such as wireless LANs, wireless PANs, wireless MANs, and cellular systems [18]. Mobility management, therefore, is one of the most challenging issues in 4G all-IP mobile networks.

Mobility management for all-IP mobile networks is originally considered *host-based mobility*, where individual or a small number of mobile users roam across heterogeneous wireless networks. This subclass is further classified as either *macromobility* or *micromobility*. Macromobility is used to manage the mobility of mobile users among wireless networks under the different authorities, and Mobile IPv4 (MIPv4) as well as Mobile IPv6 (MIPv6)[1] are probably the most well-known IP mobility support protocol. However, handoff[2] latency and signaling overhead in macromobility are the main causes of packet loss which results in performance degradation. Thus, numerous methods of minimizing handoff latency and/or signaling overhead have been proposed. In the scope of this book chapter, only IP-layer (L3) and link-layer (L2) solutions are considered, mostly in the scope of micromobility, that is, the mobility of mobile users among subnetworks managed by the same authority. Security is out of the scope of this chapter.

Micromobility solutions aim at achieving low-latency handoff mechanisms for delay-sensitive or real-time applications, with low to zero packet loss, and to ensuring that the signaling overhead is kept to a minimum. Micromobility protocols are designed for environments where mobile nodes change their point of attachment to the network so frequently that the basic MIPv4/v6 protocol tunneling mechanism introduces the unacceptable network overhead

---

[1]Though IPv6 is the de facto network protocol in the next-generation all-IP mobile networks, IPv4 and network address translation (NAT) are likely to coexist with IPv6 for a long time. Thus, mobility management protocols for both IPv4 and IPv6 mobile networks are considered in this book chapter.

[2]Handoff and handover are the interchangeable terminologies used in this book chapter. These two terms refer to the process of transferring an ongoing connection if either the source or the destination mobile node of this connection moves from one domain (e.g., home network in MIPv6) to another (e.g., foreign network in MIPv6).

in terms of increased delay, packet loss, and signaling. Micromobility protocols support the handling of local movement, for example, within a domain,
of mobile nodes without interaction with the mobile IP-enabled Internet. This
has the benefit of reducing the delay and the packet loss during the handoff
and eliminating registration between mobile nodes and possibly distant home
agents when mobile nodes remain inside their local coverage areas. Micromobility solutions can be further classified according to (1) the techniques
used to reduce the signaling overhead, for example, Hierarchical MIPv4/v6
(HMIPv4/v6) [29, 85], cellular IP (CIP) [86], handoff-aware wireless access
Internet infrastructure (HAWAII) [78]; (2) the techniques to reduce handoff
latency, for example, Fast Handover in MIPv4/v6 (FMIPv4/v6) [41, 42], or (3)
the hybrid techniques, for example, Fast Handover for HMIPv6 (FHMIPv6)
[39].

However, MIPv4/v6 and its various enhancements basically require modifications in the protocol stack of mobile nodes. On the other hand, recently,
in *network-based host mobility*, for example, Proxy MIPv6 (PMIPv6) [28, 88],
the serving network handles mobility management on behalf of mobile nodes.
Thus, the mobile node is not required to participate in any mobility-related
signaling. Therefore, the tunneling overhead and a significant amount of packet
exchanges via wireless links can be reduced, resulting in higher performance
and lower latency. In the Internet Service Provider (ISP) perspective, the deployment of mobility management in all-IP mobile networks will be faster and
more reliable.

While *host mobility* considers the mobility of individual mobile nodes, *network mobility* (NEMO) considers the mobility of a whole mobile network, for
example, mobile users within a train, an airplane, or a ship. There is recent research extending beyond host mobility to network mobility, for example, IETF
NEMO. However, the NEMO basic support protocol [17] does not address
important issues such as route optimization and handoff, causing unacceptable delay and low performance. Further improvements of route optimization,
which are border gateway protocol (BGP) based, have been implemented in
Connexion by Boeing [20], and wide-area IP network mobility (WINMO) [33].

Currently, IEEE is working on a standard, 802.21 [3], for media-
independent handover (MIH) services. IEEE 802.21 enables handover and interoperability between heterogeneous network types, including both 802 and
non-802 networks. The key ideas in this standard are (1) providing access network service to take care of link-layer triggers and allow upper layers to use
this information in a timely manner; (2) providing command service to allow
upper layers to send various instructions to the link-layer; and (3) providing
an information service as a basis for making more effective handover decisions.

While there are many approaches to mobility management for all-IP mobile
networks, there has still been the lack of a complete, up-to-date comparison
on the state of the art among these approaches, and this is the first contribution of this chapter (cf. Section 2.2). In this section, a qualitative analysis of
positive and negative points among alternative approaches will be discussed

and presented. Additionally, a major direction in reducing handoff latency in the mobility management of all-IP mobile networks is to use link-layer triggers. However, the cross L3/L2 relations to control mobility management are normally specified in a high-level abstraction. To map this high-level abstraction into a specific one, in Section 2.3, mobility management for all-IP mobile networks spanning a specific and most popular one, the WLAN IEEE 802.11 domain, is considered in both L3 and L2 layers. This is the second contribution of this book chapter.

Though WLAN 802.11 is one of the most popular wireless access networks, it only provides 1-hop coverage. In the very near future, mobile nodes will roam across multiple heterogeneous platforms while continuously maintaining connectivity. A mobile node may connect to a WLAN and then move into an area where the coverage from the WLAN does not exist. In this situation, a mobile ad-hoc network is required to extend the 1-hop to n-hop coverage. Thus, one of the wireless network types in 4G systems will be the MANET. The third contribution of this chapter (cf. Section 2.4) is mobility management for all-IP mobile networks spanning the MANET domain.

To ensure the self-configured, infrastructureless, and mobility-controlled characteristics of MANETs when accessing the Internet, in Section 2.5, different required functions need to be provided [46]. These functions include (1) MANET node location determination; (2) Internet gateway discovery, selection, and forwarding strategy; (3) auto-configuration addressing scheme, and (4) handoff control. This section will also review the existing approaches of each required function in providing Internet connectivity for MANETs and mobility management. Additionally, the connection between a MANET node and an Internet gateway (IGW) is multihop. Therefore, there is normally no direct wireless link from this MANET node to the IGW, as in WLAN. Instead, they are connected via other intermediate nodes. Thus, different problems, such as, inconsistent context, and cascading effect, can happen during the mobility of ad-hoc nodes within a MANET domain if multiple IGWs exist [47, 48] [23]. The required functions, the discovered problems, and the proposed solutions in Section 2.5 are the fourth contribution of this chapter. Section 2.6 presents open issues for IP-based mobility management and some research directions in the mobility management of next-generation all-IP mobile networks Finally, in Section 2.7, this chapter ends with conclusion.

2.2   IP Mobility Management for Mobile Networks

2.2.1   Host Mobility Management

2.2.1.1   Host-Based Mobility Management

In the IP environment, when a mobile node moves and attaches itself to another network, it needs to obtain a new IP address. This changing of IP address means that all existing IP connections to the mobile node need to be terminated and then reestablished. Thus, there is a request on how to keep on-going IP connections upon the mobility of a mobile node. This section presents protocols to ensure the ongoing connections to a mobile node when the mobile node changes its points of attachment, and does not consider further characteristics such as security and quality of service (QoS) support.

(A) **Host-Based Macromobility Management:**
Mobile IP (MIP) describes a global solution (macromobility) that overcomes this problem by using a set of network agents. It does not require any modifications to existing routers or end correspondent nodes [45]. With MIP, each mobile node is identified by an address from its home network, regardless of the point of attachment. While a mobile node is away from its home network, it obtains an IP address from the visiting network, which is called foreign agent care-of address (FA CoA) or colocated care-of address (CCoA) in MIPv4, or CCoA in MIPv6, and registers this IP address with a home agent (HA) within its home network. The HA intercepts any packets destined to the mobile node and tunnels or explicitly routes (source routing) them to the current location of the mobile node. Thus, initiating this indirection requires a timely address reconfiguration procedure and a home network registration. The time taken for a mobile node to configure a new CoA in the visiting network, and the time taken to register with the HA, together constitute the handoff latency. Figures 2.1 and 2.2 show the architectures and operations of MIPv4 and MIPv6, respectively. Figures 2.3 and 2.4 describe the packet sequences in the registration process of MIPv4, using FA CoA and CCoA, respectively. Figure 2.5 shows the packet sequences in the handoff procedure of MIPv6 with the standard IPv6 neighbor discovery [63].

Handoff latency in macromobility is the primary cause of packet loss and results in performance degradation, especially in the case of reliable end-to-end communication. As a result, numerous methods of minimizing handoff latency have been proposed in the literature. The proposed schemes can be broadly classified into (1) those that operate above the IP layer such as TCP-Migrate [84] and pTCP [30]; and (2) those that operate at the IP layer. In general, the solutions that operate at the IP layer are regarded as being more suitable as they do not violate any of the basic Internet design principles and, more importantly, because they do not require any changes to the protocols at the

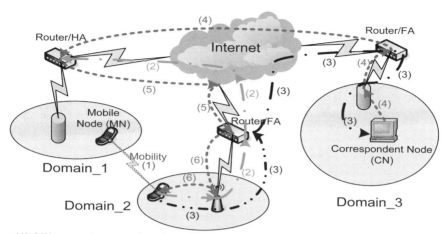

(1): MN moves to a new domain   (2): MN gets & registers its new CoA with its HA
(3): MN sends data directly to CN   (4): CN sends data to MN indirectly via MN's HA
(5): MN's HA tunnels data to MN's FA (6): MN's FA delivers data to MN

FIGURE 2.1
Architecture and operations of Mobile IPv4 (MIPv4).

correspondent nodes [12]. The IP layer solutions are presented and discussed in the next subsection, mostly in the micromobility scope.

(B) **Host-based Micro Mobility Management:**
Micromobility protocols support local movement, such as within a domain, of mobile nodes without the interaction with the MIP-enabled Internet thus reducing delay and packet loss during handoff. Several micro-mobility protocols have been proposed, which can be classified as shown in Figure 2.6. Hierarchical mobility management reduces the performance impact of mobility by handling local migrations locally and hiding them from home agents. In this case, the Internet address known by a home agent no longer reflects the exact attachment point of the mobile node. Rather, it represents the address of a gateway that is common to a potentially large numbers of network access points. When a mobile node moves from one access point to another, which is reachable through the same gateway, there is no need to inform its home agent. The role of micromobility protocols is to ensure that packets arriving at the gateway are forwarded to the appropriate access point. Two styles of hierarchical mobility management are supported by micromobility: (1) hierarchical tunneling, and (2) the mobile-specific routing.

In hierarchical tunneling approach, the location database is maintained in a distributed form by a set of foreign agents in the access network. Each foreign agent reads the original destination address of the incoming packet, searching its visitor list for a corresponding entry. If the entry exists, then it contains the address of the next lower-level foreign agent. The sequence

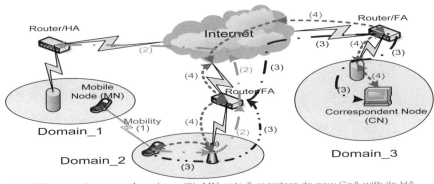

FIGURE 2.2
Architecture and operations of Mobile IPv6 (MIPv6).

of visitor list entries corresponding to a particular mobile node constitutes the location information of the mobile node and determines the route taken by its downlink packets. Entries are created and maintained by registration packets that are transmitted by mobile nodes. These proposals rely on a tree-like structure of foreign agents. Encapsulated traffic from the home agent is delivered to the root foreign agent. Each foreign agent on the tree decapsulates and then reencapsulates data packets as they are forwarded down the tree of foreign agents toward the attachment point of the mobile node. As a mobile node moves between different access points, location updates are made at the optimal point on the tree, tunneling traffic to the new access point. Examples of micromobility protocols that use hierarchical tunneling include HMIPv4 [29] and HMIPv6 [85].

Mobile-specific routing approaches avoid the overhead introduced by decapsulation and reencapsulation schemes, as is the case with hierarchical tunneling approaches. These proposals use routing to forward packets toward an attachment point of the mobile node using mobile-specific routes. These schemes typically introduce implicit signaling, for example, based on snooping data, or explicit, signaling to update mobile-specific routes, or they are aware that a routing protocol is in use. In the case of cellular IP (CIP), mobile nodes attached to an access network use the IP address of the gateway as their mobile IP care-of address. The gateway decapsulates packets and forwards them toward a base station. Inside the access network, each mobile node is identified by its home address, and data packets are routed using mobile-specific routing without tunneling. The routing protocol ensures that packets are delivered to the actual location of the mobile node. Examples of micromobility protocols that use mobile-specific routing include CIP [86] and HAWAII [78].

Typically, fixed hosts connected to the Internet remain online even though most of the time they do not communicate. Mobile nodes connected to the

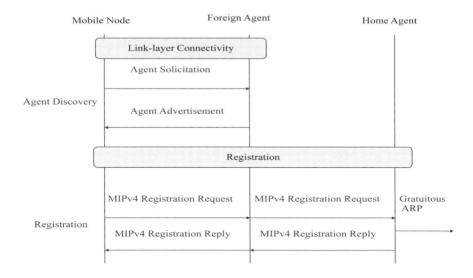

FIGURE 2.3
MIPv4 registration via foreign agent care-of address.

wireless Internet will expect the same service. In order to maintain the location information for routing support, each mobile node requires frequent location updates. Thus, the maintenance of location information consumes bandwidth and battery power. This signaling overhead can be reduced through the introduction of paging. Mobile nodes are typically powered by batteries with limited lifetime. This makes it important to save idle mobile nodes from having to transmit frequent location update packets. This requires explicit support from networking protocols, such as the ability to track location approximately and the ability to page idle mobile nodes. Idle mobile nodes do not have to register if they move within the same paging area. Rather, they only register if they change the paging area. A number of micromobility protocols, such as CIP and HAWAII, have implemented paging.

Supporting fast- and/or low-latency handover, which reduces delay and packet loss during the handoff, is another important attribute of micromobility protocols. A number of design choices influence handoff performance, including handoff control, buffering and forwarding techniques, radio behavior, movement detection and prediction, and coupling and synchronization between the IP and radio layers [12]. The low-latency handoff proposal for MIPv4 (LLMIPv4) [53] describes two methods of achieving this, namely, preregistration and post-registration. With preregistration handoff, the mobile node is assisted by the network to perform L3 handoff before it completes the L2 handoff. It uses L2 triggers, which arises as a result of beaconing signals from the network that the mobile node is about to move to, to initiate an IP layer (L3) handoff. Its design however, diverges from the clean separation

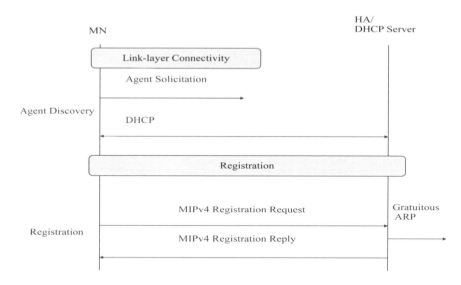

FIGURE 2.4
MIPv4 registration via colocated care-of address.

of L2 and L3 of the base MIPv4 scheme. With post-registration handoff, L2 triggers are used to setup a temporary bidirectional tunnel between the old foreign agent (oFA) and the new foreign agent (nFA). This allows the mobile node to continue using its oFA while performing the registration at the same or later time. A combined method is also possible where, if the preregistration does not complete in time, the oFA forwards traffic to the nFA using the post-registration method in parallel.

Fast handover for MIPv4 (FMIPv4) [42] and for MIPv6 (FMIPv6) [41] are similar in concept to the combined method described above and consists of three phases: (1) handover initiation (HI), (2) tunnel establishment, and (3) packet forwarding. Figures 2.7 and 2.8 show the architecture and packet sequences of the handoff procedure in FMIPv6, which is similar to those of FMIPv4 except for the name of the agents, for example, previous access router (pAR) instead of oFA, new access router (nAR) instead of nFA, and signaling packets, for example, binding update (BU) instead of registration request (RegReq).

The handover initiation (HI) is started by the L2 trigger based on certain policy rules (unspecified by IETF at the time of writing). This is done by the mobile node, which sends a router solicitation proxy (RtSolPr) packet to pAR, indicating that it wishes to perform a fast handoff to a new attachment point. The RtSolPr contains the link-layer address of the new attachment point, which is derived from the beacon packets of nAR. The mobile node will receive, in response, a proxy router advertisement (PrRtAdv) packet from the pAR, with a set of possible responses indicating that the point of attachment

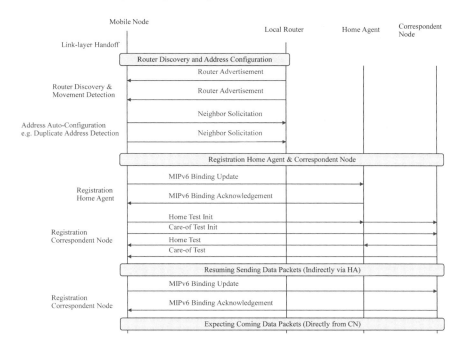

FIGURE 2.5
Handoff procedure with Mobile IPv6 (MIPv6).

is either unknown, or known but connected through the same access router, or is known and specifies the network prefix that the mobile node should use in forming the new CoA. Subsequently, the mobile node sends a fast binding update (F-BU) to the pAR using its newly formed CoA based on the prior PrRtAdv response, as the last packet before the handover is executed. The mobile node receives a fast binding acknowledgement (F-BAck) via either pAR or nAR, indicating a successful binding [31].

The tunnel establishment phase creates a tunnel between the nAR and the pAR. To establish a tunnel, the pAR sends a HI packet containing the requesting CoA and the current CoA of the mobile node to the nAR. In response, the pAR receives a handover acknowledgement (HAck) from the nAR. If the new CoA is accepted by the nAR, the pAR sets up a temporary tunnel to the new CoA. Otherwise, the pAR tunnels packets destined for the mobile node to the nAR, which will take care of forwarding packets to the mobile node temporarily. Finally, the forwarding phase of the packet is performed to smoothen the handoff until subsequent registration by the mobile node to the home agent is completed. The pAR interacts with the nAR to facilitate the forwarding of packets between them through the previously established tunnel. The initiation of forwarding is based on an anticipation timing interval

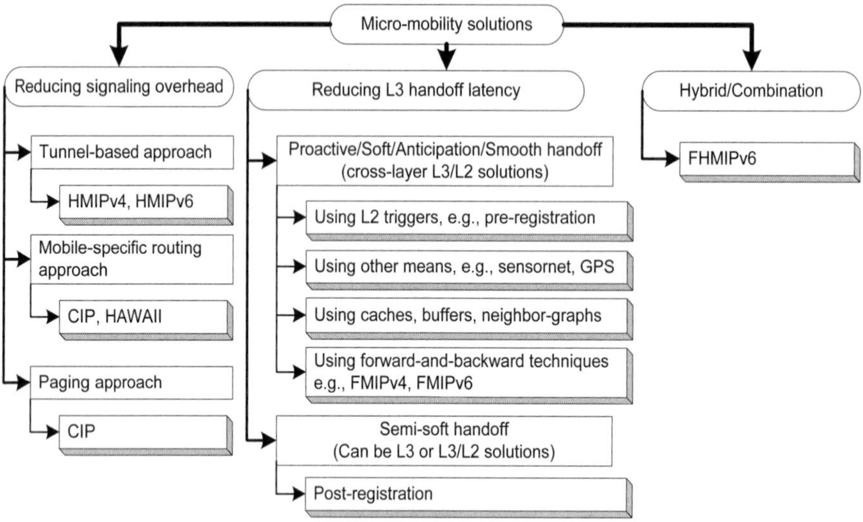

FIGURE 2.6
Classification of host-based micromobility management.

heuristic, that is the network anticipates as to when a mobile node is likely to handoff and therefore infers the appropriate packet-forwarding moment based on the anticipated timing interval. Such an interval is, however, extremely difficult to generalize, and forwarding too early or too late will result in packet losses, negating the purpose of packet forwarding. Once it arrives at the new access network, the mobile node sends the fast neighbor advertisement (F-NA) packet to initiate the flow of packets (to itself) from the nAR.

(C) **Host-based Hybrid Mobility Management:**
Hierarchical mobile IPv6 with fast handover (FHMIPv6) [39] is another attempt to further reduce the overall handoff latency from what fast handover can offer alone. By combining HMIPv6 with fast handover, the latency due to the address configuration and the subsequent home network/agent registration, can be reduced. The mobility anchor point (MAP) can be viewed as the local home agent, and in most cases, it is located closer to the mobile node than the home agent. Therefore, the signaling cost saved is the difference between the round-trip time from the mobile node to the MAP and the round-trip time between the mobile node and the home agent, assuming that the packet processing time within a network node is insignificant in comparison. This combination requires minor modification to the standard HMIPv6 protocol and the fast-handover protocol, that is relocating the forwarding anchor point from the pAR to the MAP.

An alternative to the scheme of packet forwarding has also been proposed, namely, the simultaneous bindings framework (SBF) [54]. It proposes to reduce

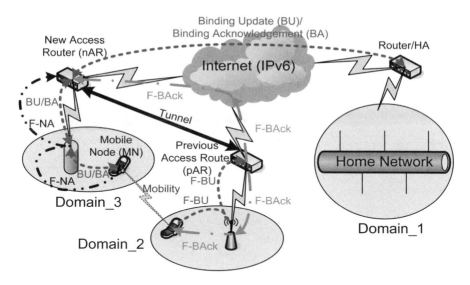

FIGURE 2.7
Architecture and operations of FMIPv6.

packet losses at the mobile node by n-casting packets for a short period to the current location of the mobile node and to n-other locations where the mobile node is expected to move to. The n-casting can be carried out by the pAR, the MAP, or the HA. The simultaneous binding scheme recognizes the problem of not knowing when the mobile node is likely to move, the timing ambiguity, and attempts to remove it by careful packet duplication to multiple access networks. It also claims to be able to address the problem associated with the rapid back-and-forth movement of mobile nodes between two access routers, called the ping-pong movement, by this packet duplication process, as it is not necessary to reconfigure the mobile node CoA during the ping-pong movement.

The seamless handoff architecture for mobile IP (SMIP) is described in [31]. SMIP is a proactive advanced configuration where a movement pattern of mobile node (linear/ping-pong) is augmented with a specialized forwarding algorithm when performing a handoff and provides the best result. Unlike FHMIPv6, which uses the L2-trigger, or SBF, which uses the L2-trigger and n-casts packets to multiple destinations, in SMIP the network uses mobile node location and movement patterns to instruct the mobile node when and how handoff should be carried out. It is a structured approach combining mobile node location tracking, movement patterning and handoff algorithms into an integral unit, providing smarter handoffs. The mobile node initiates a handoff while the network determines the handoff decision. SMIP uses signal strength together with triangulation to track the mobile node location and determine its movement pattern. The entity that stores the history of the locations and

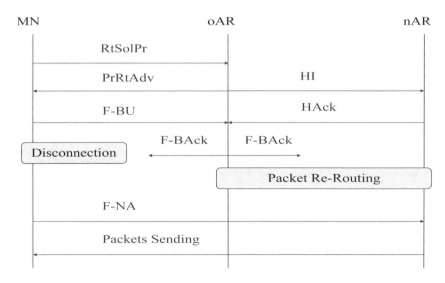

FIGURE 2.8
Predictive mode of FMIPv6 protocol.

determines the movement pattern, is referred to as the decision engine (DE), is similar to a MAP, and covers a group of access routers and the associated mobile nodes. The DE makes the handoff decision on behalf of the mobile node. Once a decision is made, it is relayed to the access routers (pAR and nAR) and eventually to the mobile node as part of the fast-handover PrRtAdv packet. Note that SMIP is built on the structure of FHMIPv6 and operates similarly to the mobile-node-initiated fast handover.

2.2.1.2   Network-Based Mobility Management

One of the challenges of host-based host mobility management in Section 2.2.1.1 is the requirement of changes in the IP stack of mobile nodes. These changes limit broad usage even if the modifications are small. This subsection describes existing and ongoing work on network-based host mobility management, which enables IP mobility for a mobile node without requiring its participation in any mobility-related signaling. The network is responsible for managing IP mobility on behalf of the mobile node. The mobility entities in the network are responsible for tracking the movements of the mobile node and initiating the required mobility signaling on behalf of the mobile node.

The scope of this section is local mobility[3], which is restricted to providing IP mobility management for mobile nodes within an access network. While some mobile node involvement is necessary and expected for generic mobility

---

[3]Local mobility has the same meaning as the micro host-based host mobility in Section 2.2.1.1. We use this terminology as it is defined in the IETF NETLMM charter.

functions such as movement detection and to inform the mobile access gateway (MAG) about mobile node movement, no specific mobile-node-to-network protocol will be required for localized mobility management itself. The mobile node stack involvement in mobility management is thereby limited to generic mobility functions at the IP layer, and no specialized localized mobility management software is required.

Proxy Mobile IPv6 (PMIPv6) [28] supports mobility for IPv6 mobile nodes without the involvement of IPv6 mobile node by extending MIPv6 signaling messages between a network node, called MAG in PMIPv6, and a home agent, called local mobility anchor (LMA) in PMIPv6. The MAG in the network performs the signaling with the LMA and does the mobility management on behalf of the mobile node attached to the network. Figures 2.9 and 2.10 show an operation overview and handoff procedure for PMIPv6, respectively. Once a mobile node (MN) enters its PMIPv6 domain (see Figure 2.9) the serving network ensures that the MN is always on its home network and can obtain its home address (HoA) on any access network. That is, the serving network assigns a unique home network prefix to each MN, and conceptually, this prefix always follows the MN wherever it moves within a PMIPv6 domain. From the perspective of the MN, the entire PMIPv6 domain appears as its home network. Accordingly, it is needless to configure the care-of-address (CoA) at the MN.

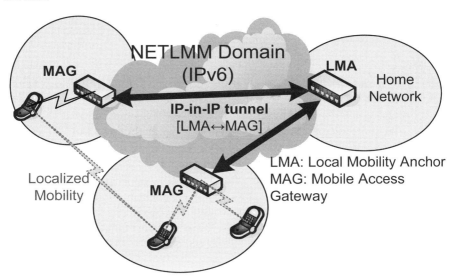

FIGURE 2.9
Proxy Mobile IPv6 (PMIPv6) architecture.

The new principal functional entities of PMIPv6 are the MAG and LMA. The MAG typically runs on the access router (AR). The main role of the MAG is to detect the MN movements and initiate mobility-related signaling with

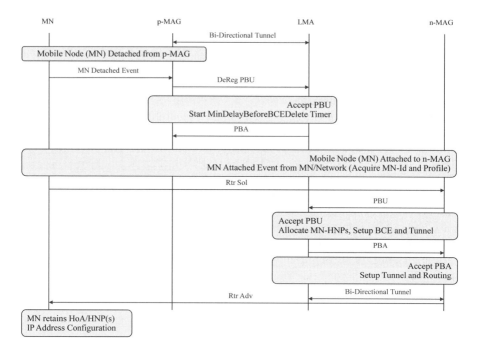

FIGURE 2.10
Handover procedure with PMIPv6.

the MN LMA on behalf of the MN. In addition, the MAG establishes a tunnel with the LMA for enabling the MN to use an address from its home network prefix and emulates the MN home network on the access network for each MN. The main role of the LMA is to maintain reachability to the MN address while it moves around within a PMIPv6 domain, and the LMA includes a binding cache entry for each currently registered MN. Figure 2.10 shows the signaling call flow for the mobile node handoff from the previously attached mobile access gateway (p-MAG) to the newly attached mobile access gateway (n-MAG). After obtaining the initial address configuration in the PMIPv6 domain, if the mobile node changes its point of attachment, the p-MAG will detect the mobile node detachment from the link. It will signal the LMA and will remove the binding and routing state for that mobile node. The LMA, upon receiving this request, will identify the corresponding mobility session for which the request was received and accepts the request, after which it waits for a certain amount of time to allow the mobile access gateway on the new link to update the binding.

However, if it does not receive any proxy binding update (PBU) message within the given amount of time, it will delete the binding cache entry. The n-MAG, upon detecting the mobile node on its access link, will signal the

LMA to update the binding state. Upon accepting this PBU message, the LMA sends a proxy binding acknowledgement (PBA) message including the MN home network prefix (MN-HNPs). It also creates the Binding Cache Entry (BCE) and sets up its endpoint of the bidirectional tunnel to the n-MAG. After completion of the signaling, the serving n-MAG will send the router advertisement (RtrAdv) containing the MN-HNPs, and this will ensure that the mobile node will not detect any change with respect to the layer-3 attachment of its interface. Note that the router solicitation (RtrSol) message from the mobile node may arrive at any time after the MN attachment and has no strict ordering relation with the other messages in the call flow. While PMIPv6 requires only IPv6 mobile nodes and IPv6 transport network between the mobility entities, it is also reasonable to expect the same mobility infrastructure in the PMIPv6 domain to provide mobility to the mobile nodes operating in IPv4, or in dual IPv4/IPv6 mode, and whether the transport network is IPv4 or IPv6. This is due to the transition from IPv4 to IPv6 being a long process, and during this period of transition, both the IPv4/IPv6 protocols will be enabled over the same network infrastructure. Reference [88] describes the IPv4 support in PMIPv6 (IPv4/PMIPv6) for two scenarios: (1) IPv4 home address mobility support, and (2) IPv4 transport network support.

The IPv4 home address mobility support essentially enables an IPv4 mobile node in a PMIPv6 domain to obtain the IPv4 home address configuration for its attached interfaces and to be able to retain that address configuration even after performing a handoff anywhere within that PMIPv6 domain. The mobile node on the access link using any of the standard IPv4 address configuration mechanisms supported on that access link, such as dynamic host configuration protocol (DHCP), will be able to obtain an IPv4 home address (IPv4-MN-HoA) for its attached interface. Although the address configuration mechanisms for delivering the address configuration to the mobile node is independent of the PMIPv6 protocol operation, there needs to be some interactions between these two protocol flows. Figures 2.11 and 2.12 illustrate, respectively, how DHCP-based address configuration support can be enabled for a mobile node in a PMIPv6 domain in two scenarios: (1) DHCP server colocated with the MAG, and (2) DHCP relay agent colocated with the MAG. The DHCP server (DHCP-S) or the DHCP relay agent (DHCP-R) configured on the MAG is required to have an IPv4 address for exchanging the DHCP messages with the mobile node. This address is the default router address of the mobile node provided by the LMA. Optionally, all the DHCP servers colocated with the MAGs in the PMIPv6 domain can be configured with a fixed IPv4 address. This fixed address can be potentially an IPv4 private address that can be used for the DHCP protocol communication on any of the access links. This address will be used as the server identifier in the DHCP messages.

The MAG may choose to ignore the DHCPDISCOVER messages till the PMIPv6 signaling is successfully completed, or it may choose to send a delayed response for reducing the additional delay waiting for a new DHCPDISCOVER message from the mobile node. For acquiring the mobile node's IPv4

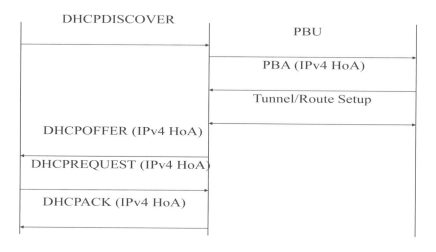

FIGURE 2.11
DHCP server colocated with MAG.

home address from the LMA, the MAG will initiate PMIPv6 signaling with the LMA. After the successful completion of PMIPv6 signaling and upon acquiring the mobile node's IPv4 home address from the LMA, the DHCP server on the MAG will send a DHCPOFFER message to the mobile node. The offered address will be the mobile node's IPv4 home address assigned by the LMA. The DHCPOFFER message will also have the subnet mask option and router option, with the values in those options set to the mobile node's IPv4 home subnet mask and default router address, respectively. Additionally, the Server Identifier option will be included, and with the value in the option set to the default router address.

In the IPv4 transport network support, the LMA and the MAG are configured and reachable using only the IPv4 addresses, and the MAG serving a mobile node can potentially send the PMIPv6 signaling messages over IPv4 transport and register its IPv4 address as the CoA in the mobile node's Binding Cache entry. An IPv4 tunnel with any of the supported encapsulation modes, for example IPv4 with UDP header, can be used for tunneling the data traffic of the mobile node. The MAG can be potentially in a private IPv4 network behind a network address translation (NAT) device, with a private IPv4 address configured on its egress interface. But the LMA must not be behind a NAT and must be using a globally routable IPv4 address. However, both the LMA and the MAG can be in the same private IPv4 routing domain, that is when both are configured with private IPv4 addresses and with no need for NAT translation between them. Note that the IPv6 address configuration requirement on the MAG does not imply that IPv6 routing enabled between

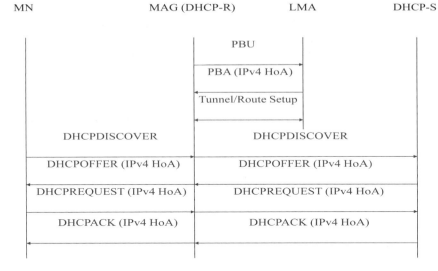

FIGURE 2.12
DHCP relay agent colocated with MAG.

the LMA and the MAG, is needed. It just requires each of the MAGs and LMAs in a PMIPv6 domain to be configured with a globally unique IPv6 address.

2.2.2 Network Mobility Management

While host mobility management deals with the mobility of individual mobile nodes, network mobility (NEMO) is concerned with managing the mobility of an entire network, which changes, as a unit, its point of attachment to the Internet. The mobile network includes one or more mobile routers (MRs) that connect the mobile network to the global Internet.

In NEMO basic support (NEMO-BS) [17], the IETF NEMO charter has adopted the methods for host mobility support used in MIPv6, without routing optimization, and has extended these methods in the simplest way possible to achieve its goals, that is allowing the session continuity for every node in the mobile network as the network moves. The basic support solution is for each MR to have a HA and use bidirectional tunneling between the MR and HA to preserve the session continuity while the MR moves. The MR acquires a CoA at its attachment point much like what is done for mobile nodes using MIPv6. This approach allows nested mobile networks since each MR will appear to its attachment point as a single node. Additionally, the NEMO basic support is transparent to mobile nodes behind the MR, and as such, does not require: (1) mobile nodes taking any actions in the mobility management, and (2) modifications to any mobile nodes other than MRs and HAs.

When a MR moves away from the home link and attaches to a new access router (AR), it acquires a CoA from the visited link. Then the MR immediately sends a binding update (BU) to its HA. When the HA receives this BU, it creates a cache entry binding the MR HoA to its CoA at the current point of attachment. The HA acknowledges the BU by sending a binding acknowledgement (BA) to the MR. Once the binding process finishes, a bidirectional tunnel is established between the HA and the MR. The tunnel end points are the MR CoA and HA address. If a packet with a source address belonging to the mobile network prefix (MNP) is received from the mobile network, the MR reverse-tunnels the packet to the HA through this tunnel. This reverse-tunneling is done by using IP-in-IP encapsulation. The HA decapsulates this packet and forwards it to the correspondent node (CN). When a CN sends a data packet to a node in the mobile network, the packet is routed to the HA that currently has the binding for the MR. The MR network prefix would be aggregated at the HA, which would advertise the resulting aggregation.

Alternatively, the HA may receive the data packets destined for the mobile network by advertising routes to the MNP. When the HA receives a data packet meant for a node in the mobile network, it tunnels the packet to the MR current CoA. The MR decapsulates the packet and forwards it onto the interface where the mobile network is connected. Before decapsulating the tunneled packet, the MR has to check whether the source address on the outer IPv6 header is the HA address. The MR also has to make sure that the destination address on the inner IPv6 header belongs to a prefix used in the mobile network before forwarding the packet to the mobile network. If it does not, the MR should drop the packet.

To illustrate the operation of NEMO basic support, Figure 2.13 depicts a scenario where both the mobile node (MN) and the MR move from their own home links and attach to visited links. The mobile network consists of a local fixed node (LFN) and a local fixed router (LFR). The MR sends a BU to its HA (HA_MR) when it attaches to a visited link and configures a CoA. HA_MR creates a binding cache entry for the MR HA and also sets up forwarding for the prefixes on the mobile network. The MN also configures a CoA from the prefix advertised on the mobile network and sends a BU to its HA (HA_MN) and to its correspondent node (CN_MN). Both HA_MN and CN_MN create binding cache entries for the MN home address. Note that an AR is used to connect the correspondent network to the Internet, and IPv6 is used for addressing.

However, in NEMO basic support, the IETF NEMO charter does not consider (1) routing issues inside the mobile network, and existing routing protocols, including MANET protocols, can be used to solve these problems; (2) routing optimization; (3) managing multiple bidirectional tunnels between the MR(s) and the corresponding HA in NEMO multihomed configurations. Figure 2.14 shows a scenario of a nested mobile network using NEMO basic support, and thus, the flow of packets between MN and CN1 would need to go through three separate tunnels. This results in the increase of: (1) length of

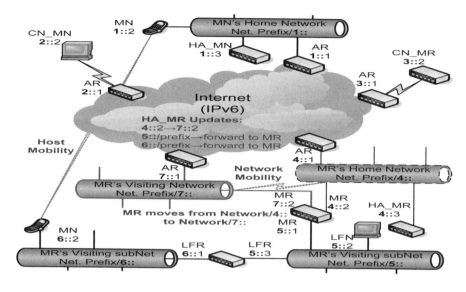

FIGURE 2.13
NEMO basic scenario operation.

packet route, (2) signaling overhead, (3) handoff delay, (4) packet transmission delay, and (5) protocol complexity and processing load.

To overcome above limitations, in [65], authors has analyzed various scenarios of route optimization in NEMO, and explores the benefits and tradeoffs in different aspects of NEMO route optimization. These scenarios of route optimization include: (1) nonnested route optimization, (2) nested mobility optimization, (3) infrastructure based optimization, and (4) intra-NEMO optimization. However, the proposed solutions are not completed, not evaluated, and require further improvements. The border gateway protocol (BGP) announcements has been applied as a mobility method in the Connexion by Boeing (CbB) network [20]. The proposed solution provides a global IP mobility architecture that does not require the use of special IP stacks on Internet hosts or mobile nodes. Virtual private networks (VPNs) and other long-lived TCP sessions can be maintained across satellite transponder handoffs. The benefit of the Boeing Connexion is that the protocol does not rely on the use of tunneling, compared to NEMO basic support, and thus, substantial latency incurred due to tunneling and route optimization problem are improved. However, the extra signaling overhead, that is, BGP routes, corresponding BGP, updates and announcements due to aircraft handoffs, also need further study and examination. Further improvements of the Boeing Connexion to reduce BGP updates and resolve the issue of route optimization, are presented in wide-area IP network mobility (WINMO) [33]. Through scoped BGP updates, route aggregation, tunneling, and mobility packet state, WINMO has achieved

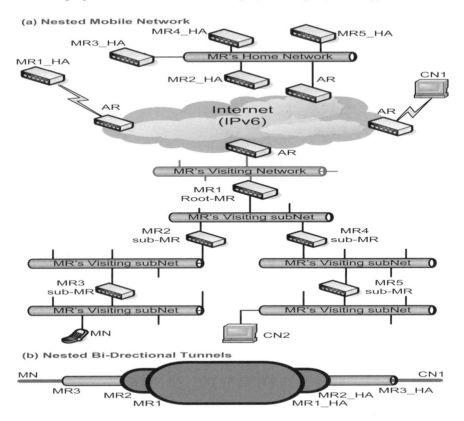

FIGURE 2.14
NEMO routing optimization problem.

low-stretch global Internet routing for mobile networks roaming across wide areas with minimal interdomain routing overhead.

2.2.3 Hybrid Mobility Management

While host mobility management (Section 2.2.1) and network mobility management (Section 2.2.2) propose different protocols and network architectures to support the mobility of individual mobile nodes and mobile networks either at a macro level or a micro level, there is still a loose coupling between these existing work toward a solution for providing global IP mobility. The Boeing Connexion can be considered as an exception that introduces a total new BGP-based, network-based NEMO solution. Next, this section continues with a new hybrid protocol, called *Identifiers Separating and Mapping Scheme* (ISMS) [19], for a network-based host mobility and NEMO management in the future Internet.

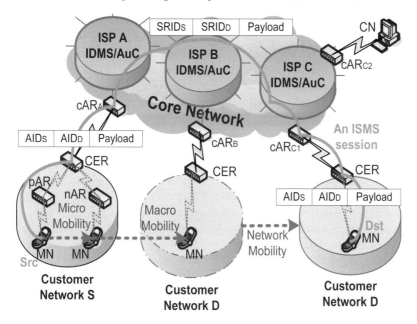

FIGURE 2.15
ISMS architecture, session, and mobility scenarios.

ISMS is designed to satisfy the requirements of the future Internet, such as (1) faster handover, (2) route optimism, (3) advanced management, and (4) location privacy and security. To achieve these requirements, ISMS separates customer networks from transit providers and decomposes Internet addresses into accessing identifiers (AIDs) and switching routing identifiers (SRIDs) to decouple the locator/identifier overload of the IP address. AIDs are network layer identifiers to be used by the transport layer. On the other hand, SRIDs serve as routing tags in the core network. For flexibility, AID-to-SRID mapping is used to bridge the customer networks and the core network. As a result, a network-based mobility management scheme is obtained, which provides lower handoff latency and highly efficient mobility management. Figure 2.15 shows the ISMS architecture, a simple ISMS session from a source MN Src to a destination mobile node Dst, under different scenarios of mobility: (1) micro host mobility, (2) macro host mobility, and (3) network mobility.

When a source mobile node Src in customer network S wants to send a packet to a destination mobile node Dst in customer network D, it forwards the packet with traffic engineering (TE) requirements to one of the S providers, say, ISP A. The core access router (cAR$_A$) contacts the identifiers mapping server (IDMS) to find the SRIDs of Dst. According to the TE requirements of Src, cARA chooses the SRID with higher preference, swaps the Src/Dst AID to Src/Dst SRID, respectively, and forwards it to cAR$_{C1}$. Upon receiving the packet, cAR_{C1} swaps the Src/Dst SRID to Src/Dst AID, respectively, and forwards it to Dst. In ISMS, the IP address is overloaded with the semantics of both who, that is end-point identifier as used by transport layer, and where, that is, locators for the routing system. The overloading is one of the main causes of the mobility problem. ISMS decomposes IP addresses into AIDs and SRIDs, and uses a separating and mapping scheme to support host mobility and network mobility.

For the micro host mobility of a MN within a single customer network S, from its previous access router (pAR) to a new AR (nAR), this MN will receive AR advertisement from the nAR and triggers the route update mechanism from nAR up to customer edge router (CER), and back to pAR. Thus, the packets to this MN will be redirected at CER and sent to nAR. Note that micro host mobility will not cause the remapping of AID and SRID. Only the customer network S where the MN Src locates is aware of the movement.

For the macro host mobility of an MN across different cARs, for example, from customer network S to customer network D via cAR_B in Figure 2.15, ISMS uses a separating and mapping scheme to support the mobility. Figure 2.16 shows the packet exchange sequences of the handoff procedure when an MN moves from customer network S to customer network D while communicating with the correspondent node (CN). Note that when the macro host mobility handoff process finishes, the packet from CN to MN will follow the route $CN \rightarrow cAR_{C2} \rightarrow cAR_A \rightarrow cAR_B \rightarrow MN$, which is not the shortest path. Route optimization is then achieved by using the traffic-driven method. After MN has moved to cAR_B, when cAR_A receives the first packet from CN to MN, it detects that MN has moved away and sends an AR mapping update message to cAR_{C2}, which includes the new mapping of MN. Alternatively, if MN sends packets to CN firstly, when cAR_B receives the first packet from CN to MN, it will inform cAR_{C2} the new mapping of MN by sending an AR-mapping update message. Once cAR_{C2} gets the new mapping of MN, the packet sent by CN to MN will be routed by the optimal route $CN \rightarrow cAR_{C2} \rightarrow cAR_B \rightarrow MN$.

For the network mobility of a mobile network, for example, a customer network D in Figure 2.15, ISMS exploits the network-based mobility scheme, that is, only the mobile router (MR) is aware of the moving when a mobile network moves to a new cAR. The MR works likes an MN and sets the new core AR as its default route. Since the MR and the core AR can communicate with each other using their AIDs, the MR needs not require a CoA. It will require the new core AR to reassign new mappings for the nodes in the mobile network. The nodes in the mobile network are no longer aware of the network they attach to. All the connections based on AIDs will not be interrupt. The benefit from the mapping service and unchanging AIDs is that the problems of nested tunnels and triangle routing will not occur.

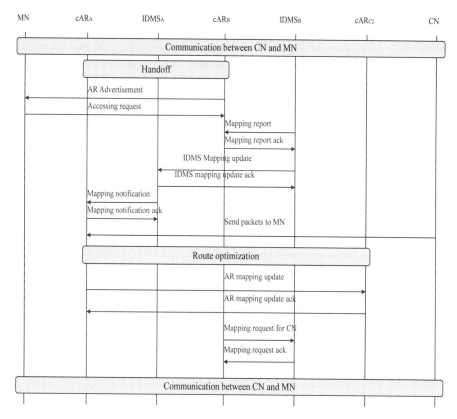

FIGURE 2.16
Macromobility management with ISMS.

2.2.4 Discussion

The operations of many IP mobility management protocols have been described in previous sections in various mobility scopes of mobile nodes and mobile networks. This section concludes with a qualitative comparison among these mobility protocols (see Table 2.1) which are based on the following metrics: (1) low handoff latency (LHL), (2) low signaling overhead (LSO), (3) routing optimization (ROP), (4) location privacy (LoP), and (5) tunneling (TUN). While the LHL metric is normally considered as a QoS metric for real-time applications, LSO, in conjunction with ROP and TUN metrics, are used to evaluate the successful ratio of packet delivery and efficient bandwidth utilization. Finally, the LoP metric ensures location privacy upon user request.

The localized or micromobility of mobile nodes in host-based host mobility (HBHM) between IP subnets under the control of the same authority achieves LHL, LSO, and ROP, and thus, these micro HBHM mobility protocols are subjected to guarantee the required QoS for time-sensitive and real-time applications, but only in the local scope, for example, within the enterprise networks. Tunneling or mobile-specific routing within IP subnets should also be considered in the forwarding packet strategies, which will affect much on the performance of wireless multihop access technologies, such as MANET, with more details in Section 2.4. For global or macro HBHM, MIPv4/v6 is the de facto protocol of mobility management due to its operational independence from underlying and above layers, though signaling overhead and latency handoff are two critical weaknesses. Thus, hybrid HBHM comes to balance for a global mobility solutions. However, the existing work in hybrid HBHM exploits L3/L2 cross-layer interaction without any standardized interface, and thus, the protocols are dependent on the underlying wireless access technologies.

While the existing protocols for HBHM can satisfy different QoS requirements on latency and packet delivery, their success is limited to only theory work, or in the network testbeds. The reason is the modification of IP stacks in mobile nodes to support mobility, and thus, the operations of these HBHM protocols are not safe and reliable from the ISPs perspective. Therefore, the real deployment of HBHM protocols are not popular in 2G/3G mobile networks, and this problem leverages the development of (1) network-based host mobility (NBHM) protocols, in which network devices will control the signaling on behalf of nobile nodes; and (2) NEMO protocols to deal with the mobility of the whole mobile network. However, existing work in NBHM and NEMO protocols provide only basic support for mobility, that is, lack of LHL, LSO, and with or without ROP, and thus, further optimization needs to be developed. In NEMO, such an optimization exploits the BGP announcements to update the mobility of mobile networks, for example, CbB-BGP and WINMO. This new concept is totally different from MIPv4/v6 for macro HBHM, avoiding the tunneling problem and improving much the handoff latency in a global perspective. The limitation is only in the use of BGP, which requires access routers on visited domains, and network routers within the Internet infrastructure uses BGP as interdomain routing protocol.

Location privacy is a new metric that is required to be supported in recent work in hybrid mobility protocols (e.g., ISMS). Since the current IP addresses of mobile nodes in visited networks also show their locations, an additional sublayer is needed for mapping between IP addresses and mobile node identifiers, and for hiding the locations of mobile nodes from upper application layers. This overlayed sub-layer can increase the signaling overhead for mapping between IP addresses and identifiers of mobile nodes.

TABLE 2.1

Comparison of IP mobility management protocols.

Metrics	Host-based Mobility									Network-based Mobility	Network Mobility (NEMO)			Hybrid Mobility
	Macro	Micro					Hybrid							
	MIPv6	HMIPv6	CIP	LLMIPv4	HAWAII	FMIPv4/v6	FHMIPv6	SBF	SMIP	PMIPv6	NEMO-BS	CbB-BGP	WINMO	ISMS
LHL	Y	N	N	N	N	N	N	N	N	Y	Y	Y	N	N
LSO	Y	N	N	N	N	N	N	N	N	Y	Y	Y	N	N
ROP	N	N	N	N	N	N	N	N	N	N	Y	N	N	N
LoP	Y	Y	Y	Y	Y	Y	Y	Y	Y	Y	Y	Y	Y	N
TUN	N	N	Y	N	Y	N	N	N	N	N	N	Y	Y	Y

2.3 IP Mobility Management in 1-Hop WLAN

The fast-handoff and low-latency schemes proposed in IP mobility management in Section 2.2 usually assume that mobile nodes can anticipate link-layer handovers and maintain connectivity with the old as well as new agents, for example, access routers and access points. That is, these optimizations are applicable mainly to soft and backward handovers. While such assumptions are valid for WLANs operating in the ad-hoc or peer-to-peer mode, they do not hold for WLANs running in the infrastructure mode [81].

In the infrastructure mode, a mobile node is associated with only one access point at a time. Although the IEEE 802.11 network interface card (NIC) on a mobile node may be able to access the signal strength information for all neighboring access points, such information is not available to mobility management software. As a result, previous proposals on fast handoff in IP mobility management in Section 2.2, which rely on the ability to anticipate an imminent link-layer handoff through signal strength comparison, cannot be applied to IEEE 802.11 networks operating in the infrastructure mode. Moreover, since the mobile nodes cannot receive packets transmitted from other access points, it is not possible to receive multiple foreign agent advertisements and maintain a list of neighboring foreign agents a priori for future use. Thus, the link-layer handovers in the WLAN infrastructure mode are hard and forward.

A handoff is hard if a mobile node can communicate with exactly one access point before and after a handoff, and it is forward if the mobile node cannot communicate with the old agent during the handoff and has to carry it out by reestablishing a connection with the new foreign agent. Both properties are detrimental to the fast mobile IP handoff. Because the link-layer handoff is hard, a mobile node cannot obtain the address of a new foreign agent before the handoff. Because the link-layer handoff is forward, a mobile node cannot contact the old agent during the handoff and request for additional buffering to minimize data loss during the handoff period [81].

Numerous schemes have been proposed to reduce the WLAN 802.11 handoff latency, which can be classified as either native (only L2) handovers between *Basic Service Sets* (BSSs) within an *Extended Service Set* (ESS), or cross-layer (L3/L2) handovers between ESSs (usually with different IP subnets). In this Section, the WLAN L2 handoff procedure is first described in Section 2.3.1, including the discovery phase and the reauthentication/reassociation phase. Then a classification on different L2 techniques to reduce the WLAN L2 handoff latency, including probe delay, authentication delay, and reassociation delay, is presented in Section 2.3.2. Finally, Section 2.3.3 shows cross-layer L3/L2 techniques to reduce WLAN 802.11 handoff latency and interworking with IP mobility management, for example, MIP. For IP mobility management in 1-hop WLAN, the existing work only focuses on host mobility management, and interworking with IP host mobility management is the scope of Section 2.3.3 in this chapter.

2.3.1 Handoff Procedure in 802.11/WLAN Networks

The handoff process in IEEE 802.11 networks has several phases, each with its own costs. First, a client must determine that it is nearing the periphery of its coverage, and thus, must find an alternative access point to continue. This involves detecting that packets are no longer received successfully. However, typical commercial implementations also monitor the current signal-to-noise ratio (SNR) or current receiving signal strength indicator (RSSI), and initiate the scanning phase when this value passes a predefined minimum threshold. Setting this threshold is something of a black art: if the mobile node waits too long to look for new access points, then it may incur additional disruption, yet if the mobile node is too eager, then it may ping-pong between access points needlessly [77].

2.3.1.1 Pure Layer-2 Handoff Algorithm

Current WLAN products support handoff at layer 2 (L2) with the aid of information from the physical layer (L1). The handoff is mobile initiated. Each mobile node is equipped with a WLAN adapter, and a serving channel will be selected for carrying all the data packets between the mobile node and the corresponding access point (AP). The mobile node, while staying in the radio coverage of the access point, will periodically check the current RSSI and calculate the current frame error rate (FER) on the serving channel. Besides, the mobile node actively scans all other channels for their receiving signal strength periodically. At any time, if the quality of the serving channel falls below certain predefined thresholds, the mobile node will decide to start a handoff process and look for a new serving channel. The pure layer-2 handoff procedure is illustrated in [92].

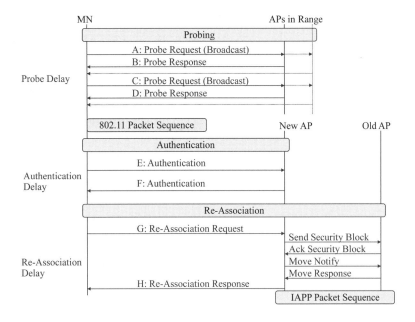

FIGURE 2.17
Handoff procedure in 802.11 networks.

Figure 2.17 shows the sequence of steps that are designed to occur during a WLAN handoff. The first step (not indicated in Figure 2.17) is the termination of a mobile node association to the current AP. Either entity can initiate a disassociation for various reasons. Due to mobility or degradation of physical connectivity (signal strength), it might not be possible for the mobile node or the access point to send an 802.11 disassociate packet. In such cases, a timeout on inactivity or communication between APs or the receipt of an interaccess point protocol (IAPP) Move-Notify packet terminates the association [60].

During the second step, the mobile node scans for access points by either sending probe request packets (in active scanning) or by listening for beacon packets (in passive scanning) broadcasted by APs on the channels of interest. Packets A, B, C, and D in Figure 2.17 show active scanning. For each channel, a probe request (packets A and C) is sent by the mobile node, and probe responses (packets B and D) are received from the APs in the vicinity of the mobile node. After scanning all intended channels, the mobile node selects the new AP based on the data rates and signal strength (see the pure layer-2 handoff procedure above as an example). Probe delay is the time spent by the mobile node in scanning and selecting the next AP.

In passive scanning, mobile nodes listen to each channel for the beacon frames (typically every 100 ms). The main inconvenience of this method is how to calculate the time to listen to each channel. This time must be longer than the beacon period, but the beacon period is unknown to the mobile node until the first beacon is received. Incidentally, the mobile node cannot switch to another channel when the first beacon arrives and has to wait for the whole beacon period because several access points of different WLANs can operate in the same channel. Since the standard mandates that the whole set of allowed channels must be scanned, mobile nodes need over a second to discover the access points in range with the default 100 ms beacon interval (e.g., there are 11 allowed channels in USA, thus it would take 1.1 seconds).

When faster scanning is needed, mobile nodes must perform active scanning. Active scanning means that mobile nodes will broadcast a probe-request management frame in each channel and wait for probe responses generated by access points. Each mobile node should wait for the responses in each channel within an interval, which is controlled by the two timers: MinChannelTime and MaxChannelTime. The first is the time to wait for the first response in an idle channel. If there is neither response nor traffic in the channel during MinChannelTime, the channel is declared empty, that is, no access point in range. The second timer, MaxChannelTime, indicates the time to wait in order to collect all responses in a used channel. This limit is used when there was activity in the channel during MinChannelTime. Both timers are measured in Time Units (TU), which the IEEE standard defines to be 1024 microseconds. The exact values of these timers are not given in the standard [87].

After the probe exchange step, the mobile node and new AP exchange 802.11 authentication packets, and the latency incurred is the authentication delay (packets E and F). After authentication, the mobile node sends an 802.11 reassociation request to the AP (packet G) and receives a reassociation response from the AP (packet H), which completes the handoff process. The latency incurred during this exchange is the reassociation delay and this process is called the reassociation process. During reassociation, the APs involved exchange mobile node context information. This is achieved via the use of the IAPP [2].

The mobile node attempts to reauthenticate to an AP according to the priority list. The reauthentication process typically involves an authentication and a re-association to the posterior AP. The reauthentication phase involves the transfer of credentials and other state information from the old-AP. As mentioned earlier, this can be achieved through a protocol such as IAPP [2]. IAPP plays a significant role during a handoff. The two main objectives achieved by interaccess point communication are

1. Maintaining a single association of a mobile node with the wireless network.

2. Securing transfer of state and context information between APs involved in a reassociation. The client context information can include, but is not limited to, IP flow context; security context; QoS information (e.g., differentiated service or integrated service, header compression) and authentication, authorization, and accounting (AAA) information [60].

Association and reassociation events change a mobile node point of access to the network. When a mobile node first associates to an AP, the AP broadcasts an Add-Notify packet notifying all APs of the mobile node association. Upon receiving an Add-Notify packet, the APs clear all stale associations and the state for the mobile node. This enforces a unique association for the mobile node with respect to the network. When a mobile node reassociates to a new-AP, it informs the old-AP of the reassociation using IAPP packets. At the beginning of a reassociation, the new-AP can optionally send a Security Block packet to the old-AP, each of which acknowledges with an Ack-Security-Block packet. This packet contains security information to establish a secure communication channel between the APs. The new-AP sends a Move-Notify packet to the old-AP requesting mobile node context information and notifying the old-AP of the reassociation. The old-AP responds by sending a Move-Response packet. Although IAPP communications serve to fulfill the mandatory distribution system functions, they invariably increase the overall handoff latency because of their reactive nature [60].

2.3.2 Native Layer-2 Handoff in WLAN

Based on the description of 802.11 operations and its handoff procedure in Section 2.3.1, different methods have been proposed to reduce the L2 handoff latency in WLAN 802.11, which can be classified as either proactive or reactive (see Figure 2.18). In [62], the authors use the global positioning system (GPS) in handoff management. The idea is to predict the next mobile node point of attachment and the associated subnetwork using the position of the mobile node. The selection of the next mobile node AP prior to the handoff allows the configuration of all the required parameters on the mobile node to reduce the L2 discovery phase and L3 new link detection.

The proactive context caching and forwarding technique [60] introduces a novel and efficient data structure, the neighbor graph, which dynamically captures the mobility topology of a wireless network through real-time examination of the handovers occurring in the network. The neighbor graph is used

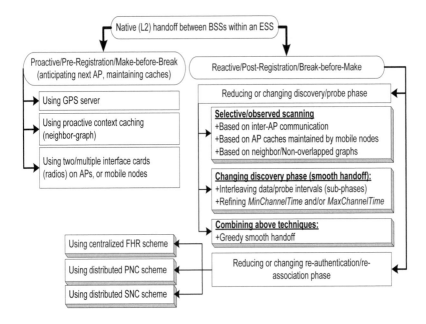

FIGURE 2.18
Classification of methods to reduce handoff latency in WLAN.

to provide the new-AP with the mobile node context prior to the handover, or proactively. Thus, this technique reduces the reassociation latency.

A novel AP with two transceivers to improve network efficiency in terms of supporting seamless handoff and balancing the traffic load is shown in [55]. In this scheme, the novel AP will use the second transceiver to scan and find neighboring mobile nodes in the transmission range and later send the results to its associate AP. This information will be useful for the AP to control its associated mobile nodes to initiate the handoff process when a neighbor AP can provide higher quality and/or share the traffic load with neighbor APs. Ramachandran *et al.* [76] introduces the novel concept of applying the make-before-break mechanism to 802.11 MAC layer handoffs. It optimizes the L2 handoff by periodically probing in the background, which uses a second radio card, to gather neighborhood information even when it is already connected to an AP. An empirical analysis of the 802.11 MAC (L2) handoff process carried out in [59] has shown that the probe phase is the significant contributor to the handoff latency. Thus, most reactive schemes concentrate on reducing the probe delay, while some focus on reducing the authentication and/or re-association delay.

Shin *et al.* [83] concentrates on reducing probe delay. First, probe delay is reduced by improving the scanning procedure using a selective scanning algorithm. Second, the number of times in the previous scanning procedure is minimized using the AP caching mechanism. Specifically, full-scan is triggered at first and the channel mask is then constructed by the information obtained in the first full-scan. In 802.11b, only three channels do not overlap among all the 11 channels. Hence, in a well-configured wireless network, almost all APs operate on channels 1, 6, and 11. Thus, the channel mask is formed by combining these three frequency channels. By using this channel mask, a mobile node can reduce the amount of unnecessary time that it spends on probing nonexistent channels among neighboring APs. To reduce further the handoff delay, a cache mechanism is also introduced. The basic idea of the cache mechanism is for each mobile node to store its handoff history. When a mobile node associates with an AP, the AP is inserted into the cache maintained at the mobile node. When a handoff is needed, the mobile node first checks whether there is an entry corresponding to the current AP MAC address in the cache. If there is a matched cache entry, the mobile node can associate with the AP without any further probing procedures [67].

In [82], the authors describe the use of a novel and efficient discovery method using neighbor graphs and non overlap graphs to reduce the total number of probed channels and the total time spent waiting on each channel. Using neighbor graph, the set of channels on which neighboring APs are currently operating and the set of neighbor APs on each channel can be learned. Based on this information, a mobile node can determine whether a channel needs to be probed or not. On the other hand, the non overlap graph abstracts the non overlapping relation among APs. Two APs are considered non overlapped if and only if the mobile node cannot communicate with both of them simultaneously with acceptable link quality. Therefore, if a mobile node has received a probe response frame from an AP_i, it implies that this mobile node cannot receive a response frame from an another AP_j if AP_i and AP_j are nonoverlapping each other. By means of the non overlap graph, a mobile node can prune some of the APs, which are non overlapped with the current APs that have already responded [67].

Reference [51] presents a smooth MAC layer handoff scheme and a greedy smooth MAC layer handoff scheme. In the former, the scan channel phase is divided into multiple subphases. The mobile node can use the interval between two consecutive sub-phases to send and receive data frames. Obviously, this can reduce the packet delay and jitter during the scanning channel phase, which is important for time-critical applications such as VoIP. The latter scheme not only scans channel smoothly but also reduces the number of channels being scanned. Reference [87] shows a tuning scheme using the packet loss distribution caused by collisions in order to determine the optimal handoff trigger timing for a new AP. To reduce the handoff detection time, the mobile node starts the channel probe procedure as soon as it deems that collision can be excluded as a reason for packet transmission failure. In other words, based on the probability distribution, if a packet and its next two consecutive retransmissions fail, the mobile node concludes that the packet failure is caused by the mobile node's movement instead of collision and therefore a further handoff process is required. In addition, they leverage the active scan mode and derive new values for MinChannelTime and MaxChannelTime from their measurement results and analytical models [67].

Even though the probe delay takes a large portion of the total handoff delay, the authentication/reassociation delay should be also reduced to achieve seamless mobile services. Actually, in public WLAN services, the authentication scheme based on the centralized authentication server is widely adopted for the sake of secure service and efficient accounting. In such an environment, the authentication/reassociation delay may be higher than that observed in the case where the open authentication procedure is employed. Different fast authentication methods in IEEE 802.11 networks are overviewed and analyzed in [67] in terms of network architectures and trust models. Reference [68] proposes a fast handoff scheme that minimizes the re-authentication latency in public wireless LANs. When a mobile node sends an authentication request, the AAA server authenticates not only the currently used AP but also multiple other APs, and sends multiple WEP keys to the mobile node. A centralized algorithm is used to select these APs. This selection algorithm is based on the frequent handoff region (FHR), traffic pattern, and user mobility characteristics within the public WLAN.

Instead of using the centralized system, a proactive scheme based on a distributed cache structure, called Proactive Neighbor Caching (PNC) [60], is introduced. This scheme uses a neighbor graph, which dynamically captures the mobility topology of the wireless network for prepositioning a mobile node context. The scheme ensures that the mobile node's context is always dispatched one-hop ahead and therefore handoff delay can be substantially reduced. The neighbor graph is constructed using the information exchanged during the mobile node's handoff, and it is maintained at each AP in a distributed manner. The propagated mobile node context is stored in the cache. Recently, this scheme has been included in the IAPP specification [2]. Reference [69] enhances the PNC scheme by adding a new concept of neighbor weight. The neighbor weight represents the handoff probability for each neighboring AP. Based on the neighbor weight, the mobile node context is propagated only to the selected neighboring APs. The neighbor graph and its neighbor weights can be constructed by monitoring the handoff patterns among the APs.

2.3.3   Cross-Layer Handoff in WLAN

The problems with mobile wireless Internet arise even at the availability of MIP and WLAN. One of them is that MIP and WLAN solve their problems independently at different layers, that is, MIP works at IP layer (L3) and does not talk to L2, and vice versa. Another drawback is that there is no standard or guidelines on cell planning for the Internet in the wireless environment. Cell planning is essential in cellular networks such as GSM. With cell planning, each base station is given a neighbor list consisting of the network candidates to which a mobile node can be handed over. Without such planning, a mobile node does not know where the neighbor is. It also does not know which radio channel to tune into at the lower layers [92].

When an IEEE 802.11-based wireless network is configured in the infrastructure mode, each mobile node is associated with the access point of the WLAN segment in which it currently resides. The link layer (L2) handoff implementation in infrastructure-mode wireless LANs poses two problems in reducing MIP handoff latency:

1. Because L2 handoff is transparent to software, it is not possible for the MIP code to detect its occurrence immediately.

2. Because a mobile node can only receive packets from one access point at a time in the infrastructure mode, the MIP code can only receive advertisements from the current foreign agent but not those from neighboring foreign agents. As a result, even if a mobile node can detect the L2 handoff immediately without advertisements from the new foreign agent, it still does not know which foreign agent to contact to facilitate the L3 handoff process.

To reduce MIP handoff latency, let us first identify the individual delay components involved. An MIP handoff period for an infrastructure-mode WLAN can be divided into four distinct subperiods [81]:

1. Time between when a link-layer handoff takes place and when it is detected by the mobile IP software.

2. Time from link-layer handoff detection to the reception of first mobile IP advertisement from the new foreign agent.

3. Time required for a mobile node to register with the new foreign agent after receiving the first advertisement.

4. Time between request sent to and response returned from the new foreign agent.

To minimize the overall handoff latency, each of these delay components needs to be reduced as much as possible. A low-latency mobile IP handoff scheme is proposed in [81] that can reduce the handoff latency of infrastructure-mode wireless LANs to less than 100 ms. This scheme expedites link-layer handoff detection based on the access-point ID probing supported by Wi-Fi cards, and speeds up network-layer handoff by replaying cached foreign agent advertisements. The proposed scheme strictly adheres to the MIP standard specification, and does not require any modifications to existing MIP implementations. A new architecture is described in [92], which is compatible with MIP and most L2 wireless networks. The solution consists of three major extensions: (1) packet buffering, (2) neighbor list update, and (3) L2 handoff notification. The effect of this enhancement provides a linkage between different layers for preventing packet loss and reducing handoff latency.

Reference [94] proposes a new low latency handoff method where access points used in a wireless LAN environment and a dedicated MAC bridge are jointly used to alleviate packet loss without altering MIP specifications. Reference [25] introduces the fast hinted cell switching movement detection method for MIP. This method assumes that the L2 is capable of delivering information, that is, the IEEE 802.11 SSID field was utilized to contain information important to MIP, such as the identity of the local mobility agent. This tends to negate the need for MIP movement detection as well as agent discovery, and leads to accelerated MIP handovers. Moreover, the existence of the L2 information renders the presence of periodic mobility agent broadcast advertisements unnecessary, which enables a more efficient utilization of network capacity.

Finally, and last but not least, reference [56] describes how an FMIPv6 could be implemented on link layers conforming to the 802.11 suite of specifications. This work which requires little effort, attempts to give a set of examples for FMIPv6 over 802.11 networks and examine how and when handover information might become available to the IP layers that implement fast handover, both in the network infrastructure and on the mobile node.

2.4   IP Mobility Management in MANET

Attempts are in progress to connect mobile ad-hoc networks (MANETs) to the Internet infrastructure to fill in the coverage gaps in the areas where the first-hop coverage, for example, WLANs or cellular networks, is not available. In the very near future, mobile nodes will roam across multiple heterogeneous platforms while continuously maintaining session connectivity. A mobile node may connect to a WLAN and then move into an area where the coverage from the WLAN does not exist. There, it may reconfigure itself into ad-hoc mode and connect to a MANET. Essential to such seamless mobility is efficient mobility management. This section focuses on IP host mobility management in n-hop MANET, in conjunction with the integration of MANETs with IP networks, and how MANET adapt to the network functionalities within IP networks.

2.4.1   Comparison of 1-Hop and n-Hop Access Networks

The location of a node in the Internet is identified by the network part of its IP address. When a packet arrives at the router connecting to the 1-hop WLAN hosting the destination, the destination IP address is converted to a MAC address. This conversion uses either the address resolution protocol (ARP) in IPv4, or the neighbor discovery protocol (NDP) in IPv6, before the packet is sent in a frame in the last hop using the MAC address as the identifier of the destination. The MAC address represents a flat address space without information about the host locality within the WLAN [5].

Considering that the n-hop MANET has a flat address space, it can be seen as either a major network or subnetwork within the Internet. This will identify a MANET connected to the Internet by its own network number. There is, however, a major difference between a WLAN and a MANET. Mobile nodes connected to a WLAN are within the same broadcast domain and are managed as 1-hop connections by the IP protocol. A packet broadcasted from a mobile node in the WLAN will reach all other mobile nodes connected to the network.

In the MANET, however, a broadcast sent by one MANET node may not reach all other MANET nodes. A broadcast needs to be retransmitted by immediate MANET nodes in the network so that it will reach all MANET nodes. A broadcast in the MANET running the IP protocol uses the time to live (TTL) value to limit the spreading of a packet. A packet TTL, when arriving at a router connected to a WLAN, requires a value of "1" to reach a node connected to the network. In the MANET, a TTL of "1", when forwarded on the MANET, will be discarded after the first hop. This behavior needs to be managed by gateways connecting MANETs with the Internet. Instead of using the ARP/NDP, the MANET routing protocol need to be used in what is defined as the last hop in the Internet. The functionality in reactive ad-hoc routing protocol maps well to the functionality in the ARP/NDP protocol, with a request for the mobile node in the last hop and a soft state table. For example, the route request (RREQ) in the reactive MANET routing protocol can be compared to the ARP request, and the soft-state MANET routing table to the ARP table [5].

A node performs a handoff if it changes its Internet Gateway (IGW) while communicating with a correspondent node (CN) on the Internet. In conventional mobile networks, for example, WLAN in Section 2.3, the quality of the wireless link between mobile node and base stations (i.e., APs) determines when to handoff from one AP to another. The performance of these types of handoffs depends on the mobility management protocol in the access network. In MANETs, the situation is more complicated. In general, some nodes do not have a direct wireless link to an AP, but they are connected via other immediate nodes. Thus, they cannot initiate handoffs that are based on the link quality to the AP. Rather, the complete multihop path to the AP, which serves as the current IGW, must be taken into consideration. A handoff can occur if the mobile node itself or any of the intermediate relay nodes moves and breaks the active path. In general, if the path between a mobile node and the IGW breaks, and there is no other path to the same IGW, the mobile node has to perform the gateway discovery to establish a new path to another IGW [61].

2.4.2   Functionalities of IP Mobility Management in MANET

In this section, the required functions of providing Internet access and IP host mobility management for MANET nodes are described first. They include: (1) MANET node location determination, (2) IGW discovery, (3) IGW selection, (4) IGW forwarding strategy, (5) address autoconfiguration, and (6) handoff-style. Next, the related work is discussed following the descriptions of the above functions.

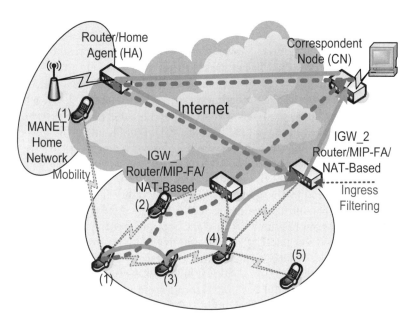

FIGURE 2.19
Overview of mobility management in MANET.

Figure 2.19 illustrates a typical mobility scenarios of a mobile node (i.e., node (1)) in MANET while connecting to the Internet. This is a scenario where the above functionalities are needed. In Figure 2.19, MIPv4 is used for macromobility management, that is, between MANET domains, while ad-hoc routing is used for micromobility management, that is, within each MANET domain. This is a popular solution for IP mobility management in MANETs [47].

2.4.2.1   Node Location Determination

MANET node location determination is the function that allows a source MANET node to determine whether a destination node is located within the same MANET domain as the source node or outside the MANET domain (e.g., on an external network, such as an Internet host). The function can be implemented by one of the following methods:

- **Network prefix**. All MANET nodes share the same network prefix. With this method, each MANET node must be assigned a global unicast IP address, both home of address (HoA) and care-of address (CoA), that is, the MANET node address is topologically correct [73].

- **Routing table** in MANET proactive ad-hoc routing protocols [16, 70]. If an entry for the destination is in the routing table of the source MANET node, the destination is either in the same MANET domain or an Internet host reachable via the IGW. Otherwise, the destination is unreachable.

- **Flooding** the route request (RREQ) and waiting for the route reply (RREP) in reactive ad-hoc routing protocols [72, 71]. If a host route is returned, the destination is in the same MANET domain. If a default route is returned, the destination is either unreachable or an Internet host reachable via the IGW.

- **Internet gateway**. The IGW responds to an RREQ, sending a proxy RREP to signal that it can route to the requested destination, that is, analogous to the functionality of a proxy ARP but over the multihops. To do this, an IGW must determine that the destination is not in the same MANET by keeping a list of currently known active nodes, called visitor list, or by pinging the destination on the IGW network interface attached to the Internet, or by flooding the whole MANET with a new RREQ [4, 73].

2.4.2.2   Internet Gateway Discovery

Internet gateway discovery is the function that allows a MANET node to discover an IGW to which traffic bound for the Internet can be forwarded, and from which traffic returned from the Internet can be received. The different discovery mechanisms can be classified into three subclasses: proactive, reactive, or hybrid [35]. In the proactive approach, each IGW periodically broadcasts an advertisement, while in the reactive approach a MANET node sends a solicitation and waits for a reply from the IGW. The former requires much overhead traffic on the MANET, while the latter entails the longer discovery delay. The hybrid approach compromises with the balance, in which each IGW periodically broadcasts the advertisement within the radius of n-hops. MANET nodes that are located further than n-hops away from the IGW must use the reactive approach to discover the IGW [80].

2.4.2.3 Metrics for Internet Gateway Selection

Internet gateway selection is a function used when a MANET node discovers multiple IGWs for accessing the Internet. A metric is normally needed in order to select the right one. Different metrics can be used:

- **Shortest hop-count**. To the nearest IGW [73].

- **Load-balancing**. For intra-MANET traffic, choose different immediate relay nodes to destination MANET nodes within the same MANET domain [79, 14], while for inter-MANET traffic, choose different IGWs for forwarding traffic from MANET to the Internet and vice verse [32].

- **Service class**. Depending on the service classes and data caches provided and managed by each IGW, respectively [13], as well as the wireless link quality among MANET nodes and IGWs [64].

- **Euclidean distance**. Spatial distance between the MANET node and the IGW [7].

- **Hybrid**. A combination of some of the above metrics [7].

2.4.2.4 Internet Gateway Forwarding Strategies

Internet gateway forwarding strategies are a function that takes the responsibility to forward traffic within the MANET, out of the MANET to the Internet, or from the Internet into the MANET. Typically, it can be classified into inter-MANET and intra-MANET forwarding strategies. The inter-MANET forwarding strategies use different approaches as follows:

- **Default routes**. Representing the default next-hop to send packets to that do not match any other explicit entry in a MANET node's routing table. Usually, the default route is used to forward packets toward an IGW, where packets are further forwarded toward the destination on the Internet [73, 9].

- **Tunneling (or encapsulation)**. Usually, the IP-in-IP encapsulation technique is used to get traffic into and out of the MANET. The outer IP header is for the tunneling connection between the source MANET node and the IGW, while the inner IP header is for the connection between the source MANET node and the destination [38].

- **Half-tunneling**. Traffic to the Internet from the MANET domain uses tunneling, while traffic from the Internet to the MANET domain uses ad-hoc forwarding without tunneling [38].

- **Source routing**. A list of all intermediate nodes between the source MANET node and the IGW are added into the IP header. At the IGW, the source routing header is removed, and the packet is forwarded further to the Internet as a normal packet [11].

- **Spanning tree rooted at the IGW**. A tree rooted at the IGW is built and maintained using the agent advertisements broadcasted periodically by the corresponding IGW [24].

The intra-MANET forwarding strategies, on the other hand, are based entirely on the operation of ad-hoc routing protocols. These can be classified as proactive, or reactive, or hybrid. In the proactive approach, each node continuously maintains up-to-date routing information to reach every other node in the network. Routing table updates are periodically transmitted throughout the network in order to maintain table consistency. Thus, the route is quickly established without any delay. However, for a highly dynamic network topology, the proactive schemes require a significant amount of resources to keep routing information up-to-date and reliable. In the reactive approach, a node initiates a route discovery throughout the network only when it wants to send packets to its destination. Thus, nodes maintain the routes to only active destinations. A route search is needed for every new destination. Therefore, the communication overhead is reduced at the expense of delay due to route discovery. Finally, in the hybrid approach, each node maintains both topology information within its zone via the proactive approach and the information regarding neighbor zones via the reactive approach.

2.4.2.5 Address Autoconfiguration for MANET

In order to enable a MANET to support IP services and internetworking with the Internet, a MANET address space based on IPv4/IPv6 is required. Moreover, the MANET addressing schemes must be autoconfigured and distributed to support the self-organized and dynamic characteristics of MANETs. Numerous addressing schemes for MANETs based on IP address autoconfiguration have been proposed in the literature. They can be classified into two approaches: conflict-detection allocation and conflict-free allocation [91].

Conflict-detection allocation mechanisms are based on picking an IP address from a pool of available addresses, configuring it as tentative address and asking the rest of the nodes of the network, checking the address uniqueness, and requesting for approval from all the nodes of the network. In case of conflict, for example, the address has been already configured by another node, the node should pick a new address and repeat the procedure (as a sort of "trial and error" method). This process is called duplicate address detection (DAD). Conflict-free allocation mechanisms, on the other hand, assume that the addresses are delegated uniquely and that they are therefore not being used by any other node in the network. This can be achieved by ensuring that the nodes that delegate the addresses, have disjointed address pools. In this way, there is no need to perform the DAD procedure.

Research has also been in progress to apply the IP address autoconfiguration scheme for the addressing of MANETs. However, only the stateless mechanism is suitable for MANETs [10]. This is because the stateful mechanism requires a centralized server to maintain a common address pool, while the stateless mechanism allows the node to construct its own address and is suitable for self-organized MANETs. However, to use the IP address stateless autoconfiguration scheme for MANET addressing, that is, a conflict-detection allocation approach, a DAD mechanism is required to assure the uniqueness of the address with multihop distance, especially to support MANET merging and partitioning.

Finally, the address allocation space is important. It must be large enough to cover the large-scale MANETs and reduce the probability of address conflicts. The following IPv4 and IPv6 addressing spaces have been proposed for MANETs [73]: 169.254.0.0/16 for IPv4, and FEC0:0:0:FFFF::/64 for IPv6 as MANET IP PREFIX.

2.4.2.6   Handover Style

A node performs a handover if it changes its IGW while communicating with a CN on the Internet. In MANETs, mobile nodes do not have a direct wireless link to an AP, and thus, they cannot initiate handoff that are based on the link quality to the AP. A handoff can occur if an ad-hoc node itself or any of the intermediate relay ad-hoc nodes moves and breaks the active path. In general, if the path between an ad-hoc node and IGW breaks, and there is no other path to the same IGW, the ad-hoc node has to perform an IGW discovery to establish a new path to another IGW.

The IGW discovery scheme and the ad-hoc routing protocol both have huge influence on multihop handoff performance. Multihop handoff schemes can be classified into forced handoff and route-optimization-based handoff. The former occurs whenever the path between the source/destination mobile node and the IGW is disrupted during data transmission due to, for example, the movement of the MANET node. Therefore, a new path to the Internet has to be set up. The following IGW discovery process may result in the detection of a new IGW that will consequently result in a handoff. The latter is a handoff that results from route optimization. If the source/destination MANET node detects that a shorter path to the Internet becomes available while communicating with a corresponding node, the active path will be optimized. In case the shorter path goes via a different IGW, a route-optimization-based handoff occurs.

2.4.3   Overview of IP Mobility Management in MANET

Table 2.3 summarizes the comparison of different approaches to providing Internet connectivity and mobility management for MANETs based on the description of the required functions in Section 2.4.2. The acronyms in Table 2.2 are used for this comparison.

- **MIPv6 + AODVv6**. Globalv4 [8] describes MIPv4 extensions for the AODV routing protocol. Destinations are first searched for in the MANET. If none is found, a host route is set up to the IGW. This

TABLE 2.2 Protocols acronyms and abbreviations.

Scheme	Description
TBBR	Tree-based bidirectional routing [24]
OLSR	Optimized link-state routing [16]
AODV	Ad-hoc on-demand distance-vector routing [72]
DSDV	Destination-sequenced distance-vector routing [70]
DSR	Dynamic source routing [36]
NAT	Network address translation
DHCP	Dynamic host-control protocol
TD/RD	Table-driven/root-driven routing [24]
MEWLANA	Mobile IP enriched wireless local area network architecture [24]

solution suffers from long route discovery delays and lack of the same route aggregation that half-tunneling provides. Similarly, Globalv6 [89] can also work with MIPv6 [37], but it is not mandatory. A node may acquire a network prefix from an IGW and construct a globally routable IP address through IPv6 stateless address autoconfiguration. Globalv6 employs a similar technique as Globalv4 to determine the locality of destinations. Routing toward the IGW is done on a hop-by-hop basis using a default route of AODVv6 [71]. Cascading effects [48], that is, all immediate MANET nodes on the chain from the source MANET node to the IGW need to flood RREQ to determine whether the destination is located in the same MANET, are avoided by requiring intermediate nodes to configure host route entries for Internet destinations, with the downside of losing route aggregation. A summary of Globalv4 and Globalv6 is also presented in [73].

- **MIPMANET** [38] studies the integration of MIPv4 in the MANET. Tunneling from ad-hoc nodes to the foreign agent (FA) is proposed as a way to achieve default route-like behavior. This is the half-tunneling approach, where the outbound traffic to the Internet from the MANET uses tunneling, and the inbound traffic from the Internet to the MANET is delivered to the corresponding destination MANET node via the host route using the AODV routing protocol. However, this work does not explore the benefits of using tunneling but studies different approaches to disseminating MIP information in the MANET instead.

- **MIP + DSR**. A technique is described in [11] to integrate MANETs with the Internet and to use MIP to support the migration of nodes between MANET and the Internet. Local delivery within a MANET subnet is accomplished using the DSR routing protocol, while standard IP routing mechanisms decide which packets should enter and leave the subnet. For sending packets from the MANET subnet to the Internet, a source MANET node uses the source routing header of DSR to forward

TABLE 2.3

IP mobility management mechanisms for MANET.

Mechanism	Location determination	IGW discovery	IGW selection metrics	IGW forwarding	Addressing	Handoff style
MIPv6+ AODVv6	Network prefix	Proactive & reactive	Not specified, implicitly shortest hop-count	Default route & AODVv6	Deriving from IPv6 stateless auto-configuration	Both
MIPMANET	Flooding RREQ	Proactive	Shortest hop-count	Half-tunneling & AODV	Not specified, but Home Address must be IP global unicast	Route optimization-based
MIP+DSR	Using IGW	Reactive	Not specified, implicitly shortest hop-count	Source routing & DSR	Home Address must be IP global unicast	Not specified, implicitly route optimization-based
MIP+OLSR	Using routing table	Proactive	Not specified, implicitly shortest hop-count	Default route & OLSR	Not specified	Forced (when a prefix change)
MEWLANA TD/RD	Using routing table (DSDV) or TBBR tree	Proactive	Shortest hop-count	Default route & DSDV or TBBR	Not specified	Forced (when a route change or node leave)
Two-tier MANET	Using routing table	Reactive	Load-balancing	Tunneling uses extra UDP/IP header & DSDV	Private address & NAT, allocating using DHCP	Not specified, implicitly route optimization-based
Hybrid MANET	Using routing table	Reactive	Hybrid: Euclidean distance & load-balancing	Default route & DSDV	Not specified	Forced (using automatic mode-detection and switching)
WLAN & MANET	Using routing table	Proactive	Not specified, implicitly shortest hop-count	Default route & OLSR	IPv6 stateless auto-configuration	Forced (using automatic mode-detection and switching)

the packet to the IGW, where the source routing header will be removed. The packet is then forwarded to the Internet. However, this technique requires that each MANET node select a single IP address (its home address) from the ones assigned to it and that it use only that address when participating in the DSR protocol.

- **MIP + OLSR**. The management of universal mobility, including both large-scale macromobility and local-scale micromobility, is the focus of [9]. A hierarchical architecture is proposed, including (1) extending micro-mobility management of a wireless access network to a MANET, (2) connecting this MANET to the Internet, and (3) integrating MIP and the OLSR routing protocol to manage universal mobility. In addition, it uses the optimal default route via the IGW to reach a host outside the MANET, that is, the Internet host. The traffic to and from the Internet is distributed between base stations (APs) of the local network.

- **MEWLANA TD/RD**. In addition to on-demand and table-driven routing protocols in MANETs, a novel ad-hoc routing type called root-driven routing is introduced [24]. Using this protocol type for networks where the intensity of inside traffic is negligible makes the protocol efficient by eliminating the routing overhead. The main idea of this routing type is formation of a tree whose root is the foreign agent and branches are mobile nodes, and periodical initiation of this tree formation procedure by the root. To take into account these different cases, two protocols called MEWLANA-TD, which uses the table-driven routing type, and MEWLANA-RD, which uses the root-driven routing type, are designed. MEWLANA-TD uses the DSDV routing protocol, in which there is a trigger updating either periodically or when there is a change in the routing table. DSDV enables each node to have an entry in their routing table for all other nodes. MEWLANA-RD uses tree-based bidirectional routing (TBBR) as the routing protocol. TBBR is a special routing protocol designed only by using MIP entities, introducing low overhead at the expense of performance degradation. Two-tier MANET [32] shows a seamless roaming and load-balancing routing capability for providing Internet connectivity to the MANET. It modifies MIP to make traversing private networks, that is, using NAT. It also proposes a load-balancing routing protocol to improve the Internet access quality by allowing mobile nodes dynamically change their IGWs, thus relieving the bottleneck problem.

- **Hybrid MANET** [7] proposes an architecture to provide MANET nodes with Internet access using fixed IGWs and exploiting the mobility capability of additional mobile nodes (mobile IGWs). Since the Internet access for MANET nodes is provided via mobile IGWs, the quality of such service depends on the selection procedure used by MANET nodes to choose the most convenient mobile IGWs and register with. This work suggests using a hybrid criterion based on the weighted sum of the Euclidean distance between MANET nodes and mobile IGWs, and the load of mobile IGWs measured as the number of MANET nodes currently registered with them. MIP and DSDV are extended to integrate the suggested hybrid criterion.

- **WLAN and MANET** [43] presents a novel approach to integrating WLAN and MANET to the Internet (IPv6). The OLSR routing protocol is used within the MANET, and IGWs are used to connect the MANET to the Internet. In addition, automatic mode-detection and switching capability are also introduced in each MANET node to facilitate hand-offs between WLAN and MANET. Mobility management across WLAN and MANET is achieved through MIPv6, which is integrated into the extended functionality of OLSR.

2.4.4   Discussion

A proactive approach to providing Internet connectivity to a MANET relies on ensuring that all nodes are registered with a foreign agent at all times. Mobile IP uses on-link layer broadcasts to provide foreign agent information to interested nodes. However, these broadcasts can prove to be extremely expensive in a MANET where a broadcast means that packets are being flooded throughout the network.

In contrast, in a purely reactive approach, mobile nodes obtain foreign agent information by sending out agent solicitations only when data needs to be sent to a node outside the MANET. To limit the amount of flooding, these solicitations might be piggybacked on RREQ packets. An expanding ring search might also be used. In addition, intermediate nodes are allowed to reply with a route to the foreign agent, which reduces the overhead further.

The hybrid approach provides Internet access to MANETs while attempting to balance the proactive and reactive approaches, and it has many benefits. A proactive solution allows mobile nodes to find the foreign agent closest to them and enables better handoffs, which in turn leads to lesser delay. Periodic registrations in such a proactive scheme help foreign agents track the mobility of the mobile node. However, if not all the nodes in the MANET require connectivity, the repeated broadcasting of agent advertisements and solicitations can have a negative impact on the MANET due to excessive flooding overhead. A hybrid approach combines the advantages of both approaches so that the required information is received in a timely fashion and the MANET's scarce resources are not further burdened with the MIP overhead.

Two strategies, default routes and tunneling, are usually used to integrate gateway forwarding with ad-hoc routing protocols. It is found that default routes that are adopted from traditional LAN settings need modifications to work in a multihop ad-hoc environment. Despite these additions, default routes have problems with multiple gateways and inconsistent routing state. Establishing tunnels to the gateways, either half-tunneling or tunneling, on the other hand, provides an architecturally appealing solution and works well with multiple Internet gateways [48, 22, 23].

2.5 Issues and Solutions for Mobility Management

2.5.1 Mobility Management Issues in MANET

In this section, we assume that MANET mobility management uses MIPv4 for macromobility and the ad-hoc routing protocol for micromobility. The data traffic sent from a MANET node to an Internet host is forwarded through IGWs using either the default route or tunneling techniques [4, 23, 38]. There are also multiple IGWs and NAT devices, together with their ingress filtering policies between the MANET domain and the Internet. Figure 2.20 shows different scenarios on IGW forwarding strategies, which are used to illustrate the next different discovered problems.

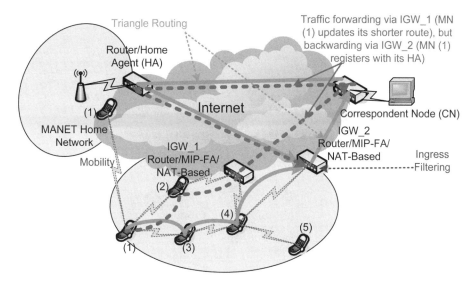

FIGURE 2.20
Issues on Internet gateway forwarding strategies.

2.5.1.1 Inconsistent Context due to Default Route Forwarding

1. **Type I**: In this forwarding mode, a MANET node (MN) sends packets to the Internet, that is, communicate with the CN, using the default route. Whenever it associates with an IGW, it sets its default route pointing to this IGW. The default route is used to forward data packets to the unknown destination. In the scope of MANET, it usually means that the destination is not in the same MANET domain and can be located in the Internet, and thus the data packet is forwarded to the IGW where it is dropped if the destination is unreachable, or continued to be forwarded to the Internet if the destination is reachable via IGW.

   The MN (5) routing table using the default route in Figure 2.21 shows an example, where MN (4) is used as the next-hop to the associated IGW_1. The problem with this default route setting is that MN (5) does not know its current associated IGW. Thus, the inconsistent context on sending packets from MN (5) to the Internet, for example, via IGW_2, and receiving packets from CN to MN (5), for example, via IGW_1, can terminate the two-way connection such as TCP. The next section shows an advanced setting to reduce this effect.

2. **Type II**: In this forwarding mode, a MANET node adds an additional entry into its routing table for the default route, indicating the current IGW that this MANET node associates to. Figure 2.22 shows the MN (5) routing table using the extended default route. With this advanced

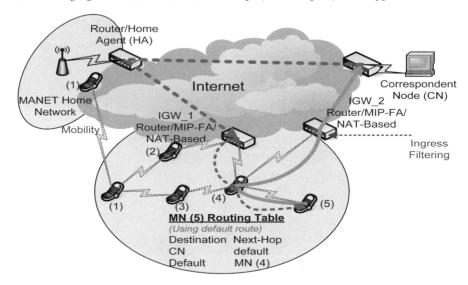

FIGURE 2.21
Inconsistent context due to default route forwarding: Type-I.

setting, the ambiguous IGW association of the MANET node is solved.
However, an additional cost on storage and access time for the additional entry is introduced. Moreover, the inconsistent context problems
still exist in other scenarios, which are presented in the next section.

3. **Type III**: Different scenarios indicating the inconsistent context are
analyzed in this section, which can be dependent on the operation of
MANET routing protocols and IGW discovery methods (proactive vs.
reactive vs. hybrid) [4, 35, 80].

   ○ **Scenario I**: A MANET node updates a shorter route to another
   IGW without reregistering the new IGW with its home agent (HA),
   and/or its foreign agent (FA), as well as the NAT device located in
   the visiting MANET domain.

   Figure 2.23 shows an example where MN (1) moves from its home
   network to the new MANET domain, registering to its home agent
   (HA) via IGW_2 through MN (3) as the next-hop to IGW_2. The
   distance from MN (1) to IGW_2 is 3-hop. Later, MN (1) finds a
   shorter route to the Internet via IGW_1. MN (1) chooses MN (2)
   as the next-hop to IGW_1, and the hop-count is 2. In this scenario,
   traffic from MN (1) to the CN is forwarded via IGW_1. However,
   traffic from the CN to MN (1) is still forwarded via IGW_2 since it
   is registered by MN (1) to MN (1) HA. This creates the inconsistent
   context.

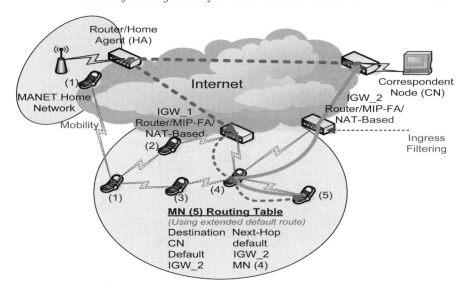

FIGURE 2.22
Inconsistent context due to default route forwarding: Type-II.

○ **Scenario II**: A MANET node associated with an IGW, for example, IGW_1, forwards the agent advertisement packet for another IGW, for example, IGW_2. As a result, its downstream nodes can associated to IGW_2 using it as the next-hop. However, the traffic is actually forwarded through IGW_1.

As an example, in Figure 2.24, suppose that MN (4) is currently associated with IGW_1, and MN (5) is not associated with any IGW. In the proactive IGW discovery, the IGW will broadcast periodically its agent advertisement packets. When MN (4) receives IGW_2 agent advertisement packets, MN (5) will set MN (4) as the next-hop to IGW_2 in its default route if MN (4) continues forwarding IGW_2's agent advertisement packets. However, whenever the traffic to CN is generated by MN (5) and then forwarded to IGW_2 via MN (4), this traffic is actually forwarded via IGW_1 by MN (4). This creates the inconsistent context. Note that the decision of MN (4) whether or not to continue forwarding other IGWs packets, for example, agent advertisement, is dependent on the corresponding operation and implementation of the MANET routing protocol and IGW discovery method.

○ **Scenario III**: A MANET node loses its association to the current IGW, for example, on detection of a broken link, and reassociates to another IGW. As a result, traffic to the Internet from its down-

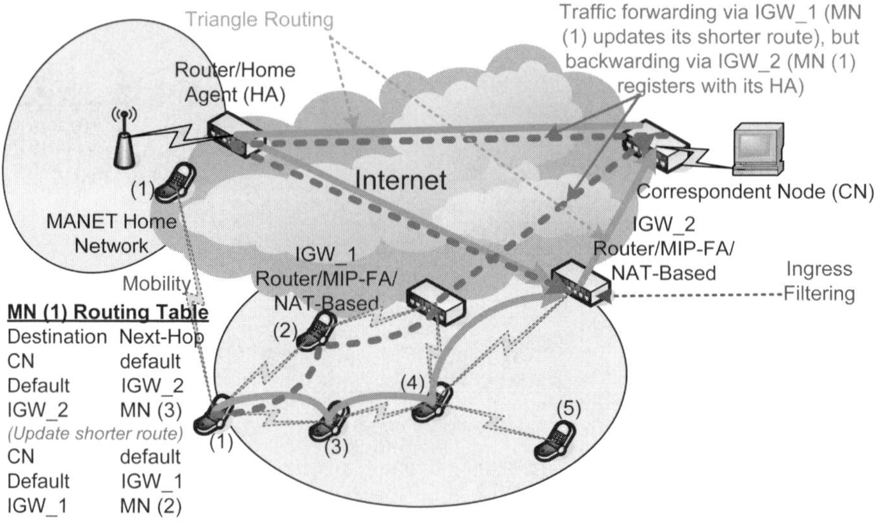

FIGURE 2.23
Inconsistent context due to default route forwarding: Type-III (scenario I).

stream nodes choosing it as the next-hop to the current IGW will be forwarded via another IGW instead.

As an example, in Figure 2.24, suppose MN (3) chooses MN (4) as its next-hop to the IGW_1, and MN (4) is currently associated to the IGW_1. Later, if the link between MN (4) and IGW_1 is broken, MN (4) reassociates to IGW_2. Thus, traffic to the Internet from MN (3) will be forwarded via IGW_2, though MN (3) thinks it is forwarded via IGW_1. This creates the inconsistent context.

2.5.1.2 Inconsistent Context in MIP-FA Triangle Routing

The use of MIPv4 can easily lead to the triangle routing problem (see Figure 2.24), that is, traffic from a MANET node to the correspondent node (CN) is sent directly, while return traffic is sent to the MANET node home agent (HA) and then is tunneled to the MANET node's foreign agent (FA) and delivered to the MANET node.

In providing Internet connectivity for MANET nodes, this also means that traffic to the Internet from a MANET node can be forwarded to one IGW, for example, IGW_1 in Figure 2.24, for a shorter route, while return traffic is forwarded through registered IGW, for example, IGW_2 in Figure 2.24. This creates the inconsistent context. In the case where IGW_2 is behind a stateful firewall, that is, the first outbound to the Internet packet sets the soft state in the firewall for the return packets to enter the MANET domain, the connection will be terminated. Note that the triangle routing problem can be

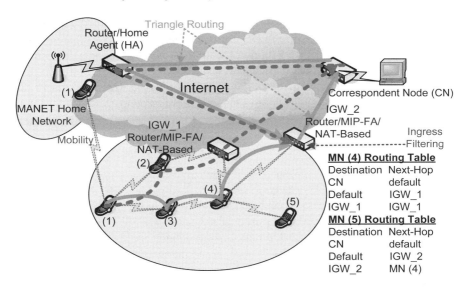

FIGURE 2.24
Inconsistent context due to default route forwarding: Type-III (scenarios II-III).

considered as a consequence of Scenario I, and thus solution to this problem can be referred to in Scenario I, Section 2.5.2.

2.5.1.3 Inconsistent Context in MIP-FA Ingress Filtering

Ingress filtering means that a router/firewall will not accept on its ingress interface packets with a source IP address that is not topologically correct for that interface. The motivation is to prevent IP address spoofing. If the ingress filtering is integrated into IGWs, the traffic to the Internet from a MANET node ends up at another IGW and not the IGW it is registered with, that IGW can drop its packet.

As an example, in Figure 2.24, MN (1) moves to a new MANET domain, getting a new topologically correct care-of address (CoA) and registering it with its HA via IGW_2. If the MANET node later updates its shorter route through IGW_1, for example, using another IP subnet, its outbound packets can be dropped at IGW_1 due to ingress filtering. Note that the ingress filtering problem can be also considered as a consequence of Scenario I, and hence, the solution to this problem can be referred to in Scenario I, Section 2.5.2.

2.5.1.4 Inconsistent Context in MIP-FA NAT Traversing

If the IGWs connecting MANET domains with the Internet are either behind the NAT devices, or integrating the NAT function, and private IP subnets are used with MANET domains, a MANET node communicates with an Internet host through the private IP address of its associated IGW, which will be mapped into another public IP address. Thus, with a NAT-based IGW, the connection of a MANET node is bound to the NAT that the connection passes through. However, whenever the MANET node updates its new NAT-based IGW, the return traffic can be dropped at this one due to the stateful router/firewall. This creates the inconsistent context. The solution to this problem is shown in Section 2.5.2.3.

2.5.1.5   Cascading Effect for Node Location Determination

In the reactive ad-hoc routing protocol, whenever a route to a destination is needed, the source MANET node needs to perform a route discovery. Usually, it broadcasts a route request (RREQ) packet and waits for a route reply (RREP) if the destination is located in the same MANET. In providing Internet connectivity for MANET nodes using default routes, no RREP will be sent back to the source MANET node if the destination is an Internet host. Note that, for some mechanisms, an IGW is allowed to send an RREP to the source MANET node to indicate that the destination is an Internet host and can be reached through this IGW, called proxy RREP.

If the destination is an Internet host, the source MANET node will send packets to the destination using its default route setting in its routing table. However, when its upstream MANET node receives its packets, it again performs another route discovery for the destination. This route discovery is repeated at each upstream MANET node on the chain from the source MANET node to the corresponding IGW, creating a problem called the cascading effect. The advantage of the cascading effect on MANET reactive routing protocols is that the shortest host route to the destination is found if the destination is located within the same MANET domain. However, this mechanism is redundant and tolerates much overhead if the destination is an Internet host.

Therefore, the cascading effect problem should be removed on providing Internet connectivity for MANET nodes. One solution is to insert directly in the routing table of each MANET node an entry for each destination Internet host. When a relay MANET node receives data packets from its downstream nodes, it continues forwarding to the destination Internet host without any route discovery if an entry for that destination Internet host is available in its routing table.

2.5.2   Mobility Management Solutions in MANET

Since the inconsistent context problems due to MIPv4-FA triangle routing and ingress filtering are consequences of Scenario I (see Section 2.5.1), and the solution for the cascading effect problem has been already presented, this section continues with only solutions to the inconsistent context due to default route forwarding. The solution to the inconsistent context due to MIPv4-FA NAT traversing is presented in Section 2.5.2.3.

2.5.2.1 Reducing Inconsistent Context Using Default Route

- **Scenario I:** A MANET node is not allowed to update its shorter route to another IGW unless its current transmissions on any two-way connections to the Internet hosts are finished and it has already reregistered this new IGW with its home agent. This reregistration can be prepared in advance, for example, during the data transmissions on the current connections via the old IGW. Clearly, this rule removes completely the inconsistent context in Scenario I since the inbound/outbound traffic to/from the Internet is always forwarded via the same IGW. However, its disadvantage is that the intra-MANET route from the source MANET node to the registered IGW is not an optimal one, for example, the shortest route.

  This rule also reduces the inconsistent context in Scenario II and Scenario III. The reason is that the less changes on the IGW reassociation of MANET nodes, the less inconsistent context problems appear on their downstream MANET nodes.

- **Scenario II:** On proactive IGW discovery, a MANET node that does not register to any IGW is allowed to rebroadcast the received agent advertisement if it decides to register with this IGW. Otherwise, the rebroadcasting of the agent advertisement is prohibited. On reactive IGW discovery, a MANET node is allowed to generate/forward an agent advertisement in one of the following three cases:

  1. It does not register to any IGW, registering it to the IGW of which the agent advertisement it receives, then forwarding the agent advertisement to the source MANET node.
  2. It has already registered to an IGW, receiving the agent advertisement generated by the same IGW, forwarding this agent advertisement to the source MANET node.
  3. It has already registered to an IGW, receiving the agent solicitation from the source MANET node, generating an agent advertisement to the source MANET node.
  4. On hybrid IGW discovery, the above rules are applied whenever an agent advertisement or an agent solicitation is received.

  Clearly, the above rules ensure that all the MANET nodes on the chain from any source MANET node to their registered IGWs use the same IGW for inbound/outbound traffic to/from the Internet. However, on proactive IGW discovery, the applied rule creates nonoverlapped MANET domains, each domain associating to only one IGW. Thus, it does not have the advantage of using multiple IGWs for load balancing and fault tolerance. Moreover, there can be the appearance of orphan MANET nodes due to collision or high mobility.

On reactive or hybrid IGW discovery, the generation of the agent advertisement of an IGW or an immediate MANET node, called gratuitous agent advertisement or proxy route reply (RREP), if this agent advertisement is piggybacked on the RREP of the MANET reactive routing protocol, to the source MANET introduces non-optimal route. This happens whenever the destination is located in the same MANET domain of the source MANET node, and the source MANET node receives a proxy RREP before the normal RREP sent by the destination, for example, due to collision or longer hop-count. This problem can be further reduced by the use of the destination sequence number which, in this field on the proxy RREP packet, is always set to a small value, for example, "0". Thus, when the source MANET node receives the RREP sent by the destination later, it will update the host route to the destination instead of the default route via the IGW, since the destination sequence number sent by the destination is greater than that of the proxy RREP sent by the IGW.

- **Scenario III:** There should be a mechanism for the MANET node to detect the broken link, sending this information to its downstream MANET nodes so that these MANET nodes can reregister their new IGWs with their home agents. This mechanism is dependent on the MANET routing protocol.

  As an example, in the AODV routing protocol, a MANET node keeps a list of other neighbor MANET nodes using it as the next-hop to a set of destinations called the precursor list. Whenever this MANET node detects a broken link to any destination, it searches its precursor list for that destination and sends a route error (RERR) to all the nodes on this list. This process is propagated to the MANET nodes in its precursor list.

  In the scope of this scenario, a MANET node detecting a broken link to its registered IGW will integrate this information to the RERR and send it to all nodes in its precursor list. However, to reduce also the inconsistent context in Scenario II, the detected MANET node also sends the RERR to all the precursor lists associated to all IGWs as the destinations. This is because some MANET nodes can be associated with an IGW but their traffic can be forwarded to another IGW instead.

2.5.2.2   Removing Inconsistent Context Using Tunneling

Clearly, in IGW forwarding strategies using the default route between a MANET domain and the Internet connecting through multiple IGWs, the inconsistent context is always a problem. This is because the traffic from the MANET to the Internet and the return traffic can be forwarded through different IGWs, taking to the termination of two-way connections.

With the described solutions in the previous section, the problems of default route are reduced. However, depending on the implementation of MANET routing protocols, the IGW discovery, whether they are implemented independently or dependently, there can be the appearance of another scenarios of inconsistent context. Thus, it is the purpose of this part to introduce another approach to removing completely the effect of inconsistent context.

The idea is that a MANET node always sends its outbound traffic to the Internet via its registered IGW irrespective its immediate MANET nodes updating the shorter routes to other IGWs. This is achieved via the use of tunneling, for example, IP-in-IP tunneling [48, 23]. In this scheme, the source MANET node encapsulates the original IP packet to another IP packet, in which its registered IGW is the destination IP address in the outer IP header. The encapsulated IP packet is then forwarded to its registered IGW using the MANET routing protocol. When the encapsulated IP packet arrives to the IGW, it is decapsulated, and the inner original IP packet is forwarded to the destination Internet host by IGW.

Figure 2.25 shows an example, where an IP packet to the Internet from the source MANET node MN (1) is first tunneled, that is, encapsulated, to its registered IGW_2, and then decapsulated and the original IP packet forwarded to the CN. In the reversed direction using tunneling through MN (1) HA in this scenario, CN sends its packet to the MN (1) home address (HAddr) via MN (1) HA, where the packet is encapsulated and forwarded to MN (1) FA (IGW_2). At the IGW_2, this packet is decapsulated, then encapsulated, and forwarded to the MN (1).

With the tunneling, the inconsistent context problems are completely removed. However, an additional cost of adding an outer IP header is introduced.

2.5.2.3 Solutions on Passing the NAT Device

Mobile IP with NAT traversal can be used to pass the NAT-based IGW. The NAT traversal is based on the MIP UDP tunneling mechanism [49]. In MIP UDP tunneling, the mobile node may use an extension in its registration request to indicate that it is able to use MIP UDP tunneling instead of standard MIP tunneling if the home agent sees that the registration request seems to have passed through an NAT. The home agent may then send a registration reply with an extension indicating acceptance or denial. After assent from the home agent, MIP UDP tunneling will be available for use for both forward and reverse tunneling. UDP-tunneled packets sent by the MANET node use the same ports as the registration request message. In particular, the source port may vary between new registrations but remains the same for all tunneled data and reregistrations. The destination port is always 434. UDP-tunneled packets sent by the home agent uses the same ports but in reverse.

To reduce the inconsistent context problems, this solution requires an additional IP-in-IP tunneling from the source MANET node to its registered IGW to ensure that the source MANET IP address and source MANET UDP port number are translated consistently by the same NAT-based IGW.

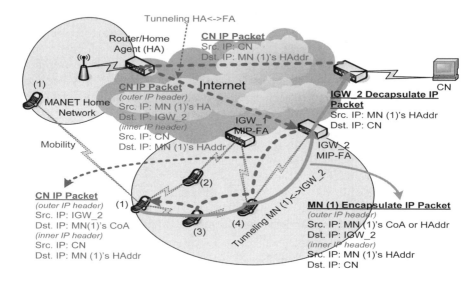

FIGURE 2.25
Packet header passing MIP-FA and IGW using tunneling.

Figure 2.26 shows an example where the tunneling is established between the MN (1) and its registered IGW_2, and between the MN (1) MIP-FA (mapping to publish IP address of IGW_3) and MN (1) HA. MIP UDP tunneling is used to transmit the packet through the NAT-based IGW_3, while the tunneling between MN (1) and IGW_2 is used to remove the inconsistent context problems. Note that, in Figure 2.26, the MN (1) IP packet is used to indicate the direction of sending data packets from the source MANET node to the destination CN, while the CN IP packet is used for indicating the packet transmissions in the reverse direction.

2.6 Mobility Management Open Issues

Major problems of existing work in IP mobility management for all-IP mobile networks are either due to the lack of, or little focus on:

1. a standardized interface for cross-layer, especially L3/L2, interaction to shorten the handoff latency and reduce the signaling overhead;

2. the convergence toward an open architecture to support different mobil-

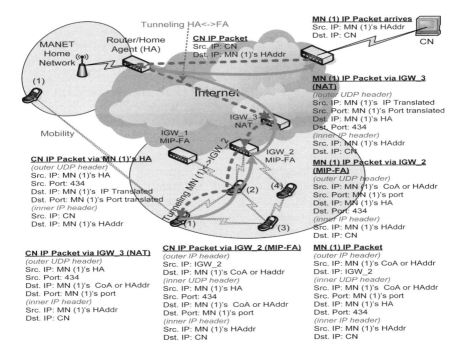

FIGURE 2.26
Packet header passing both MIP-FA and NAT-based IGWs.

ity scenarios under both horizontal and vertical handovers[4] in heterogeneous infrastructures and wireless access technologies;

3. interoperation between existing mobility protocols and new developed protocols to either improve further the performance (e.g., handoff latency, packet delivery) or reduce signaling overhead;

4. location privacy;

5. implementation of protocols, network architectures, and testbeds for critical evaluations of different mobility metrics.

These problems are research challenges in the mobility management for next-generation all-IP mobile networks and will be discussed next, together with pointing in possible directions toward solutions. Security is also a critical issue and should be included in mobility management, but it is out of the scope of this chapter.

---

[4]A horizontal handover is a handoff between two network access points that use the same network technology and interface. A vertical handover is a handoff between two network access points, that are usually using different network connection technologies.

2.6.1 Standardized Interface for Cross-Layer Interaction

To fulfill the required quality of service (QoS) in mobility, for example, handover latency, packet delivery ratio, signaling overhead, cross-layer interaction, especially at L3/L2, is needed. However, the existing work in general IP mobility management (Section 2.2) does not define any standardized interface or API for cross-layer interaction, and thus, functionalities for cross-layer interaction developed in WLAN for faster handoff (Section 2.3) are not compatible with those functionalities provided by other work. Therefore, the lack of a standardized interface for cross-layer interaction decreases the portability of existing mobility management protocols, especially for vertical handoff. Among on-going work to fill in this gap, IEEE 802.21 or *Media-Independent Handover* (MIH) services [3, 74, 58, 44] is the best candidate.

The IEEE 802.21 MIH services aims at providing functionalities to assist with seamless handoffs between heterogeneous link-layer technologies and across IP subnet boundaries. MIH services can be delivered through L2 specific solutions and/or through an L3 or above protocol. The IEEE 802.21/MIH advantages for mobility are multifold. First, it enables seamless handovers across different access technologies. Second, it optimizes L3 handover, for example, MIP or PMIP, by using L2 triggers associated with events to provide early warning of impending handovers and thereby decrease handover latency. Third, it provides QoS continuity across different technologies and minimizes service interruption. Finally, it provides ease of implementation through (1) thin software client on terminals, (2) no radio access network modifications required, and (3) support either network or client-controlled handovers.

The IEEE 802.21 architecture consists of MIH users located above L4 that use MIH service access points (SAPs) to communicate with MIH services at lower layers. MIH users are the initiation and termination points for MIH signaling sessions. The MIH Function (MIHF), located between L2 and L3, provides handover services, including the event service (ES), information service (IS), and command service (CS) through SAPs that are defined by the IEEE 802.21 Working Group. Figure 2.27 presents the standardized interfaces for cross-layer interaction in IEEE 802.21 MIH. The scope of IEEE 802.21 includes only the operation of MIHF and the primitives associated with the interfaces between MIHF and other entities. A single media-independent interface between MIHF and MIH user (MIH_SAP) is sufficient. On the other hand, there is a need for defining a separate technology-dependent interface that is specific to the corresponding media type supported, between the MIHF and the lower layers (MIH_LINK_SAP). The primitives associated with the MIH_LINK_SAP enable MIHF to receive timely and consistent link information and control link operation during handovers. Besides these, IEEE 802.21 specifies a media-independent SAP (MIH_NET_SAP) taht provides transport services for L2 and L3 MIH message exchange with remote MIHFs. Functions over the LLC_SAP are not specified in IEEE 802.21.

However, there are still open problems in IEEE 802.21 MIH services. First,

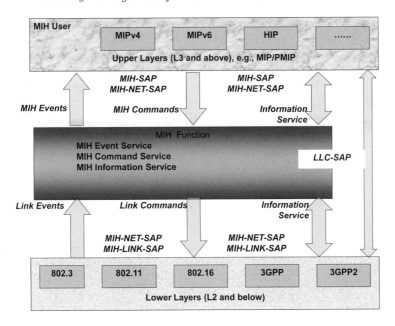

FIGURE 2.27
Media-Independent Handover (MIH) reference model.

reliable delivery is one of the key requirements to be imposed by MIH users, and thus, MIH signaling messages should be carried over a reliable transport protocol. Second, the scope of current specifications of IEEE 802.21 MIH is restricted to access technology-independent handovers. Intra-technology handovers, handover policies, security mechanisms, media-specific L2 enhancements to support IEEE 802.21, and L3 or upper-layer enhancements are outside the scope of IEEE 802.21.

Some ongoing work has recently focused on the reliable delivery of IEEE 802.21 MIH signaling messages. The IETF mobility for IP performance, signaling and handoff optimization (MIPSHOP) has been working on developing a protocol for transport of MIH services information and mechanisms for discovering the MIH server. In particular, a mobility services framework design (MSFD) is described in [58] for the IEEE 802.21 MIH protocol, which addresses mechanisms for mobility service discovery, and transport layer mechanisms for the reliable delivery of MIH messages.

2.6.2 Convergence Toward an Open Architecture

Up to now, the existing solutions of mobility management have mainly dealt with user handovers between wireless access points in an operator-controlled infrastructure. In the emerging wireless network scenarios with heterogeneous user devices and wireless technologies, the term mobility should have a wider sense and involve system responses to any changes in the user and network environments, including changes in radio and network resources as well as commercial conditions. Furthermore, mobility solutions need to support a simple-to-use, affordable, anytime-anywhere network access so that mobile users can soon take for granted a rich set of services regardless of the underlying connectivity, in the forthcoming eSociety [21]. Therefore, a single mobility paradigm cannot cover all this diverse set of requirements. Instead, a set of solutions, that can be flexibly combined and integrated on demand into an open architecture is needed.

2.6.3   Ambient Networks

The ambient networks project [6] is one of a few works that addresses the above challenges by developing innovative network solutions for increased competition and cooperation in an environment with a multitude of access technologies, operators, and business actors. The project aims at offering a complete, coherent open architecture solution based on dynamic composition, which provides access through the instant establishment of internetwork agreements using common control functions to a wide range of different applications and access technologies, enabling integrated, scalable, and transparent control of network capabilities.

In particular, three new concepts have been introduced in [66]. The mobility toolbox is the first of these concepts, where different both legacy and new mobility components can be slotted together as needed. Components can be optimized to suit particular handover scenarios, such as the movement of applications between devices, mobility of networks, and handover in more complex, multidomain environments between different administrative domains. The toolbox allows for new mobility services, which, to a large extent, rely on the ambient network composition concept. The second concept is the separation of mobility-relevant triggers from the actual mobility management mechanism, which allows processing of a wide range of mobility-relevant triggers generated by virtually any part of the system. The third concept is the development of mobility functions for groups of nodes, which provide new services for network mobility and optimize existing solutions in terms of consumed resources and increased handover performance.

2.6.4   Interoperation

Another research approach is to exploit the advantages of existing mobility management protocols to either improve or optimize further the protocol operations for different objectives: (1) shorter handoff latency, (2) higher packet delivery ratio, and (3) lower signaling overhead, to list a few. Such an interoperation work includes: (1) PMIPv6/MIPv6 [26] for a global IP mobility management solution, (2) fast handovers for PMIPv6 [93] adapts

FMIPv6 for PMIPv6 to shorten the handoff latency and reduce the packet loss, (3) NEMO/MANET [57] use NEMO to support the global reachability of MANET nodes, and (4) IEEE 802.21/MANET [50] is based on the IEEE 802.21 MIH services to develop an integrated framework for seamless soft handoff in ad-hoc networks.

2.6.5   Location Privacy

While many IP mobility management protocols have been developed for wireless mobile networks, location privacy has not been addressed much. The reason is that IP addresses of mobile nodes and related entities, for example, home agents or access routers or foreign agents, are often used as topological contact identifiers to establish and maintain ongoing connections. Since IP routing is based on the IP network prefix, which also shows the network location, an IP mobility management protocol that exploited IP addresses as identifiers usually does not provide location privacy. However, location privacy is one of the user requests in 4G mobile networks. A current approach to location privacy is to enable a separation between the end-point identifier and locator role of IP address and thus overcome one of paradigms of the current Internet architecture – the association of a (possible temporary) IP address with a host end-point identifier. Such examples in this approach include identifiers separating and mapping scheme (ISMS) [19], and the locator/ID separation protocol [15].

The IETF is also working on the host identity protocol (HIP), in which all occurrences of IP addresses in applications are eliminated and replaced with cryptographic host identifiers. However, this approach to location privacy still has several unsolved aspects in the area of mobility management: (1) a NAT traversal solution, (2) a description of the interactions between legacy location-unaware vs. location-aware applications, (3) how to build location-aware-based overlays, (4) how to match location privacy properties to the requirements of existing standards, and (5) support from existing infrastructure such as DNS. A possible solution to some of these problems can be exploited from the Internet indirection infrastructure [34].

2.6.6   Implementation and Testbed

Many IP mobility management protocols are proposed without evaluation, and thus, there has still been the lack of a wireless mobile network testbed, together with the qualitative/quantitative analysis and implementation of protocols for evaluation, testing, and further improvements in solutions for problems. Recently, a few work have focused on these issues. These works include: (1) quantitative analysis modeling for more comparisons and evaluations [40, 27, 52]; (2) GNU/Linux implementation of IEEE 802.21 [75]; and (3) the ongoing WEIRD project [90], which aims at implementing the European national research network testbed using WiMAX technology.

2.7   Conclusion

Realizing all-IP mobile networks has been an ongoing process for many years. This trend is fueled by the demand for Internet applications and services. Furthermore, network operators and administrators also expect that all-IP networking will lower the operating costs of their mobile networks. IP technology is abundant and designed after a philosophy of simplicity, strict layering, and modularity. This makes IP technology an interesting solution for 2G/3G operators, where it can provide a cost advantage relative to their legacy networking solutions. Indeed, some consider all-IP a disruptive technical solution to traditional 2G/3G networking.

However, the transition to all-IP networking is a technical hurdle. One of the most challenging issues of all-IP networking is mobility management. This chapter has surveyed these challenges and the technical solutions developed to handle them, mainly focusing on the network and link layers.

This chapter provides a complete up-to-date comparison of the state of the art of different approaches to IP mobility management. The main distinctions were made between host mobility, network mobility (e.g., NEMO) and hybrid mobility (e.g., ISMS). The former distinguishes between network-based mobility (e.g., Proxy MIP) and host-based mobility (e.g., MIP for macromobility, HMIP/CellularIP/Hawaii for micromobility and FHMIP for a hybrid approach.)

When surveying the state of the art of mobility management for all-IP mobile networks, we observed that alternative approaches have qualitatively different pros and cons. Thus, there is no one-size-fits-all solution. Instead, when implementing and introducing all-IP networking, one has to carefully make a selection of a suitable technical solution.

As pointed out, one of the drivers of transition to all-IP is lowering costs, by the simple, layered, and modular approach of IP. In this respect, all-IP fits well with a number of IEEE link-layer solutions, such as IEEE 802.3 Ethernet and IEEE 802.11 Wireless LAN (WLAN) solutions, both in terms of reasonable costs, in terms of abundance/popularity and in terms of a matching design philosophy. As far as mobility management is concerned, the IEEE 802.11 WLAN technology is of particular interest. Thus, the chapter considered network layer and link layer mobility management for all-IP mobile networks that are spanning WLAN networks. The WLAN 802.11 handoff procedure was discussed, as well as the native link-layer handoff between different BSSs (Basic Service Sets) in an ESS (Extended Service Set). The cross-layer handoff between different ESSs was also described.

Finally, mobility management in all-IP mobile networks that span mobile ad-hoc networks (MANETs) was also considered. The reason is that WLANs are typically one-hop access networks (except for the 802.11s technology, which can easily be analyzed similar to MANETs). MANETs, on the other hand, are multihop access networks. Mobility management in MANETs calls for solutions to issues such as node location determination, gateway discovery, gateway selection, gateway forwarding strategy, autoconfiguration and the actual handoff style.

We observed that, due to the mobility of MANET nodes and the multihop connectivity in MANETs, different problems appear. A number of issues were raised, including the inconsistent context in default route forwarding, the cascading effect in node localization determination, the triangle routing problem, the ingress filtering problem, and the NAT traversal problem. The chapter proposed solutions to either reduce partly, or remove completely, these discovered problems.

A number of open issues for further research have been outlined. First, there is a need to establish a standardized cross-layer interface for interaction between L2 and L3 during a handover in order to reduce latency and overhead. Furthermore, there is a need to establish an open architecture that is able to span a number of different mobility paradigms, and this in an interesting topic for further research. Moreover, the interoperation between existing and newly developed mobility protocols is a promising area for further research. Location privacy is another topic that has become relevant since many IP mobility protocols were not designed with this in mind in the first place. The demand for location privacy has become evident later, especially when IP mobility protocols have become realistic alternatives in 4G. Finally, many protocols and architectures for all-IP mobility management need further evaluation and testing in realistic testbed implementations.

While these research directions are followed, traditional operators of 2G and 3G networks will gradually be making a transition toward 4G and all-IP networks. Getting mobility management solutions implemented and tested in a real setting will be a part of this work. Interoperation will also be relevant, and will provide further insights. This trend might lead much of the research focus in a practical direction. However, at the same time, the more fundamental research challenges mentioned earlier (e.g., open architectures) will still persist.

Bibliography

[1] 3GPP. Setting Standards for Mobile Broadband, http://www.3gpp.org/.

[2] IEEE Std 802.1f. Recommended practice for multi-vendor access point interoperability via an inter-access point protocol across distribution systems supporting IEEE 802.11 operation. *IEEE Draft 802.1f/D3*, 2002.

[3] IEEE Std 802.21. IEEE standard for local and metropolitan area networks – Part 21: media independent handover services. *IEEE Std 802.21-2008*, 2009.

[4] F. M. Abduljalil and S. K. Bodhe. A survey of integrating IP mobility protocols and mobile ad hoc networks. *IEEE Communication Surveys and Tutorials*, pages 14–30, 2007.

[5] C. Ählund and A. B. Zaslavsky. Software solutions to Internet connectivity in mobile ad hoc networks. *Proc. of 4th International Conference on Product Focused Software Process Improvement*, pages 559–571, 2002.

[6] Ambient Network. `http://www.ambient-networks.org/index.html`.

[7] H. Ammari and H. E. Rewini. Using hybrid selection schemes to support QoS when providing multihop wireless Internet access to mobile ad hoc networks. *Proc. of 1st International Conference on Quality of Service in Heterogeneous Wired/Wireless Networks*, pages 148–155, 2004.

[8] E.M. Belding-Royer, Y. Sun, and C.E. Perkins. Global connectivity for IPv4 mobile ad hoc networks. IETF Internet Draft draft-royer-manet-globalv4-00.txt, 2001.

[9] M. Benzaid, P. Minet, K. A. Agha, C. Adjih, and G. Allard. Integration of mobile-IP and OLSR for a universal mobility. *Wireless Networks*, pages 377–388, 2004.

[10] C. Bernardos and M. Calderon. Survey of IP address autoconfiguration mechanisms for MANET. IETF Internet Draft draft-bernardos-manet-autoconf-survey-00.txt, 2005.

[11] J. Broch, D. A. Maltz, and D. B. Johnson. Supporting hierarchy and heterogeneous interfaces in multi-hop wireless ad hoc networks. *Proc. of the International Symposium on Parallel Architectures, Algorithms and Networks*, pages 370–375, 1999.

[12] A. T. Campbell and J-G. Castellanos. IP micro-mobility protocols. *ACM Mobile Computing and Communications Review*, pages 45–53, 2000.

[13] N. Chand, R. C. Joshi, and M. Misra. Cooperative caching in mobile ad-hoc networks based on data utility. *Mobile Information Systems*, pages 19–37, 2007.

[14] C. Y. Chiu and C. Gen-Huey. A stability aware cluster routing protocol for mobile ad hoc networks. *Wireless Communications and Mobile Computing*, pages 503–515, 2003.

[15] N. Choi, T. You, J. Park, T. Kwon, and Y. Choi. ID/LOC separation network architecture for mobility support in future Internet. *Proc. of the 11th International Conference on Advanced Communication Technology*, pages 78–82, 2009.

[16] T. Clausen, C. Dearlove, and P. Jacket. The optimized link-state routing protocol version 2. IETF Internet Draft draft-ietf-manet-olsrv2-02.txt, 2006.

[17] V. Devarapalli, R. Wakikawa, A. Petrescu, and P. Thubert. Network mobility NEMO support protocol. IETF RFC 3963, 2005.

[18] S. Ding. A survey on integrating MANETs with the Internet: challenges and designs. *Computer Communications*, pages 3537–3551, 2008.

[19] P. Dong, H. Zhang, H. Luo, T-Y. Chi, and S-Y. Kuo. A network-based mobility management scheme for future Internet. *Computers and Electrical Engineering*, pages 291–302, 2009.

[20] A.L. Dul. Global IP network mobility using border gateway protocol. *Connexion by Boeing Company, http://www.quark.net/docs/Global_IP_Network_Mobility_using_BGP.pdf*, March 2006.

[21] e-Society. http://www.york.ac.uk/res/e-society/.

[22] P. E. Engelstad and G. Egeland. NAT-based Internet connectivity for on-demand ad hoc networks. *Proc. of the 1st IFIP TC6 Working Conference on Wireless On-Demand Network Systems*, pages 344–358, 2004.

[23] P. E. Engelstad, A. Tonnesen, A. Hafslund, and G. Egeland. Internet connectivity for multi-homed proactive ad hoc networks. *Proc. of International Conference on Communications (ICC)*, pages 4050–4056, 2004.

[24] M. Ergen and A. Puri. MEWLANA-mobile IP enriched wireless local area network architecture. *Proc. of the 56th IEEE Vehicular Technology Conference (VTC)*, pages 2449–2453, 2002.

[25] N. A. Fikouras, A. J. Könsgen, and C. Görg. Accelerating mobile IP hand-offs through link-layer information, an experimental investigation with 802.11b and Internet audio. *Proc. of International Multiconference on Measurement, Modelling, and Evaluation of Computer-Communication Systems*, 2001.

[26] G. Giaretta. Interactions between PMIPv6 and MIPv6: scenarios and related issues. IETF Internet Draft draft-ietf-netlmm-mip-interactions-04.txt, 2009.

[27] D. Griffith, R. Rouil, and N. Golmie. Performance metrics for IEEE 802.21 media independent handover MIH signaling. *Wireless Personal Communications*, pages 537–567, 2008.

[28] S. Gundavelli, K. Leung, V. Devarapalli, K. Chowdhury, and B. Patil. Proxy Mobile IPv6. IETF RFC 5213, 2008.

[29] E. Gustafsson, A. Jonsson, and C. Perkins. An end-to-end approach for transparent mobility across heterogeneous wireless networks. IETF Internet Draft draft-ietf-mobileip-reg-tunnel-04.txt, 2001.

[30] R. Hsieh and A. Seneviratne. A comparison of mechanisms for improving mobile IP handoff latency for end-to-end TCP. *Proc. of 9th International Conference on Mobile Computing and Networking*, pages 29–41, 2003.

[31] R. Hsieh, Z.G. Zhou, and A. Seneviratne. S-MIP: a seamless handoff architecture for mobile IP. *Proc. of the 22nd IEEE INFOCOM*, pages 1774–1784, 2003.

[32] Y.Y. Hsu, Y.C. Tseng, C.C. Tseng, C.F. Huang, J.H. Fan, and H.L. Wu. Design and implementation of two-tier mobile ad hoc networks with seamless roaming and load-balancing routing capability. *Proc. of 1st International Conference on Quality of Service in Heterogeneous Wired/Wireless Networks*, pages 52–58, 2004.

[33] X. Hu, L. Li, Z.M. Mao, and Y.R. Yang. Wide-area IP network mobility. *Proc. of the 27th IEEE INFOCOM*, pages 1624–1632, 2008.

[34] Internet Indirection Infrastructure. http://i3.cs.berkeley.edu/.

[35] X. Jin and B. Christian. Wireless multihop Internet access: gateway discovery, routing, and addressing. *Proc. of the 3rd Generation Wireless and Beyond*, 2002.

[36] D. Johnson, Y. Hu, and D. Maltz. The dynamic source routing (DSR) protocol for mobile ad hoc networks for IPv4. IETF RFC 4728, 2007.

[37] D. Johnson, C. Perkins, and J. Arkko. Mobility Support in IPv6. IETF RFC 3775, 2004.

[38] U. Jönsson, F. Alriksson, T. Larsson, P. Johansson, and G. Q. Maguire. MIPMANET mobile IP for mobile ad hoc networks. *Proc. of the 1st ACM Interational Symposium on Mobile Ad Hoc Networking and Computing*, pages 75–85, 2000.

[39] H.Y. Jung, S.J. Koh, and J.Y. Lee. Fast Handover for Hierarchical MIPv6 (FHMIPv6). IETF Internet Draft draft-jung-mobopts-fhmipv6-00.txt, 2005.

[40] K-S. Kong, Y-H. Han, M-K. Shin, and You H. Mobility management for all-IP mobile Networks: mobile IPv6 vs. proxy mobile IPv6. *IEEE Wireless Communications*, pages 36–54, 2008.

[41] R. Koodli. Fast Handovers for Mobile IPv6. IETF RFC 4068, 2005.

[42] R. Koodli and C. Perkins. Mobile IPv4 fast handovers. IETF Internet Draft draft-ietf-mip4-fmipv4-07.txt, 2007.

[43] L. Lamont, M. Wang, L. Villasenor, T. Randhawa, and S. Hardy. Integrating WLANs & MANETs to the IPv6 based Internet. *Proc. of the IEEE International Conference on Communications (ICC)*, pages 75–85, 2003.

[44] G. Lampropoulos, A. K. Salkintzis, and N. Passas. Media-independent handover for seamless service provision in heterogeneous networks. *IEEE Communications Magazine*, pages 64–71, 2008.

[45] D. Le, X. Fu, and D. Hogrefe. A review of mobility support paradigms for the Internet. *IEEE Communication Surveys & Tutorials*, pages 38–51, 2006.

[46] Q. Le-Trung, P. E. Engelstad, V. Pham, T. Skeie, A. Taherkordi, and F. Eliassen. Providing internet connectivity and mobility management for MANETs. *International Journal of Web Information Systems*, pages 239–263, 2009.

[47] Q. Le-Trung, P. E. Engelstad, T. Skeie, and A. Taherkordi. Load-balance of intra/inter-MANET traffic over multiple Internet gateways. *Proc. of the 6th International Conference on Advances in Mobile Computing and Multimedia*, pages 50–57, 2008.

[48] Q. Le-Trung and G. Kotsis. Reducing problems in providing Internet connectivity for mobile ad hoc networks. *Proc. of the 4th International Workshop of the EuroNGI/EuroFGI Network of Excellence on Wireless Systems and Mobility in Next Generation Internet*, pages 113–127, 2008.

[49] H. Levkowetz and S. Vaarala. Mobile IP NAT/NAPT traversal using UDP tunneling. IETF Internet Draft draft-ietf-mobileip-nat-traversal-07.txt, 2007.

[50] J.H. Li, S. Luo, and S. Das. An integrated framework for seamless soft handoff in ad hoc networks. *Proc. of the IEEE Military Communications Conference (MILCOM)*, pages 1–7, 2006.

[51] Y. Liao and L. Gao. Practical schemes for smooth MAC layer handoff in 802.11 wireless networks. *Proc. of the International Symposium on World of Wireless, Mobile and Multimedia Networks*, pages 181–190, 2006.

[52] C. Makaya and S. Pierre. An Analytical Framework for Performance Evaluation of IPv6-based Mobility Management Protocols. *IEEE Transactions on Wireless Communications*, 7(2):972–983, 2008.

[53] K.E. Malki. Low Latency Handoffs in Mobile IPv4. IETF Internet Draft draft-ietf-mobileip-lowlatency-handoffs-v4-11.txt, 2005.

[54] K.E. Malki and H. Soliman. Simultaneous bindings for mobile IPv6 fast handoffs. IETF Internet Draft draft-elmalki-mobileip-bicasting-v6-00.txt, 2001.

[55] T. Manodham and T. Miki. A novel AP for improving the performance of wireless LANs supporitng VoIP. *Journal of Networks*, pages 41–48, 2006.

[56] P. McCann. Mobile IPv6 fast handovers for 802.11 networks. IETF RFC 4260, 2005.

[57] B. McCarthy, C. Edwards, and M. Dunmore. Using NEMO to support the global reachability of MANET nodes. *Proc. of IEEE INFOCOM*, pages 2097–2105, 2009.

[58] T. Melia, G. Bajkp, S. Das, N. Golmie, and J.C. Zuniga. IEEE 802.21 mobility services framework design (MSFD). IETF Internet Draft draft-ietf-mipshop-mstp-solution-12.txt, 2009.

[59] A. Mishra, M. Shin, and W. Arbaugh. An empirical analysis of the IEEE 802.11 MAC layer handoff process. *ACM SIGCOMM Computer Communication Review*, pages 93–102, 2003.

[60] A. Mishra, M. Shin, and W. Arbaush. Context caching using neighbor graphs for fast handoffs in a wireless network. *Proc. of IEEE INFOCOM*, pages 351–361, 2004.

[61] G. Mona, H. Philipp, P. Christian, F. Vasilis, and A. Hamid. Performance analysis of Internet gateway discovery protocols in ad hoc networks. *Proc. of IEEE Wireless Communications and Networking Conference (WCNC)*, pages 120–125, 2004.

[62] J. Montavont and T. Nöel. IEEE 802.11 handovers assisted by GPS information. *Proc. of the 2006 IEEE International Conference on Wireless and Mobile Computing, Networking and Communications (WiMob)*, pages 166–172, 2006.

[63] T. Narten, E. Nordmark, W. Simpson, and H. Soliman. Neighbour discovery for IP version 6 (IPv6). IETF Internet Draft draft-ietf-ipv6-2461bis-02.txt, 2005.

[64] E. Natsheh and T. C. Wan. Adaptive and fuzzy approaches for nodes affinity management in wireless ad-hoc networks. *Mobile Information Systems*, pages 273–295, 2008.

[65] C. Ng, F. Zhao, U.C. Davis, M. Watari, and P. Thubert. Network mobility route optimization solution space analysis. IETF Internet Draft draft-ietf-nemo-ro-space-analysis-03.txt, 2006.

[66] N. Niebert, A. Schieder, J. Zander, and R. Hancock. *Ambient Networks: Cooperative Mobile Networking for the Wireless World*. John Wiley & Sons, New York, 2007.

[67] S. Pack, J. Choi, T. Kwon, and Y. Choi. Fast-handoff support in IEEE 802.11 wireless networks. *IEEE Communication Surveys & Tutorials*, pages 2–12, 2007.

[68] S. Pack and Y. Choi. Fast inter-AP handoff using predictive authentication scheme in a public wireless LAN. *Proc. of IEEE Networks Conference*, pages 15–26, 2002.

[69] S. Pack, H. Jung, T. Kwon, and Y. Choi. SNC: a selective neighbor caching scheme for fast handoff in IEEE 802.11 wireless networks. *ACM Mobile Computing and Communications Review*, pages 39–49, 2005.

[70] C. Perkins and P. Bhagwat. Highly dynamic destination-sequenced distance-vector routing (DSDV) for mobile computers. *Proc. of the SIG-COMM Conference on Communications Architectures, Protocols and Applications*, pages 234–244, 1994.

[71] C. Perkins, Belding-Royer E., and S. Das. Ad hoc on-demand distance vector (AODV) routing for IP version 6. IETF Internet Draft draft-perkins-manet-aodv6-01.txt, 2000.

[72] C. Perkins, Belding-Royer E., and S. Das. Ad hoc on-demand distance vector (AODV) routing. IETF RFC 3561, 2003.

[73] C. Perkins, J.T. Malinen, R. Wakikawa, A. Nilsson, and A.J. Tuominen. Internet connectivity for mobile ad hoc networks. *Wireless Communications and Mobile Computing*, pages 465–482, 2002.

[74] E. Piri and K. Pentikousis. IEEE 802.21: Media-independent handover services. *Internet Protocol Journal*, pages 7–27, 2009.

[75] E. Piri and K. Pentikousis. Towards a GNU/Linux IEEE 802.21 implementation. *Proc. of IEEE International Conference on Communications (ICC)*, pages 1–5, 2009.

[76] K. Ramachandran, S. Rangarajan, and J. C. Lin. Make-before-break MAC layer handoff in 802.11 wireless networks. *Proc. of IEEE International Conference on Communications (ICC)*, pages 4818–4823, 2006.

[77] I. Ramani and S. Savage. SyncScan: practical fast handoff for 802.11 infrastructure networks. *Proc. of IEEE INFOCOM*, pages 675–684, 2005.

[78] R. Ramjee, T. La Porta, S. Thuel, K. Varadhan, and S.Y. Wang. HAWAII: a domain-based approach for supporting mobility in wide-area wireless networks. *IEEE/ACM Transactions on Networking*, pages 396–410, 2002.

[79] E. M. Royer, P. M. Melliar-Smith, and L. E. Moser. An analysis of the optimal node density for ad hoc mobile networks. *Proc. of IEEE International Conference on Communications (ICC)*, pages 857–861, 2001.

[80] P. M. Ruiz and A. F. Gomez-Skarmeta. Adaptive gateway discovery mechanisms to enhance Internet connectivity for mobile ad hoc networks. *Ad Hoc & Sensor Wireless Networks*, pages 159–177, 2005.

[81] S. Sharma, N. Zhu, and T-C. Chiueh. Low-latency mobile IP handoff for infrastructure-mode wireless LANs. *IEEE Journal on Selected Areas in Communications (JSAC)*, pages 643–652, 2004.

[82] M. Shin, A. Mishra, and W. Arbaugh. Improving the latency of 802.11 hand-offs using neighbor graphs. *Proc. of the 2nd International Conference on Mobile Systems, Applications, and Services*, pages 70–83, 2004.

[83] S. Shin, A. G. Forte, A. S. Rawat, and H. Schulzrinne. Reducing MAC layer handoff latency in IEEE 802.11 wireless LANs. *Proc. of ACM International Conference on Mobile Computing and Networking (MOBICOM)*, pages 19–26, 2004.

[84] A. C. Snoeren and H. Balakrishnan. An end-to-end approach to host mobility. *Proc. of ACM International Conference on Mobile Computing and Networking (MOBICOM)*, pages 155–166, 2000.

[85] H. Soliman, C. Castelluccia, K. E. Malki, and L. Bellier. Hierarchical mobile IPv6 mobility management (HMIPv6). IETF RFC 4140, 2005.

[86] A. G. Valko. Cellular IP: a new approach to Internet host mobility. *ACM SIGCOMM Computer Communications Review*, pages 50–65, 1999.

[87] H. Velayos and G. Karlsson. Techniques to reduce the IEEE 802.11b handoff time. *Proc. of IEEE International Conference on Communications (ICC)*, pages 3844–3848, 2004.

[88] R. Wakikawa and S. Gundavelli. IPv4 support for proxy mobile IPv6. IETF Internet Draft draft-ietf-netlmm-pmip6-ipv4-support-13.txt, 2009.

[89] R. Wakikawa, J. T. Malinen, C. E. Perkins, A. Nilsson, and A. J. Tuominen. Global connectivity for IPv6 mobile ad hoc networks. IETF Internet Draft draft-wakikawa-manet-globalv6-00.txt, 2001.

[90] WEIRD. WiMAX Extension to Isolated Research Data Network Project, http://www.ist-weird.eu/.

[91] K. Weniger and M. Zitterbart. Address autoconfiguration in mobile ad hoc networks: current approaches and future directions. *IEEE Networks*, pages 6–11, 2004.

[92] J-C-S. Wu, C-W. Cheng, N-F. Huang, and G-K. Ma. Intelligent handoff for mobile wireless Internet. *Mobile Networks and Applications*, pages 67–79, 2001.

[93] H. Yokota, K. Chowdhury, R. Koodli, B. Patil, and F. Xia. Fast handovers for proxy mobile IPv6. IETF Internet Draft draft-ietf-mipshop-pfmipv6-09.txt, 2009.

[94] H. Yokota, A. Idoue, T. Hasegawa, and T. Kato. Link layer assisted mobile IP fast handoff method over wireless LAN networks. *Proc. of the ACM International Conference on Mobile Computing and Networking (MOBICOM)*, pages 131–139, 2002.

3

Integrated Network Architecture Design for Next-Generation Wireless Systems

Li Jun Zhang

Geninov Inc., Research and Development Division, Montreal, Canada
Email: lijun.zhang@polymtl.ca

Liyan Zhang

Dalian Jiaotong University, School of Electronics & Information Engineering, P. R. China

Laurent Marchand

Ericsson Research, Montreal, Canada

Samuel Pierre

Ecole Polytechnique de Montreal, Department of Computer Engineering, Canada

CONTENTS

Advancement in wireless networking and mobile computing enables users to benefit from diverse mobile communications systems such as wireless local, metropolitan, and wide area networks. On the other hand, subscribers are intensifying demands for seamless roaming across different wireless systems while profiting services anytime, anywhere, and from any device. Under the circumstance, disparate networks are expected to integrate with each other to provide ubiquitous and high data-rate services to roaming users. However, integration of such systems brings about a variety of challenges because of the heterogeneities in security protocol, quality of service provisioning, mobility management, architecture, radio access technology, etc. Therefore designing new interworking and integrated architecture is a crucial task during wireless system evolution. This chapter provides a comprehensive technical guide and references for telecom engineers, researchers, and scientists while planning and designing interworking architectures for the emerging new generation wireless system.

3.1   Introduction

On March 2009, Comsys Mobile launched a multimode smartphone reference design platform that supports different standards like mobile WiMAX, GSM, and EDGE. Such platform is also capable of integrating the WiFi system, Bluetooth, FM radio, and GPS connectivity, and allows the coexistence of radios for all usage scenarios [45]. The new platform supports call continuity between WiMAX voice over IP (VoIP) and GSM circuit-switched voice, which is based on a standard IP Multimedia Subsystem (IMS) architecture.

It is evident that with the advent of the multimode smartphone, subscribers can benefit from a number of services from wireless local, metropolitan and wide area networks. These communication systems are expected to integrate with each other to provide ubiquitous and high-speed voice and data services to roaming users. However, interworking and integrating of distinct wireless networks brings about new challenges due to their heterogeneities in terms of mobility and security protocol, system architecture, provisioning services, and radio access technology, as well as quality of service requirements. Under the circumstances, it is crucial to take into account of all the heterogeneities while designing new interworking architecture for next-generation wireless system (NGWS). In addition, such architecture should provide appropriate mobility management, fast authentication for users on the move, guaranteed quality of service for multimedia applications that are executed on the mobile device, and sufficient security protection.

Then, again, new interworking architecture must preserve the advantages of the individual system and at the same time, eliminate the weaknesses of each network [6] while allowing mobile users always to be connected [16] to the best available network. Therefore, such new integrated architecture should have the following features [6, 25]:

- **Economical**: Using as much existing infrastructure as possible, rather than putting more efforts into developing new radio interfaces and access technologies [18]. In the meantime, minimizing the use of new infrastructure is also important to ensure rapid deployment.

- **Scalable**: The new architecture should be able to integrate any number of wireless systems of the same or different service providers; in particular, these different service providers may have no direct roaming agreements or service level agreements among them [6].

- **Transparency of radio access technology**: Underlying radio access technology should be transparent and imperceptible to mobile users. This is essential for roamers with multimode mobile devices like iPad, iPhone, Droid, smartphone.

- **Seamless mobility**: Network connectivity and session continuity should be guaranteed when mobile nodes perform any kind of roaming, such as intra and interdomain, intra and intertechnology handovers.

- **Security**: Providing a level of security and privacy that is at least equivalent to the existing wireless and wired networks is essential in the design of new architecture for next-generation wireless systems.

- **Improved quality of service provisioning**: It is envisioned that next generation wireless networks will support a wide variety of multimedia services such as multimedia web browsing, video and news on demand, mobile office system, stock market information, and so on, to roaming users anywhere, anytime in a seamless way [23]. The characteristics of different wireless links, as well as the desire to maintain network connectivity while on the move, offer significant challenges to provisioning quality of service and the related performance is of great interest in next generation wireless systems.

The remainder of this chapter is organized as follows: Section 3.2 provides a comprehensive literature review of existing interworking architectures for emerging wireless networks. Thus, integration between the third-generation (3G) mobile communication system and wireless metropolitan area network (WMAN), between wireless local area networks (WLAN) and WMAN, as well as interworking of 3G system with WLAN. The latter can be implemented in different ways: tight coupling, loose coupling, no coupling, hybrid coupling as well. Each coupling method is described in details in Section 3.2.3. After the survey, Sections 3.2.4 elaborates two well-cited architectures: the Architecture for Ubiquitous Mobile Communications (AMC) and Integrated Intersystem Architecture (IISA), respectively. Section 3.3 then proposes a novel unified integrated Architecture that supports Fast authentication and Seamless roaming (AFS). Section 3.4 discusses open research issues about designing interworking architecture for next-generation wireless system. Section 3.5 then concludes this chapter with brief outline of future work.

3.2 Existing Integrated and Interworking Architecture

It is evident that next-generation wireless systems include not only the cutting-edge Beyond 3rd Generation (B3G) mobile communication networks [10], but also other type of networks, like mobile ad-hoc networks, wireless sensor networks, wireless embedded Internet, Internet of Things, etc. The latter involves connecting embedded devices such as home electrical appliances, sensors, weather stations, and toys to IP-enabled networks [38].

On the other hand, integration of different wireless networks can be done by gateway approach, modified TCP/IP, or networks overlaying. The latter can be carried out by the Delay Tolerant Networking (DTN) [13], Internet Indirection Infrastructure (i3) [43], [44], etc. This chapter only focuses on the gateway approach and interconnection between 3GPP (or 3GPP2) wireless system and WiFi (or WiMAX) networks.

So far, numerous solutions have been proposed to integrate heterogeneous wireless and mobile communication systems. Yet, most of research activities focus on design architecture for integrating the 3rd Generation (3G) mobile system with wireless local area networks (WLANs) [6, 26, 42, 19]. Among all, the 3rd Generation Partnership Project (3GPP) has introduced six scenarios for interworking between a UMTS and WLAN [2, 1], whereas the 3GPP2 working groups put their efforts on integration of WLAN with CDMA2000 network [3, 4]. Still, discussions about the functionality, architecture, and feasibility of such interworking are working in progress. However, only a few solutions are proposed to interconnect WLAN with wireless metropolitan area network (WMAN) [29], as well as integration of the 3G system with WMAN [21, 28].

3.2.1 Interworking Architecture for 3G/WMAN

The increasing demand for broadband wireless packet data access presents a fundamental technology challenge for next generation mobile communication system. Addressing such challenge requires the development of technologies and solutions to achieve higher packet data throughput over wireless link to support a greater number of users within individual cells, and to deliver a significantly enhanced experience to users.

It is envisioned that the 3G mobile communication system will provide broadband-like capabilities. In comparison to the precedent generations, such a system allows simultaneous use of voice and data services with higher data rates. It can offer roaming subscribers with downlink speeds up to 14.0 Mbps with the protocol High-Speed Downlink Packet Access (HSDPA) and with uplink speeds up to 5.76 Mbps with the High-Speed Uplink Packet Access (HSUPA). Further speed increases are available with the Evolved High-Speed Packet Access (HSPA+), which provides data rates up to 56 Mbps on the downlink and 22 Mbps on the uplink with *Multiple Input Multiple Output* (MIMO) technologies and higher order modulation like 64 QAM (Quadrature Amplitude Modulation). In summary, the 3G wireless system enables network operators to offer subscribers a wider range of value-added services while achieving greater network capacity through improved spectral efficiency.

On the other hand, the wireless metropolitan area network, also known as WiMAX system within the industry alliance WiMAX Forum, provides up to 10 Mbps broadband speed with a larger geographical area than a wireless local area network (WLAN). Typically, a wireless metropolitan area network usually covers several blocks of buildings to an entire city or a large campus. However, its geographic scope falls between a wide area network and a local area network (LAN). Such network can provide broadband wireless access up to 50 Km for fixed stations, and 5 to 15 Km for mobile stations. In contrast, the coverage area of a WLAN is limited to only 30 to 100 m, in most cases. In general, a metropolitan area network provides LAN users with Internet connectivity in a metropolitan region. In addition to offering higher data rates over longer distances, a wireless metropolitan area network allows for more efficient bandwidth use and less signal interference due to the fact that it employs the Orthogonal Frequency Division Multiplexing (OFDM) technique for transmission. Such multiplexing method makes the signal in RF systems less likely to be affected by fading.

Based on the IEEE 802.16 standards, wireless metropolitan area networks are gaining in popularity due to their capability of offering higher data rates, compared to the 3G mobile system, but still limited in global mobility support. In this context, integration of the 3G system with the wireless metropolitan area network (WMAN) is crucial to providing subscribers with wider coverage area, higher data rates, and stronger mobility support. It is expected that through integration, users can experience seamless and transparent wireless connectivity while on the move, as well as receive better communication services with guaranteed performance. However, as the 3G mobile communication system and WMAN have different characteristics such as radio coverage, data rate, network protocol, and quality of service aspect, it is a great challenge for seamless integration and interworking of such two disparate wireless systems.

Literature review shows that most of research activities focus on design interworking architecture for integration of the 3G mobile system with the WLAN and only a few solutions are designed for integration, and interworking of the 3G communication system with WMAN. This section presents a loosely-coupled integrated architecture of 3G/WMAN, together with quality of service support and Session Initiation Protocol (SIP) [37] based mobility management. Figure 3.1 illustrates the integrated network architecture between a 3G system and a WMAN [21].

In order to provide mobility support, the gateway (GW) in the WMAN is added with the functionality of the foreign agent (FA) [21]. Note that a foreign agent is a router that stores information about mobile nodes visiting its network. Such router broadcasts router advertisement messages periodically. Upon receiving such a message, mobile nodes can formulate their care-of addresses that are used within the visiting network via *Mobile IP* operation [31]. The gateway provides inter-domain and intra-domain mobility support. In addition, an authentication, authorization, accounting (AAA) server is also added into the WMAN coverage area, shown in the shaded zone on Figure 3.1. Besides, another AAA server is also added into the 3G mobile system. The AAA server handles user requests for access to network resources and deals with authentication, authorization, and billing issues. When they are implemented in data networks, they are similar to the Home Location Register (HLR) in the circuit-switched voice network. The home AAA (HAAA) server stores user profile information, responds to authentication requests, and collects accounting information. These network elements are shown in Figure 3.1.

To guarantee session continuity, two mobility management schemes are designed [21]: one for non-real-time session and another for real-time session. For non-real-time traffic, Mobile IP (MIP) protocol [30] is recommended for operation. Thus, session mobility is controlled at the network layer. However, for real-time traffic, SIP [37] is proposed to avoid a triangular routing problem and the latency caused by agent discovery procedure. Consequently, session mobility is controlled at the application layer, instead of at the network layer, for real-time traffic.

Figure 3.2 illustrates mobility management procedure for real-time sessions using the protocol SIP. When mobile nodes perform intersystem handoff, the old real-time session is recommended to be terminated. Mobile nodes (MNs) then interact with the integrated network for new link activation and quality of service negotiation. Upon attaching on the new link, they send a re-INVITE message to their peer, a corresponding node (CN) by executing the protocol SIP. Upon success, the new real-time session is activated again.

To offer users with suitable quality of service in a visiting system, new quality of service (QoS) control mechanisms are designed [21]. For example, process used to activate a QoS session, mechanism of QoS classes mapping between the UMTS and WMAN [21]. Besides, schemes for network and session layer quality of service support are also proposed [21]. However, Figure 3.1 shows that ongoing real-time sessions inevitably will be interrupted because of link activation and quality of service negotiation. When the switching latency is too long, it is impossible to support delay-sensitive multimedia applications.

On the other hand, the proposal [21] also shows that using loosely-coupled way to integrate disparate wireless systems is a better choice than the tightly-coupled solution. This is because network elements do not require modification

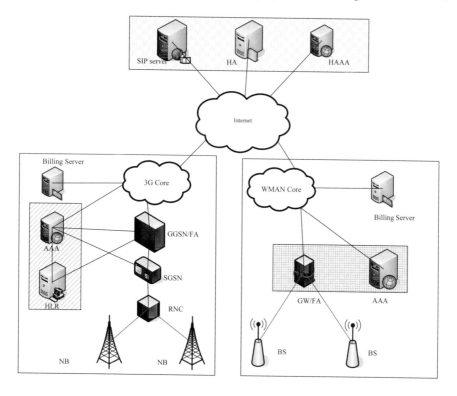

FIGURE 3.1
Integrated architecture for 3G and WMANs.

during interworking, and the interworking architecture can be decoupled from specific standards and protocols. Loose coupling also enables each network to operate in their own way with little or few intervention between them. Different coupling techniques (e.g., tight coupling, loose coupling, no coupling, hybrid coupling) are provided in detail in Section 3.2.3.

3.2.2   Integrated Architecture for WLAN/WMAN

Typically, WLANs are connected to the Internet through a wired infrastructure in hotspot area, such as airport, cafés, shopping malls, etc. However, such connection may be unavailable in remote rural or suburban areas. On the other hand, the WiMAX technology is an important choice to provide backhaul support for WLAN hotspots.

WiMAX, which stands for worldwide interoperability for microwave access, is a telecommunications technology that enables wireless transmission of data from point-to-multipoint links to portable and fully mobile Internet access. Such technology is based on the IEEE 802.16 standards suite, also known as

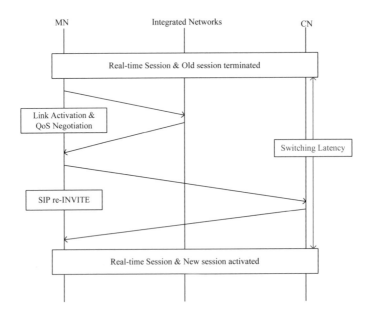

FIGURE 3.2
Real-time session mobility management using SIP

broadband wireless access (BWA). It supports point-to-multipoint single-hop transmission between a base station (BS) and multiple mobile nodes. The base station may act as mesh router, while mobile nodes can work as mesh clients, thus forming a multihop wireless mesh network (WMN). Note that a mesh router is a router forwarding traffic to and from the gateways of a WMAN, while mesh clients are often laptops, cell phones, and other wireless devices.

The complementary nature of wireless metropolitan and local area networks has attracted industry, academia, and standardization organizations for their integration, because such integration can extend the radio coverage area of wireless local area network and increase service availability for mobile Internet applications. However, as WLAN and WMAN present different characteristics, such integration presents a difficult issue [29]. For example, a Mobile Station (MS) with intersystem handover capability must equip with dual radio interfaces: one used for establishing connectivity with the WMAN, the other for the WLAN. In addition, the base station in the WMAN must be capable to relay WLAN traffic to the Internet. To do so, it must be capable of operating on mesh mode, acting as mesh router in the integrated environment. Some other research issues also arise when integrating WLAN with WMAN. The challenges will be elaborated in Section 3.4. Just to name a few, radio resource management, topology management, error control, end-to-end quality of service control aspects are issues that need to be well handled during integration.

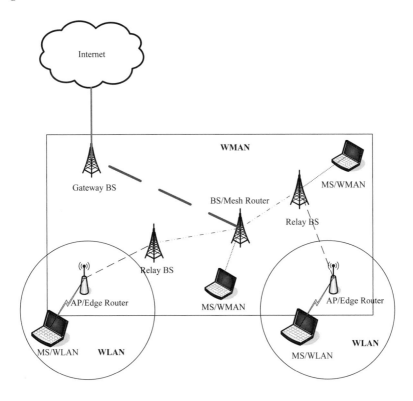

FIGURE 3.3
Mesh topology WLAN/WMAN integrated architecture.

Figure 3.3 shows an integration of WLAN with WMAN using mesh topology [29]. The access point (AP) in a WLAN acts as an edge router connected to the WMAN via a relay base station. The base station (BS) typically takes charge of traffic from and to mobile stations (MS) at its radio range. The relay base stations then connect to a base station that works as a mesh router. Such mesh router enables broadband Internet access through a backhaul connection with a gateway base station. Note that the edge router must be equipped with a dual radio interface (IEEE 802.11 and 802.16). In addition, any BS in the WMAN should provide services to stand-alone subscriber stations, called mobile stations (MS/WMAN), and WLAN access points that connect to it. The Grant Per Subscriber Station (GPSS) service class is recommended for use for every stand-alone mobile station and access point. So the base station allocates bandwidth to each subscriber mobile station and associated edge routers separately [29]. To avoid cochannel interference, adjacent base stations in the WMAN must operate on different frequency bands.

In general, integration of WLAN/WMAN can actually extend the coverage area of WLANs and augment service availability for mobile Internet applications in case where a wired infrastructure is unavailable (e.g., in remote rural or suburban areas). To guarantee success communications between WLAN and WMAN, between mobile stations in the WLAN (MS/WLAN) and the Internet, integration is done at an Edge Router, which also functions as a WLAN access point. Such router possesses two radio interfaces using different frequency bands, and two queues to separate local and Internet traffic. Local traffic corresponds to the traffic within the WLAN coverage, whereas Internet traffic is due to communications traversing the wireless mesh network via base stations of the WMAN. Packets from and to the Internet must be reformatted by the edge router and gateway base station. However, such fact adds more complexity into the implementation of interworking. On the other hand, Ethernet-based bridging technique, instead of using mesh topology, may also be used for interworking of WLANs with WMAN [14].

3.2.3 Interworking Architecture for 3G/WLAN

Integration of wireless local area networks (WLANs) and cellular telecommunication networks can enable cellular operators to benefit from the rapidly evolved WLAN technology and provide advanced high-speed data services to subscribers with one subscription, one bill, and one set of services [36]. On the other hand, effectively combining the 3rd generation (3G) systems and WLANs into an integrated wireless data access environment empowers mobile users with ubiquitous connectivity and high-speed services, especially in hotspot areas where bandwidth-sensitive applications are most demanded. However, interworking between WLANs and the 3G mobile systems can be implemented in different ways; for example, the wireless local area network could be an integral part of the 3G mobile system or the two wireless systems could be separating [2, 1].

To succeed in integration and interworking between wireless local area networks and 3GPP/3GPP2 mobile communication systems, the general requirements are described as follows [2]:

- Clearly specify the way of splitting the functionalities of wireless local area network and the 3G mobile systems.

- Wireless local area networks should be posed as little as possible during interworking.

- Support many-to-many relationship between the two systems, that is, allow that one wireless local area network can interwork with several 3G mobile systems, and vice versa.

Besides, six interworking scenarios are identified by the 3G Partnership Project (3GPP) initiative, which are listed as follows [2]:

- **Scenario 1 – Common billing and customer care**: Under this scenario, there are no requirements affecting 3GPP specifications.

- **Scenario 2 – Common access control and charging**: It consists of network selection, system recognition, access control, security, roaming and terminal aspects, as well as naming and addressing, charging and billing aspects.

- **Scenario 3 – Access to 3GPP system packet-switched based services**: It includes service access control, security, and quality of service provisioning and roaming aspects.

- **Scenario 4 – Service continuity**: The services available will survive the handover process while changing access network technology between WLAN and a 3GPP system.

- **Scenario 5 – Seamless services**: The continuity of services when performing intertechnology handover between the 3GPP and WLAN access technologies should be seamless, that is, the user experiences no interruption of the available services during the handoff.

- **Scenario 6 – Not available**: So far, no use cases have been identified for this scenario.

Generally, the way of integration between 3GPP/3GPP2 mobile system with wireless local area networks can be classified into tight coupling [1, 36, 5, 32, 8], loose coupling [35, 8], no coupling [47, 48, 49], and hybrid coupling [40, 41]. The following subsections will elaborate each coupling technique in detail.

3.2.3.1 Tight Coupling

Tight coupling (or emulator approach) allows a wireless local area network (WLAN) to appear to the 3G core network as either a GPRS radio access network [1, 36], or as a Packet Control Function (PCF) in case of CDMA2000 network [7]. Figure 3.4 illustrates an example of integration between WLAN and GPRS using tight coupling technique.

Figure 3.4 shows that the interworking of WLAN and GPRS is realized through a GPRS Interworking Function (GIF). Such node connects the WLAN's Distribution System to a Serving GPRS Support Node (SGSN) through the interface Gb, which is a standard interface to the GPRS core network. The GIF virtually hides the particularity of the wireless local area network makes the node SGSN to take for grant that the wireless local area network is a regular Routing Area (RA) to be handled in the same way as other routing areas in the GSM/GPRS mobile system. Figure 3.5 illustrates an example of integration of WLAN with CDMA2000 network using tight coupling technique.

Figure 3.5 shows that the interworking of WLAN and CDMA2000 network is done through a gateway with the Packet Control Function (PCF). Such

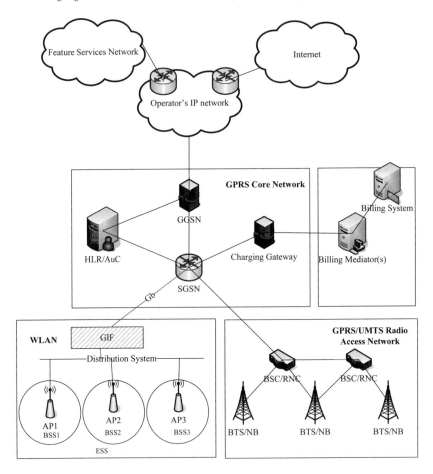

FIGURE 3.4
Tight-coupling WLAN/GPRS integrated architecture.

node connects to a Packet Data Serving Node (PDSN) that resides in the 3G core network. The gateway hides the details of the wireless local area network to the 3G core network. To achieve this, it implements all the 3G protocols, for mobility management, authentication, etc. On the other hand, to be capable of performing intersystem handover, mobile stations are also required implementing the 3G protocol stack on top of their WLAN interface cards. All the traffic from the wireless local area network are passed through the gateway/PCF, reformatted and injected into the 3G core network [7].

Using tight-coupling, the cellular radio is simply replaced by the WLAN radio that provides equivalent functionality of a 3G radio access network. Hence, such coupling technique enables the fully reuse of the 3G mobile system protocols and existing network infrastructures. Moreover, a tight-coupled approach

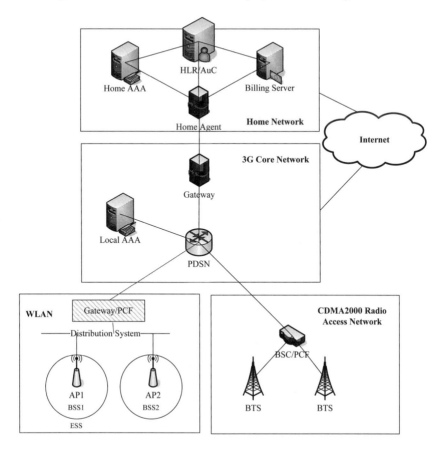

FIGURE 3.5
Tight coupling WLAN and CDMA2000 integration.

enables WLAN's users to access to the 3G mobile system services with guaranteed quality of service support and seamless mobility [40]. In particular, such interworking method allows low handover delay and reduced packet loss [26].

However, the two figures illustrate that, using tight-coupled architecture, a new interface is necessary between WLAN and the 3G core network. Such a fact implies that the involved interworking wireless systems must belong to the same network operator. Consequently, independently operated WLANs cannot be integrated into the 3G mobile cellular networks [47, 8, 7]; this results in less flexibility of integration. Furthermore, mobile nodes must implement the corresponding 3G protocol stack on top of the standard 802.11 [8]. This adds design complexity on the user's device. Moreover, as all the packets from incorporated WLANs go through the 3G core network, integrating points that connects to the 3G core network via either an SGSN or a PDSN become traffic bottlenecks [47]. It is also hard for nodes like SGSN and PDSN to accommodate bulky traffic. To fix this, modification or upgrading is required for the 3G core network to sustain the increased load from integrated WLANs.

3.2.3.2 Loose Coupling

Loose coupling, also known as mobile IP approach, allows a WLAN to connect to the 3G core network indirectly via an external IP network such as the Internet, any Packet Data Networks (PDNs) [42, 36, 8].

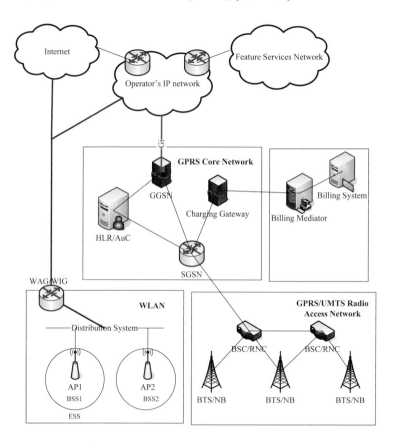

FIGURE 3.6
Loose-coupling WLAN and GPRS integration.

Figure 3.6 illustrates an example of integration of WLAN with General Packet Radio Service (GPRS) network using loose coupling [36]. A WLAN is coupled with the GPRS core network via the interface Gi, which is a standard interface that connects a Gateway GPRS Support Node (GGSN) with external PDNs, for example, an operator's IP network. A Wireless Access Gateway

(WAG), sometimes also called Wireless Interworking Gateway (WIG), is implemented within the wireless local area network to connect to external IP network. In contrast to tight coupling, shown in Figure 3.4, the WLAN data traffic does not pass through the GPRS core network, but goes directly either to an operator's IP network or to the Internet. To gain access to any services offered by the Feature Services Network, authentication that is based on the Subscriber Identity Module (SIM) should be supported by both WLAN and GPRS network [36]. In addition, such integration and interworking architecture supports integrated billing through billing mediators (or brokers). Generally, loose coupling uses the protocols that are designed by working groups within the Internet Engineering Task Force (IETF) for authentication, accounting, and mobility management.

Figure 3.7 shows an example of interworking between WLAN and CDMA2000 network with loose coupling. The loose coupling approach calls for the introduction of a new network element into a wireless local area network: a Wireless Access Gateway (WAG) with Packet Control Function (PCF). Such node connects to the Internet without any direct link to the 3G mobile system. As a result, the high-speed data traffic from wireless local area network is never injected into the 3G core network using loosely coupled architecture.

In summary, loose coupling allows independent deployment and traffic engineering of WLAN and 3G systems. With roaming agreements established among service providers, mobile nodes can obtain services offered by other communication systems, that is, loose-coupled architecture provides opportunities for wireless Internet service providers (WISPs) to extend their service scopes/areas via roaming agreements with other wireless local area networks and 3G network operators. Besides, using loose coupling, the data paths of a wireless local area network are completely separated from those of the 3G mobile system [8]. Different protocols are adopted to handle authentication, mobility management, and billing.

However, to achieve seamless integration, these protocols must interoperate with each other. For example, in case of WLAN coupled with CDMA2000 network, the WLAN must support MIP and proxy AAA (P-AAA or AAA proxy) functionalities for mobility management and billing issues. The major weakness of this approach is excessive handoff latency, because mobility signaling has to traverse a relatively long path due to the separation between the 3G radio access network and WLAN domains [42]. Furthermore, high handoff delay leads to unacceptable packet loss, traffic congestion, and handoff failure; this makes loose coupling unable to support service continuity during intersystem handoff [40]. Nonetheless, loose-coupled architecture is the most preferred solution for seamless integration of wireless local area networks and the 3G mobile communication system [42, 8].

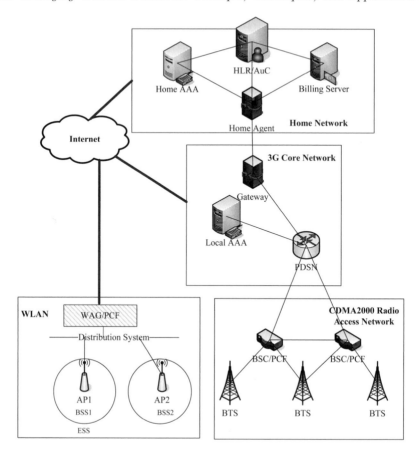

FIGURE 3.7
Loose-coupling WLAN and CDMA2000 interworking.

3.2.3.3 No Coupling

No coupling, also named the gateway approach, treats the integrated wireless local area network (WLAN) and 3G mobile system as peer-to-peer (P2P) networks [26]. The principal idea of this coupling technique is exploiting legacy mobility management protocols to handle intrasystem (or horizontal) handoff and using Roaming Agreements (RAs) to handle intersystem (or vertical) handoff via a logical node: a gateway (GW). The gateway exchanges necessary information between the WLAN and 3G communication system, converts signals and routes packets for roaming users [47, 48, 49]. Such a node can be named as a Virtual GPRS Support Node (VGSN) [48], or Interworking Gateway (IG) [6, 26, 27], or Interworking Decision Engine (IDE) [24, 25].

To enable the utilization of the no-coupling technique, several requirements are imposed for its implementation [48]. First of all, roaming users must have a dual mode device, that is, mobile devices possess two radio interfaces, just as other interworking solutions like tight and loose coupling. In addition, the two involved wireless networks should have roaming agreements. As a result, subscribers of one network are permitted to use another network for packet data services. Furthermore, the two wireless communication systems are peer-to-peer networks that are managed and operated separately. All traffic from and to the Internet are handled by each individual network without involving the other system. Besides, applications executed on a mobile device should use a unique IP address as its identifier for access to the Internet. And service continuity and quality of session should be guaranteed during handoff.

To realize no-coupling, a logical network entity, Virtual GPRS Support Node (VGSN), is introduced into the interworking environment. Such node connects to a WLAN through either a WAG or an Access Router (AR), at the same time, connecting to the UMTS (Universal Mobile Telecommunications System) network via either a SGSN or a GGSN. The VGSN is responsible for intersystem handoff between WLAN and UMTS, that is, incoming and outgoing packets must go through the VGSN while mobile nodes move their connectivity from one radio access network to another. To facilitate mobility management, a VGSN address resolution procedure is designed for inter-system handoff. However, such facts may lead to additional handoff delays and more packet losses. Figure 3.8 shows an example of interworking of WLAN and UMTS using the no-coupling technique.

The advantage of the no-coupling approach is that the integrating networks can operate independently. Compared to the loose-coupling approach, Mobile IP functionality is not mandatory, thus results in less handoff delays and lower packet losses [47]. But this coupling technique still has problems such as scalability caused by the imposed roaming agreements between involved wireless systems.

3.2.3.4 Hybrid Coupling

Tight coupling has the advantages of low handoff delay and reduced packet loss, but with the problem of flexibility, scalability. In addition, the inter-working point is a bottleneck due to the fact that the 3G core network nodes are designed to handle small-sized data: circuit-switched voice calls or short messages. On the other hand, loose coupling enables independent data paths for traffic from WLAN and the 3G mobile system, as well as independent deployment and traffic engineering. As a result, each individual network does not need to change their network architecture, neither the protocol stack. However, loosely coupled architecture suffers from long handoff latency, high packet loss, thus it is impossible to support service continuity while mobile stations perform intersystem movements. Under the circumstances, a hybrid coupling technique is proposed to interwork WLAN with the UMTS and allows the UMTS core network to efficiently accommodate traffic from the integrated WLAN [40].

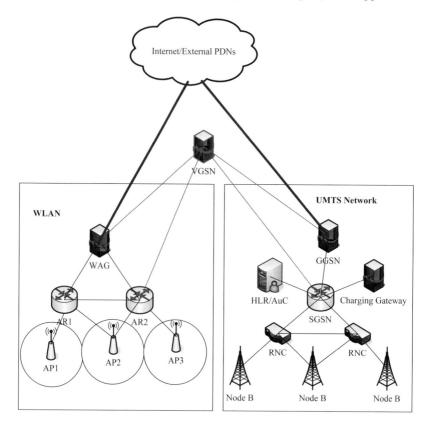

FIGURE 3.8
WLAN and UMTS integration using no-coupling.

The main idea of the hybrid coupling approach is that data paths are differentiated according to traffic type. For real-time traffic, tight coupling approach is exploited, so WLAN data traffic traverses the UMTS core network. For non-real-time traffic, loose coupling is necessary, which forces WLAN data packets go through an external IP network.

To achieve hybrid coupling, a new network entity, called access point gateway (APGW) is added into the integrated environment [41, 40]. Such a node is responsible for differentiating real-time and non-real-time traffic generated within the WLAN, and forwarding packets of real-time traffic to an SGSN that resides in a 3G core network, and forwarding non-real-time traffic to an access router, which then connects the WLAN to the Internet or external PDNs. In other words, the APGW must be able to differentiate service types, according to which, it decides the traffic path through either an SGSN or an access router. In a meantime, the APGW is also responsible for radio resource management within the WLAN and mapping the WLAN's radio resources to those of the 3G mobile network. Figure 3.9 illustrates an example of interworking of WLAN and UMTS using hybrid coupling.

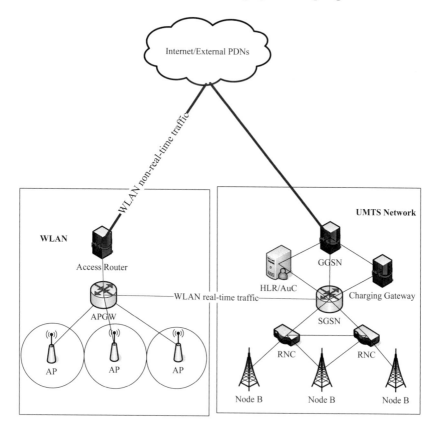

FIGURE 3.9
Hybrid coupling for WLAN and UMTS integration

Tightly coupled architecture enables mobility management by using either Mobile IP (MIP) protocol or any tunneling protocols. However, with hybrid coupling, the protocol MIP and the simultaneous binding option are recommended to support mobility management for real-time and non-real-time traffic. Such a solution better addresses the intersystem handoff while minimizing handoff latency and packet loss.

3.2.4 Interworking in Heterogeneous Wireless Systems

This section presents two typical integrated architectures in heterogeneous wireless systems: the Architecture for ubiquitous Mobile Communications (AMC) and Integrated InterSystem Architecture (IISA).

3.2.4.1 Architecture for Ubiquitous Mobile Communications

Rapid progress in wireless networking and mobile communications makes it possible for subscribers to benefit from a number of mobile communication systems, such as WiFi network, WiMAX system, 2G, 3G, as well as next-generation (NG) mobile communication networks. The latter is also known as beyond 3G or 4G wireless and mobile networks. These different wireless systems are expected to coordinate with each other to offer ubiquitous high-speed services to roaming users. Under the circumstance, the Architecture for ubiquitous Mobile Communications (AMC) is proposed [6, 26, 27], which allows the integration and interworking of disparate wireless systems including CDMA2000 network, GPRS, satellite network, and WLAN.

Disparate wireless networks connect to the Internet or other external PDNs using gateways. For example, PDSN for CDMA2000 network; Gateway Station (GS) for satellite network; GGSN for UMTS/GPRS network; WAG or an AR for WLAN. To realize the integration of these different communication systems, two network elements are introduced: Network Interoperating Agent (NIA) and Interworking Gateway (IG) [6]. The NIA functions as the third party with whom each individual wireless system establishes a direct Service Level Agreement (SLA). Its utilization can eliminate the needs of separate SLAs with all other operators. Besides, the NIA is responsible for authentication, authorization, billing, and mobility management for intersystem roaming [6]. The IG works as a gateway between a particular system and the NIA; it is responsible for mobility and traffic management, authentication, and accounting issues. The functionality of an IG can also be integrated into an interworking gateway, such as PDSN, GGSN, GS, WAG, or AR. The NIA resides in the Internet while the IG is located within each network. Figure 3.10 illustrates AMC.

Using the Internet Protocol (IP) as the interconnection protocol, AMC is capable of integrating any number of systems of different service providers. On the other hand, with the introduction of new network entities NIA and IG, the need for establishing direct Service Level Agreements (SLAs) among service providers can be eliminated. Furthermore, third-party-based authentication, billing algorithms, and mobility management protocols are developed to facilitate intersystem roaming.

However, the utilization of the NIA leads to an implicit problem of scalability. In a case where a service provider cannot establish an SLA with the designated third party, who hosts the functionality of the NIA, the corresponding wireless system cannot be integrated into AMC. Under this circumstance, a hybrid solution is preferable, which allows each individual system to establish an SLA with either the specific third party or a partner who has a direct SLA with the NIA. On the other hand, a new handoff decision mechanism is necessary for integration of disparate wireless systems. Such mechanism should take the heterogeneity into consideration. However, AMC does not design any ap-

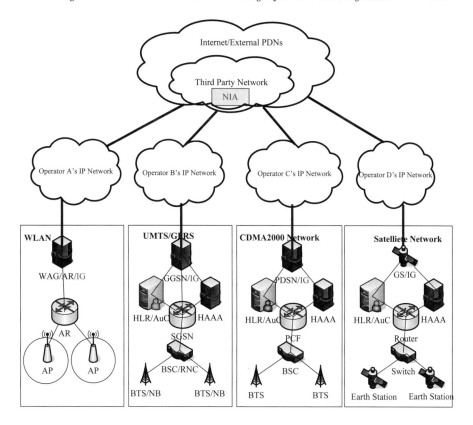

FIGURE 3.10
Architecture for Ubiquitous Mobile Communications (AMC)

propriate handoff decision algorithm [25]. Such fact hinders seamless mobility management when mobile users perform intersystem handover.

3.2.4.2   Integrated Intersystem Architecture

Coexistence of diverse but complementary mobile systems and wireless access technologies make it possible for mobile users to take benefit from each individual system and to move freely from one network to another [24, 25]. A typical next-generation wireless system consists of many different wireless networks overlaid with each other while formulating a wireless overlay network. The introduction of next generation wireless system aims to enable subscribers to access to services anytime, anywhere. Such concept is known as "always best connected" [16] to the most appropriate wireless access network. Under the circumstances, an Integrated Intersystem Architecture (IISA) [24, 25] is designed to enable the integration and interworking of existing wireless systems, which include wireless local area network (WLAN), 3GPP, and 3GPP2 packet-switched data networks. Figure 3.11 illustrates IISA.

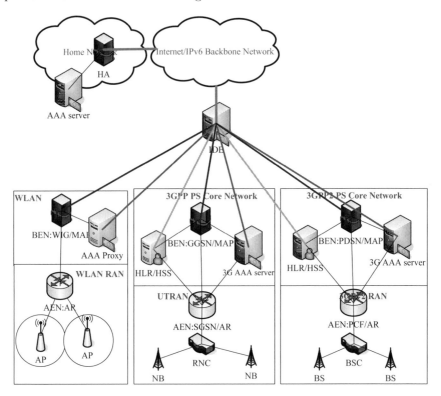

FIGURE 3.11
Integrated Intersystem Architecture (IISA).

The architecture IISA is based on 3GPP/3GPP2 packet-switched networks interworking with wireless local area networks. This architecture attempts to fulfill the requirements of designing new integrated architecture for next generation wireless system, such as scalability, transparency to underlying radio access technologies, economics and security. A new network element is added into the interworking architecture, which is called the Interworking Decision Engine (IDE). Such a node connects each individual communication system with the Internet and IPv6 backbone network, shown on Figure 3.11. It is responsible for handling signaling traffic instead of data packets, as well as mobility management, authentication, accounting, and resource management. In particular, a handoff decision module is implemented at the IDE, which assists mobile nodes to be "always best connected."

To facilitate IPv6-based mobility support, some network elements of 3G mobile system are added with new functionalities. For example, the SGSN in a 3GPP system and the PCF in a 3GPP2 network, are enhanced with AR functionalities [20] to facilitate IP mobility support. Such nodes are called Access Edge Nodes (AEN). In the same vein, acting as an interworking gateway, the GGSN, PDSN, and WIG or WAG, are also extended with Mobility Anchor Point (MAP) [39] and/or Home Agent (HA) [20] functions. Such nodes are called Border Edge Nodes (BENs). The WIG is particularly responsible for policy routing, message format conversion, and quality of service mapping between WLAN and the 3G mobile systems. Another network entity, P-AAA or AAA proxy, is also added into the WLAN which is responsible for authentication and billing issues when mobile nodes perform intersystem (or vertical) handoff.

Generally, the IISA presents almost similar idea as the AMC [6, 26, 27] with some enhancements such as the handoff decision algorithm and new Handoff Protocol for Integrated Networks (HPIN). The latter is based on localized mobility management, access network discovery and fast handoff concepts to provide efficient handover management. However, it still cannot support seamless mobility due to the fact that the signaling overheads are too expensive while executing the HPIN. On the other hand, fast authentication is not taken into account simultaneously with mobility management. Under the circumstances, a new Architecture to support Fast authentication and Seamless roaming (AFS) is proposed for next-generation heterogeneous wireless networks, which will be elaborated in the following section.

3.3   New Unified Integrated Architecture

Mobile subscribers are increasing their demands for personalized multimedia services, anytime, anywhere, and on any wireless device. This makes the integration of disparate wireless communication networks is indispensable to address users' needs. However, provisioning of value-added services imposes different quality of service requirements over the communication system, in terms of latency, bandwidth, packet loss rate, jitter, error rate, etc. Thus, new challenges are brought about while designing the next-generation wireless system.

There are two possible solutions for integrating different communication systems: either developing novel wireless systems with radio interfaces and technologies to satisfy the requirements of the services demanded by future mobile users or integrating existing wireless systems [6]. Since the former solution is expensive and impractical, we advocate using as many existing network architectures as possible, the same idea as that presented in [6, 26, 27, 24, 25].

3.3.1 Architecture for Fast Authentication and Seamless Roaming

The novel architecture to support fast authentication and seamless roaming (AFS) integrates almost all the existing communication systems, such as the 2nd and 3rd generations (2G/3G) mobile cellular networks, WLANs, satellite networks, and WMANs. The latter is also known as the WiMAX system.

The 2G mobile system to be integrated into the new unified integrated architecture is the GSM with enhanced data delivery support, that is, GSM/GPRS and GSM/EDGE. In addition, two major 3G mobile systems are selected for integration: UMTS and CDMA2000 (or EV-DO) system. The reason for this selection is that both of them will be upgraded into the fourth generation (4G) wireless network, and they are well-specified by 3GPP and 3GPP2 initiatives.

To realize seamless roaming and fast authentication, mobile service area is partitioned into mobility regions, which are further divided into mobility domains. Each administrative domain (AD) is controlled by a new network entity called enhanced Mobility Anchor Point (eMAP), which is also called Regional Intelligent Node (RIN). Another new network element, the enhanced Home Authentication, Authorization and Accounting (eHAAA) server, is also introduced into the novel AFS. Such a node resides in the mobile nodes' home network and is responsible for fast authentication. The mobility regions and domains overlap with one another, and formulate a wireless overlaid multitier network.

To achieve transparency to underlying heterogeneous radio access technologies, the novel unified integrated architecture AFS exploits the IPv6 [11] as the interconnection protocol. As a result, such architecture is capable of integrating any number of systems of different service providers, same as AMC and IISA.

Figure 3.12 illustrates AFS. Mobile service areas are partitioned into mobility regions (Mobility Region 1 and 2, shown on the figure), which are further divided into administrative domains (WLAN domain, 3GPP network domain, 3GPP2 network domain, satellite network domain and WiMAX domain). Each domain may use heterogeneous or homogeneous radio access technology. The mobility regions and domains overlap with each other and formulate a wireless overlaid multitier network.

Each mobility domain is managed by an eMAP. Such node aggregates key functionalities of the MAP [39], the new Authentication Center (nAuC), and the interworking gateway. For example, an eMAP implemented at a WLAN contain the functionality of a WAG for interworking purpose. Once deployed in a 3GPP/3GPP2 network, it works as a GGSN or a PDSN. In a WiMAX network domain, the eMAP may function as a Connectivity Service Network Gateway (CSN-GW) or a Gateway base station using mesh topology [29]. Once installed in a satellite network, the eMAP acts as a standard Gateway Station (GS) [6]. The generic functionality of an interworking gateway, such

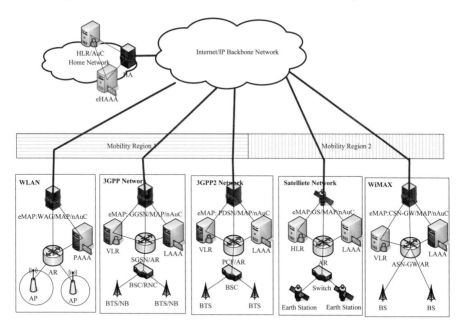

FIGURE 3.12

Architecture for fast authentication and seamless roaming (AFS).

as GGSN, WAG/WIG, PDSN, GS, and Gateway BS is to perform IP Network Address and Port Translation (NAPT) to enable the access network operator to use private-space IP addresses inside its domain, as well as simultaneously provide external IP network access. It is also possible for the interworking gateway to manage SLAs among service providers; in this case, the interworking gateway in the proposed novel architecture can work as the Network Interoperating Agent (NIA) and Interworking Gateway (IG) that are introduced in [6].

Containing the functionality of the MAP makes the node eMAP support IPv6-based mobility management. Additionally, with the new Authentication Center (nAuC), the eMAP can authenticate each mobile subscriber that attempts to access services provided by any packet-switched core network. Another option enables an eMAP to combine the functionality of the Local Authentication, Authorization and Accounting (LAAA) server [9], and/or Visitor Location Register (VLR) for authentication and billing issues. Or, the typical functions of the LAAA server can be merged into the nAuC.

Another new network entity, the enhanced Home AAA (eHAAA) server, is also introduced into the novel architecture AFS. Such a node resides in a mobile node's home network. It contains the key functionality of a standard HAAA server [9] with a new Central Database (CD). Such a database holds detailed information about each mobility region such as the identifier of the region, IP address of each associated eMAP in that region, neighborhood of each eMAP, identifier of each Regional Edge Node (REN) in that region, etc. It is also possible for the eHAAA server to implement the functionality of a Home Location Register (HLR), and/or a home agent (HA). Once their functionalities are combined together, they formulate a new network entity, called Home Intelligent Node (HIN).

In brief, the new unified integrated architecture AFS interworks disparate wireless systems including WLAN, WMAN, satellite network and 3GPP/3GPP2 mobile cellular networks. Unlike AMC [6] and IISA [24, 25], AFS does not imply any preexistence of roaming agreements (RAs) or SLAs among service providers. This is because any mobile user can obtain services from a visiting network by either direct predefined RA or SLA or by dynamically establishing indirect RA or SLA via a trusted partner. Such process is accomplished by executing fast authentication and seamless roaming schemes, which will be described in Section 3.3.3. A new access network selection method is also proposed in the following section.

3.3.2   New Access Network Selection Method

Selection of a new access network is very important to assist mobile users to make a good handoff decision. In return, a good handoff decision can result in establishing connections with the best available network. Under the circumstance, a new access network selection method is proposed for the novel unified integrated architecture. The main idea is to avoid vertical (or intersystem) handoff as much as possible, if applicable. Otherwise, a network should be selected with a direct RA or SLA; if the former two requirements cannot be met, then mobile nodes should select a network with which their home service provider could establish a temporary virtual indirect RA/SLA via a trusted third party.

Entering into a new administrative domain, a mobile node needs to select the best connected network before breaking its current connection. Therefore, it executes the following network selection method:

1. Analyze wireless environment within range by listening to broadcast channels and examining the received broadcasting messages from the nearby base stations.

2. Select a network with the same radio access technology and direct RA/SLA, that is, the home network service provider has already established a RA/SLA with the service provider of the visiting domain.

3. Otherwise, select a network with the same radio access technology and indirect RA/SLA. The latter means the involved network service providers can establish an indirect temporary virtual RA/SLA via a trusted third party.

4. Select a network with integration capability and direct RA/SLA with the mobile node's home network service provider.

5. Otherwise, select a network with integration capability and indirect RA/SLA via a trusted third-party.

To facilitate access network selection, the access point (AP) in a WLAN should insert some necessary information into its broadcasting message, such as WLAN's name, integration capability, name list of the 3GPP/3GPP2, WMAN, and satellite network partners, mobile country code (MCC)/mobile network code (MNC). The latter is also known as a "MCC/MNC tuple," used to uniquely identify a mobile phone operator/carrier using the GSM, CDMA, UMTS, satellite mobile networks, etc. Here we advocate adding such pair for WLANs and WMANs to facilitate future integration and interworking.

The name list of different partners indicates network service providers whose network is already integrated into the unified interworking architecture: AFS. It enables mobile nodes to make intelligent decisions about network selection. It shows mobile users the list of service providers with which their home network service provider can establish direct RAs or SLAs, no matter which radio access technology is deployed in their administrative domain. For example, an AP in a WLAN may broadcast a beacon with information like "butterfly: integration capability/yes: 3G Telus: 302/220." This means the WLAN's name is a butterfly; such network has the capability to integrate with a 3G mobile system, controlled by Telus Mobility. The MCC of 320 indicates the country of Canada, whereas the mobile network code 220 indicates using the technology UMTS.

The same principle is applicable for WMAN, satellite, and 3GPP/3GPP2 mobile cellular networks. The information broadcasted by a base station can help mobile nodes to first select a trusted domain with the same radio access technology, and then handoff to that domain. Otherwise, select a temporary domain with integration capability, with which a temporary trust relationship can be established via a trusted third party that resides near the visiting domain. By this means, after successful authentication with a trusted third party, all outgoing data traffic from the mobile node will be tunneled to a neighboring trusted domain, and then forwarded to the destination node. At the same time, all incoming traffic will be intercepted by a previous trusted domain and tunneled to the new domain, then de-tunneled and forwarded to the mobile node.

An example of new access network selection is given here for mobile nodes before performing a handoff. It is assumed that each mobile node should have pre-knowledge about the partners of its home service provider. For example, supposing that the mobile node's home service provider is A, a network operator with UMTS technology. B and D are A's trusted partners. When this mobile user moves into a new visiting domain, controlled by the network operator C, it receives beacons from base stations within its radio range with the following information:

- WLAN/butterfly: integration capability/yes: 3G D: 302/220

- WLAN/supercool: integration capability/yes: WMAN F: 302/999

From the received messages, mobile users know that the butterfly network is a WLAN that can integrate with a 3G mobile system, owned by the network operator D, that exploits the UMTS technology, whereas another WLAN is called, supercool, which can integrate with a WMAN, owned by the network operator F, using the WiMAX technology. Since A and D have already some RAs/SLAs between them, the mobile node must connect to the butterfly network for an appropriate handoff decision. This is because an indirect temporary RA/SLA can be established between A and C, via a trusted third party: D.

3.3.3 Fast Authentication and Seamless Roaming Schemes

Solutions currently found in the literature attempt to reduce authentication delays in a single administrative domain (AD), where centralized control is preferred [17]. However, as the demands for seamless roaming across different wireless systems (or domains) increase, new authentication solution is required for interdomain (and intersystem) handoff. Since each domain is under its own authority and administration [17], each authority exploits its own authentication mechanism that is unavailable to others. This situation makes third-party-based authentication solutions turn out to be infeasible. In this context, a trust-based fast authentication scheme is proposed for the unified integrated AFS, which replaces, third-party-based solution that was presented in AMC [6, 26, 27] and IISA [24, 25].

To achieve fast authentication and seamless roaming support, a mobile service area is partitioned into mobility regions, which are further divided into mobility domains. A mobility region consists of two or more Autonomous Systems (AS). In general, an AS is under a common administration and using same routing policies. Under the circumstance, the Border Gateway Protocol (BGP) [34] is used to exchange routing information between access routers within a radio access network. Each administrative domain is controlled by an eMAP. The eMAP combines the functionality of a standard MAP for hierarchical mobility management, an nAuC, and an interworking gateway, as well as a local proxy for the central database (CD) that is introduced into the eHAAA server. Figure 3.13 illustrates the main functional modules that an eMAP possesses.

The main idea of our proposed trust-based fast authentication scheme is that the neighboring eMAPs that trust each other share authentication material (e.g., a security key) for a visiting mobile node. By this means, lengthy end-to-end authentication can be replaced by local authentication when mobile nodes switch their network connectivity, that is, authentication is always done at the eMAP level without further propagating the signaling to mobile nodes' home network. In case where the new selected eMAP has no trust

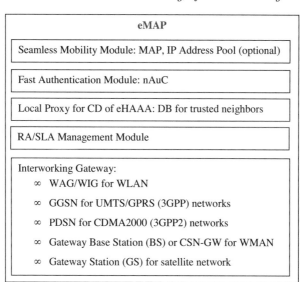

FIGURE 3.13
Functional modules of enhanced MAP (eMAP).

relationship with the mobile node's home service provider, the new eMAP may request its trusted neighbors for either authenticating that mobile or forwarding the corresponding authentication materials to it. As a result, the authentication delay can be minimized during mobility management. The details of fast authentication and seamless roaming procedure are elaborated as follows.

When entering into a new eMAP domain, a Mobile Node (MN) receives router advertisements from ARs within radio range. It then selects a trusted eMAP (teMAP) domain using the above-mentioned access network selection method. Note that here it is assumed that the integrated system supports host-based mobility management, thus network-based mobility management using the protocol Proxy Mobile IPv6 (PMIPv6) [15] is out of the scope of this chapter. However, such issue will be discussed in the near future. In addition, the router advertisement message may be modified by adding information that assists mobile nodes to select a trusted eMAP. In case where the mobile node cannot select a trusted eMAP, it may select an eMAP the same way as it does when executing the protocol Hierarchical Mobile IPv6 [39]. A trusted eMAP domain implies that the proprietor of the concerning domain/network has established direct security agreements or associations with the mobile node's home network service provider.

Upon selection of a new eMAP, the mobile node configures two care-of addresses (CoAs): a regional CoA (RCoA) within the visiting domain and an on-link CoA (LCoA). Both addresses can be formulated either in a stateless manner [46] or using IPv6 stateful address auto-configuration [12]. Then the mobile node executes duplicate address detection for its new on-link CoA. Such procedure is a mandatory for verifying the uniqueness of the addresses on a link [46]. Upon completion, the mobile node sends a local binding update to the selected eMAP in order to establish a binding between the RCoA and LCoA. Note that the selected eMAP may be trustable or untrustworthy.

After receiving the local binding update message, the eMAP executes the duplicate address detection for the specific RCoA. If failed, the eMAP searches its IP address pool and selects a new RCoA (nRCoA) for the mobile. In this case, there is no need to check the uniqueness of the nRCoA, thus minimizes the delays caused by duplicate address detection, which usually lasts for about 1 second [33]. The eMAP then binds the RCoA to the LCoA, and stores such binding in its binding cache. Afterwards, the eMAP returns a binding acknowledgement (BA) to the mobile node. As a result, the eMAP intercepts all packets addressed to registered mobile nodes and tunnels them to the corresponding on-link CoA.

Upon a successful registration with the eMAP, the mobile node sends a binding update (BU) to its home agent (HA) in order to bind the mobile's home address (HoA) to its regional CoA (RCoA). The home agent then forwards the mobile's RCoA to the eHAAA server, which resides in the same network. To reduce the delays for such transmission, the functionalities of the HA and eHAAA can be combined together in a manner that formulates a new network element, called Home Intelligent Node (HIN), for the unified integrated architecture AFS.

The eHAAA extracts the network prefix from the received RCoA, and then searches the IP address of the selected eMAP in its new central database (CD). Subsequently, the eHAAA verifies whether the concerning eMAP is an REN or not. If so, the authentication materials of the mobile are forwarded to all trusted eMAPs located in the same mobility region as the selected eMAP, as well as to those in the eMAP's neighboring regions. Otherwise, the eHAAA only sends the authentication materials to all trusted eMAPs that are located in the same region as the selected eMAP. The RCoA keeps unchanged as long as the mobile node moves within an eMAP domain. Fast authentication takes place when mobile nodes perform interdomain (or intersystem) handoff. In case where a mobile node selects a trustable domain to handoff, it only needs to send an Authentication Request to the teMAP, which then verifies its nAuC to authenticate the mobile. The nAuC stores information such as subscriber identity, user mobility pattern, user profile and preferences, terminal mobility pattern, terminal identity, security association parameters, etc. If successful, the teMAP returns an authentication response to the mobile node. As a result, authentication is always carried out locally at the level of eMAP, instead of in an end-to-end way. Figure 3.14 illustrates the fast authentication and seamless roaming procedure using trusted eMAP.

In case where the mobile node selects an untrustworthy eMAP domain to handoff, the new untrusted eMAP (ueMAP) multicasts a Further Authentica-

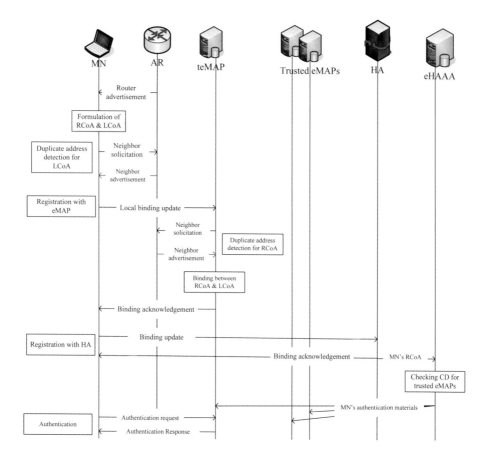

FIGURE 3.14
Fast authentication and seamless roaming with trusted eMAP.

tion Request to its trusted neighborhood. Accordingly, those trusted eMAPs respond to this message by either requesting the ueMAP to forward the received Authentication Request to it, or it simply transmits the mobile node's authentication materials to the new eMAP. The former implies additional delays for authentication, but more secure than the latter. This is because only trusted eMAP can authenticate mobile nodes. Figure 3.15 shows the case that a neighboring trusted eMAP is delegated for authentication, whereas Figure 3.16 illustrates the case that ueMAP performs authentication by itself.

Once a mobile node (MN) moves into a new trusted administrative domain, end-to-end authentication is always replaced by local authentication with the new selected trustable eMAP. However, when the mobile node moves into an eMAP domain without trust confidence, the new eMAP can request its trusted neighbors for assistance in authenticating the mobile. Accordingly, a temporary virtual trust relationship can be established via a trusted third party. Such fact results in reduced authentication latency during handoff. Figure 3.15 illustrates the process of an authentication that is delegated to a neighboring trusted eMAP.

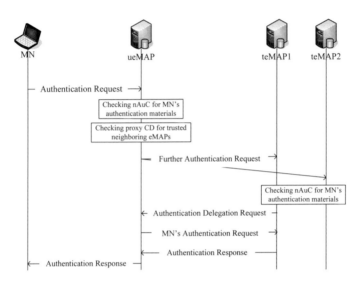

FIGURE 3.15
Authentication delegated to trusted eMAP.

Figure 3.15 shows that when an untrusted eMAP (ueMAP) receives an Authentication Request from an MN, it first checks its nAuC for the mobile's authentication materials. Since the nAuC does not hold the materials, the ueMAP then searches for its neighboring trusted eMAPs. Such information is stored in the proxy for the Central Database (CD) of the eHAAA. Further Authentication Request messages are then multicasted to all neighboring trusty eMAPs. Only those who hold the mobile's authentication materials respond. A trusted eMAP then sends an Authentication Delegation Request to the ueMAP, which then returns the mobile node's Authentication Request. Afterwards, the trusted eMAP authenticates the mobile and sends an Authentication Response to the mobile through the ueMAP. Figure 3.16 shows the case where an ueMAP performs authentication by itself.

Different from the authentication delegation procedure, shown on Figure 3.15, an ueMAP can also perform authentication by itself after receiving the authentication materials of the mobile from one trusted neighbor. However, for security concerns, the decision about whether using delegated authentication or not is made by a teMAP. Forwarding a mobile node's authentication

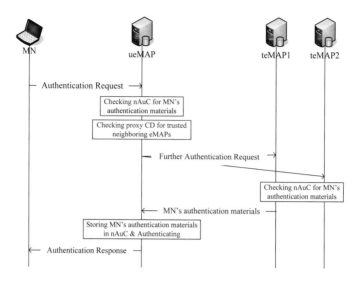

FIGURE 3.16
Fast authentication by untrusted eMAP.

materials to the ueMAP may be used to reduce authentication delay but at the price of security.

To further reduce the authentication delay, a trusted eMAP can distribute registered mobile nodes' authentication materials to all associated access routers (ARs) within its mobility domain. As a result, authentication can be carried out between mobile node and new access router, thus improves system performance.

Once attaching on a new link, an MN sends an Authentication Request to the new access router (NAR). The NAR then checks its nAuC for the mobile's authentication. If successful, it replies an Authentication Response to the mobile node. Otherwise, the NAR sends a Further Authentication Request to either its associated eMAP or neighboring access routers, or both for demanding the corresponding authentication materials. Either the eMAP or any adjacent access router who holds the materials may response to such request. Or a delegation may be exploited using the same way that described above. Consequently, end-to-end authentication, and regional authentication can be further replaced by local authentication with a NAR. Figure 3.17 provides an example on how the eHAAA distributes the authentication materials of a mobile node, as described below.

Authentication materials of a mobile node are distributed by the eHAAA to corresponding trusted eMAPs after successful home registration between the mobile and its home agent. For example, a mobile node enters into the mobility region 1 (Reg_1) associating with eMAP1, the eHAAA server multi-

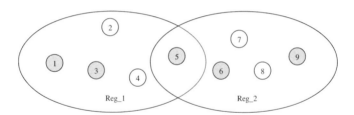

FIGURE 3.17
Authentication materials distributed by eHAAA.

casts the concerning authentication materials to eMAP3 and eMAP5. If the mobile node first associates with eMAP5, its authentication materials are distributed to eMAP1, eMAP3, eMAP5, eMAP6 and eMAP9. The circles with shadow indicate those trusted eMAPs whereas those without shadow illustrate untrusted eMAPs.

For billing management, only trusted eMAPs are permitted to generate and send billing reports to mobile nodes' home network. For those eMAPs with temporary trust relationship, the outgoing traffic from a mobile node is recommended to be tunneled to a trusted eMAP, which then generates a billing report. And the incoming traffic from a correspondent node is also recommended to be intercepted by the previously associated trusted eMAP, and to be tunneled to the new eMAP, which then forwards data to the mobile node.

In contrast to third-party-based authentication that presented in [6, 26, 27] and [24, 25], we propose trust-based fast authentication method, which allows end-to-end authentication to be replaced by either regional or local authentication. To remedy the case where no trust relationship existing between a mobile's home network and visiting domain/network, a temporary virtual trust relationship is established via a trusted neighboring third party. By this means, fast authentication is always available for mobile nodes.

3.3.4 Features of the New Unified Integrated Architecture

This section qualitatively evaluates the new unified integrated architecture to support fast authentication and seamless roaming, in the context of the design objectives, its advantages and disadvantages [6].

- **Economical**: The new unified integrated architecture, AFS, uses as much existing infrastructure of disparate wireless systems as possible. It does not imply any change to the existing infrastructure during integration. AFS achieves seamless integration by introducing new network entities: eMAP and eHAAA. The former extends the functionalities of a standard MAP [39] with additional functional modules, such as nAuC, local proxy for the Central Database (CD) of the eHAAA, roaming agreement (RA) and/or SLAs management unit, as well as interworking gateway. The eMAP is responsible for mobility management, security association, authentication, billing as well as interworking between different wireless systems. The eHAAA is responsible for fast authentication, mobility management and billing.

- **Scalability**: The architecture AFS can integrate any number of existing or future wireless systems of the same or different network operators. In addition, these operators may not have direct RAs or SLAs among them. Since the AFS does not impose any pre-existing roaming agreements between each pair of mobile systems, thus it is completely scalable.

- **Transparency to heterogeneous radio access technologies**: The Internet Protocol (IP) is employed as the common interconnection protocol. Such IP-layer based solution allows mobile nodes to roam among multiple wireless systems, regardless of underlying different radio access technologies.

- **Security**: AFS adopts existing security mechanisms to provide security and privacy equivalent to existing wireless networks.

- **Seamless mobility**: Seamless intrasystem and intersystem mobility is achieved by using the protocol Hierarchical Mobile IPv6 (HMIPv6) [39]. Furthermore, new access network selection method is proposed that allows mobile nodes to intelligently make handoff decisions while avoiding vertical handoff as much as possible. To improve the handoff latency and to reduce packet dropping, IP layer (also called layer three – L3) and data link layer (or layer two) cross-layer interaction can be carried out along with the implementation of the protocol Mobile IPv6 Fast Handover (FMIPv6) [22]. This issue will be handled within the future works.

- **Fast authentication**: Fast authentication is supported by the novel architecture AFS with introduction of new network elements: eHAAA and eMAP. End-to-end authentication is always eliminated by allowing an eHAAA server to distribute mobile nodes' authentication materials to trusted eMAPs even before roaming users move to that domain. As a result, authentication is always done either regionally (at the eMAP level) or locally (at the level of the access router). Therefore, authentication delay is reduced significantly during handoff. On the other hand, the functionality of an eHAAA server can be integrated into the Home Agent (HA) to formulate a Home Intelligent Node (HIN). Such configuration enables further minimization of authentication delay.

The limitation of the new unified integrated architecture AFS is that mobile nodes need to be modified or upgraded with the capability of executing the access network selection method. This may add design complexity on mobile device side. In the meantime, the eHAAA server should have a global vision of mobility regions and domains, that is, it should have pre-knowledge about IP address of each trusted eMAP in a region, the neighborhood of each trusted eMAP, region related information, etc. This adds new complexity to traditional AAA mechanism. Furthermore, multicasting the authentication materials of mobile nodes to trusted neighboring eMAPs may lead to high signaling overloads over the network. As the new network element eMAP is an all-functionalities-in-one gateway, it may become a bottleneck point in the integrated system.

However, as authentication is combined with mobility management, handoff delays can be significantly reduced. With the proposed fast authentication methods, end-to-end authentication is always replaced by local authentication with either an eMAP or an access router, thus authentication latencies are largely minimized. Moreover, the novel access network selection method enables mobile nodes to intelligently choose the best connected wireless network while avoiding vertical (or intersystem) handoff as much as possible.

3.4 Open Research Challenges

Next-generation wireless systems are expected to combine a number of wireless and mobile communication networks. However, these networks exhibit a variety of heterogeneity in terms of radio access technology, network architecture, mobility and security protocols, quality of service control mechanism, quality of service requirements, etc. This makes integration and interworking of existing disparate wireless systems a greatly challenging issue. On the other hand, subscribers are intensifying demands for seamless roaming across different wireless networks with guaranteed quality of service. This adds new complexity into the integrated architecture design for next-generation mobile communication system. To enable subscribers to profit from telecommunication services from different wireless system with guaranteed performance, to the extent of anytime and anywhere, from any mobile device, is really one big research issue.

For example, wireless local and metropolitan area networks possess different bandwidth, that is, their network capacity varies quite a lot. Their heterogeneities are also presented by radio coverage, network components, radio access technology, mobility protocol, security mechanism, etc. How to hide such heterogeneity and seamless integrate these two networks together is a hard task for both industrial and academic researchers. The research issues to integrate wireless local area network with wireless metropolitan area network can be listed as follows [29]:

- **Topology management using the mesh infrastructure**: Efficient solutions are required to minimize network deployment cost while satisfying the quality of service requirements for local and relay connections using mesh topology. In addition, traffic load at hotspots and user mobility patterns need to be taken into account for optimal topology design for an integrated architecture with mesh infrastructure.

- **Radio resource management**: Relay base station must possess an efficient radio resource management module while mesh topology is adopted for integration. Such a module enables intelligent bandwidth allocation, channel assignment, and admission control, as well as load balancing aspects when different types of connections are handled. Moreover, distinction between local and relay traffic, and how to schedule them with prioritization according to their quality of service requirements must be well dealt with. As wireless local and metropolitan area networks may operate on overlapped frequency spectrum (i.e., 2–11 GHz), interference is inevitable to take place. Therefore, dynamic adaptation for frequency allocation is also necessary to minimize such interference.

- **End-to-end quality of service control**: To offer users guaranteed quality of service, new radio link control mechanism is necessary. Integration of link-layer error control and error recovery is also important to improve the radio link performance. On the other hand, packet scheduling and new routing techniques with cooperative diversity are required to improve end-to-end performance.

- **Routing algorithm**: In an integrated environment, an ideal routing mechanism should take into account the quality of diverse wireless link and quality of service requirements for each corresponding connection.

- **Integration point**: Generally, integration and interworking of different wireless system is realized at an edge router. Such node performs protocol adaptation and quality of service support, so that the interworking system is enabled to support real-time traffic such as voice, video, and interactive multimedia applications.

Challenges are also omnipresent while designing the 4G mobile communication system, which is based on the Long Term Evolution (LTE) and Service Architecture Evolution (SAE) architecture. The key element of LTE/SAE architecture is the Evolved Packet Core (EPC), which radically changes key networking paradigms, compared to previous generations of mobile systems. The introduction of EPC must successfully address a number of technological challenges. For example, significant technological advances must be made on the radio interface of an evolved Node B (eNodeB). This makes the LTE system capable of efficiently using the spectrum with wider spectral bands. As a result, system capacity and performance can be improved.

On the other hand, the EPC network needs to be changed or upgraded so as to provide higher throughput and low latency; this comes from the simplified, flat all-IP network architecture. In order to design an efficient solution, the EPC must address some key aspects before deployment, such as:

- routing and network addressing aspects, as well as session management;
- the best choice between centralized or distributed interworking architecture;
- interconnection protocol, either using IPv6 or IPv4;
- a mechanism to guarantee end-to-end quality of service;
- efficient mobility management at the link-layer, the IP-layer, as well as the transport-layer that is, both network connectivity and session continuity need to be well addressed;
- design of end-to-end security protocol to protect both data and control signaling;
- interconnectivity to external PDNs and any other Virtual Private Networks (VPN);
- design for lawful Interception and deep packet inspection, etc.

Subscribers expect to benefit from a reliable, scalable, high-performance next-generation wireless system. However, it is difficult to fulfill all these requirements due to the dynamic nature of user mobility in an integrated environment. To meet the new demands for high-speed wireless Internet access and interactive multimedia application while on the move, next-generation wireless system must interwork in full harmony with all existing mobile communication systems while tackling the problems that were listed.

3.5 Conclusion

This chapter provides a comprehensive literature review about interworking architecture design for next-generation wireless systems. Interworking architectures that integrate the 3G mobile system with wireless local and metropolitan area networks are elaborated, as well as those that couple WiFi network with WiMAX system described in detail. After the survey, a new unified integrated architecture to support fast authentication and seamless roaming (AFS) is proposed. Such interworking architecture presents the features such as economy, scalability, adequate security and privacy, fast authentication, seamless mobility, and using the IP as interconnection protocol to hide the heterogeneity of different wireless access technology. Moreover, the novel architecture integrates all existing wireless systems including WLANs, WMANs, satellite networks, 3GPP (e.g., GSM, GPRS, EDGE, UMTS, LTE), and 3GPP2 (e.g., CDMA2000, EV-DO) networks.

After presenting the new unified integrated architecture, new access network selection, fast authentication, and seamless roaming mechanisms are designed. The former allows mobile nodes to make intelligent handoff decision while avoiding intersystem handoff as much as possible. As to the latter, it enables handling authentication and mobility management together, and replace, end-to-end authentication by local authentication with either an eMAP or an access router. As a result, authentication and handoff delays are significantly minimized when mobile nodes move from one radio access network to another. To achieve fast authentication, authentication materials of mobile nodes are distributed to trusted eMAPs by an eHAAA that resides at the mobile nodes' home network. Such materials can be further distributed to access routers within an eMAP domain. By this means, authentication is guaranteed to be done always locally, thus resulting in less authentication delay, compared with traditional authentication and mobility management. On the other hand, new seamless mobility protocol, called Seamless Mobile IPv6 (SMIPv6) [50], may also be used for seamless mobility management.

Following the proposals, open research issues are discussed. Examples include the challenges to interwork wireless local and metropolitan area networks, as well as issues for designing the 4G network architecture for LTE and SAE. In particularly, the aspects concerning the EPC network design is argued. In the near future, performance evaluation for the new unified integrated architecture will be done through analytical models and simulations.

Bibliography

[1] 3GPP. 3GPP system to wireless local area network (WLAN) interworking; System Description (Release 8). 3GPP TS 23.234 v8.0.0, 3rd Generation Partnership Project, 2008.

[2] 3GPP. Feasibility study on 3GPP system to wireless local area network (WLAN) interworking (Release 8). 3GPP TR 22.934 v8.0.0, 3rd Generation Partnership Project, 2008.

[3] 3GPP2. 3GPP2 – WLAN Interworking – Stage 1 Requirements. 3GPP2 TSG-S S.R0087-0 v1.0, 3rd Generation Partnership Project 2, 2004.

[4] 3GPP2. CDMA2000 – WLAN Interworking. 3GPP2 TSG-S S.R0087-A v1.0, 3rd Generation Partnership Project 2, 2006.

[5] K. Ahmavaara, H. Haverinen, and R. Pichna. Interworking architecture between 3GPP and WLAN systems. *IEEE Communications Magazine*, 41(11):74–81, Nov. 2003.

[6] I.F. Akyildiz, S. Mohanty, and J. Xie. A ubiquitous mobile communication architecture for next-generation heterogeneous wireless systems. *IEEE Communications Magazine*, 43(6):529–536, June 2005.

[7] M.M. Buddhikot, G. Chandranmenon, S. Han, Y-W. Lee, S. Miller, and L. Salgarelli. Design and implementation of a WLAN/cdma2000 interworking architecture. *IEEE Communications Magazine*, 41(11):90–100, Nov. 2003.

[8] M.M. Buddhikot, G. Chandranmenon, S. Han, Y-W. Lee, S. Miller, and L. Salgarelli. Integration of 802.11 and third-generation wireless data networks. In *Proc. of the 22nd IEEE INFOCOM*, pages 503–512, San Francisco, CA, March 30–April 3, 2003.

[9] P.R. Calhoun, J. Loughney, E. Guttman, G. Zorn, and J. Arkko. Diameter Base Protocol. IETF RFC 3588, 2004.

[10] H-H. Chen and M. Guizani. *Next Generation Wireless Systems and Networks*. John Wiley & Sons, New York, 2006.

[11] S.E. Deering and R.M. Hinden. Internet Protocol version 6 (IPv6) Specification. IETF RFC 2460, 1998.

[12] R. Droms, J. Bound, B. Volz, T. Lemon, C.E. Perkins, and M. Carney. Dynamic Host Configuration Protocol for IPv6 (DHCPv6). IETF RFC 3315, 2003.

[13] K. Fall. A delay-tolerant network architecture for challenged Internets. In *Proc. of the Annual Conference of the ACM Special Interest Group on Data Communication (SIGCOMM'03)*, pages 27–34, Karlsruhe, Germany, Aug., 25–29 2003.

[14] S. Frattasi, E. Cianca, and R. Prasad. Interworking between WLAN and WMAN: an Ethernet-based integrated device. In *Proc. of the 6th International Symposium on Wireless Personal Multimedia Communications (WPMC'03)*, Yokosuka, Japan, Oct., 19–22, 2003.

[15] S. Gundavelli, K. Leung, V. Devarapalli, K. Chowdhury, and B. Patil. Proxy Mobile IPv6. IETF RFC 5213, 2008.

[16] E. Gustafsson and A. Jonsson. Always Best Connected. *IEEE Wireless Communications*, 10(1):49–55, Feb. 2003.

[17] J. Hassan, H. Sirisena, and B. Landfeldt. Trust-based fast authentication for multiowner wireless networks. *IEEE Transactions on Mobile Computing*, 7(2):247–261, Feb. 2008.

[18] S.Y. Hui and K.H. Yeung. Challenges in the migration to 4G mobile systems. *IEEE Communications Magazine*, 41(12):54–59, Dec. 2003.

[19] M. Jaseemuddin. An architecture for integrating UMTS and 802.11 WLAN networks. In *Proc. of the 8th IEEE International Symposium on Computers and Communication (ISCC'03)*, pages 716–723, Antalya, Turkey, June 30–July 3, 2003.

[20] D.B. Johnson, C.E. Perkins, and J. Arkko. Mobility Support in IPv6 (MIPv6). IETF RFC 3775, 2004.

[21] D. Kim and A. Ganz. Architecture for 3G and 802.16 wireless networks integration with QoS support. In *Proc. of the 2nd International Conference on Quality of Service in Heterogeneous Wired/Wireless Networks (QShine'05)*, Orlando, FL, Aug. 22–24, 2005.

[22] R. Koodli. Mobile IPv6 Fast Handovers (FMIPv6). IETF RFC 5568, 2009.

[23] W. Li and Y. Pan. *Resource Allocation in Next Generation Wireless Networks*. Nova Science Publishers, New York, 2006.

[24] C. Makaya and S. Pierre. Reliable integrated architecture for heterogeneous mobile and wireless networks. *Journal of Networks*, 2(6):24–32, Dec. 2007.

[25] C. Makaya and S. Pierre. An architecture for seamless mobility support in IP-based next-generation wireless networks. *IEEE Transactions on Vehicular Technology*, 57(2):1209–1225, Mar. 2008.

[26] S. Mohanty. A new architecture for 3G and WLAN integration and inter-system handover management. *Wireless Networks*, 12(6):733–745, Nov. 2006.

[27] S. Mohanty and J. Xie. Performance analysis of a novel architecture to integrate heterogeneous wireless systems. *Computer Networks*, 51(4):1095–1105, Mar. 2007.

[28] Q.-T. Nguyen-Vuong, L. Fiat, and N. Agoulmine. An architecture for UMTS–WIMAX interworking. In *Proc. of the 1st International Workshop on Broadband Convergence Networks (BCN'06)*, pages 1–10, Vancouver, Canada, April 7, 2006.

[29] D. Niyato and E. Hossain. Integration of IEEE 802.11 WLANs with IEEE 802.16-based multihop infrastructure mesh/relay networks: A game-theoretic approach to radio resource management. *IEEE Network*, 21(3):6–14, May–June 2007.

[30] C.E. Perkins. IP Mobility Support. IETF RFC 2002, 1996.

[31] C.E. Perkins. IP Mobility Support for IPv4. IETF RFC 3344, 2002.

[32] F.A. Phiri and M.B.R. Murthy. WLAN-GPRS tight coupling based interworking architecture with vertical handoff support. *Wireless Personal Communications*, 40(2):137–144, Jan. 2007.

[33] P. Pongpaibool, P. Sotthivirat, S.I. Kitisin, and C. Srisathapornphat. Fast duplicate address detection for mobile IPv6. In *Proc. of the 15th IEEE International Conference on Networks (ICON'07)*, pages 224–229, Adelaide, Australia, Nov. 19–21, 2007.

[34] Y. Rekhter, T. Li, and S. Hares. A Border Gateway Protocol 4 (BGPv4). IETF RFC 4271, 2006.

[35] A.K. Salkintzis. Interworking techniques and architectures for WLAN/3G integration toward 4G mobile data networks. *IEEE Wireless Communications*, 11(3):50–61, June 2004.

[36] A.K. Salkintzis, C. Fors, and R. Pazhyannur. WLAN-GPRS integration for next-generation mobile data networks. *IEEE Wireless Communications*, 9(5):112–124, Oct. 2002.

[37] R. Shacham, H. Schulzrinne, S. Thakolsri, and W. Kellerer. Session Initiation Protocol (SIP) Session Mobility. IETF RFC 5631, 2009.

[38] Z. Shelby and C. Bormann. *6LoWPAN: The Wireless Embedded Internet*. John Wiley & Sons, 2009.

[39] H. Soliman, C. Castelluccia, K. El-Malki, and L. Bellier. Hierarchical Mobile IPv6 (HMIPv6) Mobility Management. IETF RFC 5380, 2008.

[40] J.-Y. Song, H.J. Lee, S.-H. Lee, S.-W. Lee, and D.-H. Cho. Hybrid coupling scheme for UMTS and wireless LAN interworking. *AEU – International Journal of Electronics and Communications*, 61(5):329–336, May 2007.

[41] J.-Y. Song, S.-W. Lee, and D.-H. Cho. Hybrid coupling scheme for UMTS and wireless LAN interworking. In *Proc. of the 58th IEEE Vehicular Technology Conference (VTC-Fall)*, pages 2247–2251, Orlando, FL, Oct., 6–9, 2003.

[42] W. Song, W. Zhuang, and A. Saleh. Interworking of 3G Cellular Networks and Wireless LANs. *International Journal of Wireless and Mobile Computing*, 2(4):237–247, Jan. 2008.

[43] I. Stoica, D. Adkins, S. Zhuang, S. Shenker, and S. Surana. Internet indirection infrastructure. In *Proc. of the Annual Conference of the ACM Special Interest Group on Data Communication (SIGCOMM'02)*, pages 73–86, Pittsburgh, PA, Aug., 19–23, 2002.

[44] I. Stoica, D. Adkins, S. Zhuang, S. Shenker, and S. Surana. Internet indirection infrastructure. *IEEE/ACM Transactions on Networking*, 12(2):205–218, April 2004.

[45] Comsys Enabling the Mobile Highway. News & Events - Press releases: Unique multimode mobile WiMAX/GSM-EDGE smartphone reference design platform by Comsys Mobile.

[46] S. Thomson, T. Narten, and T. Jinmei. IPv6 Stateless Address Autoconfiguration. IETF RFC 4862, 2007.

[47] S.-L. Tsao and C.-C. Lin. Design and evaluation of UMTS-WLAN interworking strategies. In *Proc. of the 56th IEEE Vehicular Technology Conference (VTC-Fall)*, pages 777–781, Vancouver, Canada, Sept. 24–28, 2002.

[48] S.-L. Tsao and C.-C. Lin. VGSN: a gateway approach to interconnect UMTS/WLAN networks. In *Proc. of 13th IEEE International Symposium on Personal, Indoor and Mobile Radio Communications (PIMRC'02)*, pages 275–279, Lisboa, Portugal, Sept. 15–18, 2002.

[49] D. Wisely and E. Mitjana. Paving the road to systems beyond 3G - the IST BRAIN and MIND projects. *Journal of Communications and Networks*, 4(4):292–301, Dec. 2002.

[50] L.J. Zhang. *Fast and seamless mobility management in IPv6-based next-generation wireless networks*. PhD thesis, Ecole Polytechnique de Montreal, University of Montreal, Montreal, Canada, 2008.

4

WLAN/3G Convergence and Advanced Mobility Features

Mehdi Mani

Department of Wireless Networks and Multimedia Services
Institut TELECOM SudParis
9 Rue Charles Fourier, 91011, Evry, France
Email: mehdi.mani@it-sudparis.eu

Noel Crespi

Department of Wireless Networks and Multimedia Services
Institut TELECOM SudParis
9 Rue Charles Fourier, 91011, Evry, France
Email: noel.crespi@it-sudparis.eu

CONTENTS

In this chapter we define new strategies for WLAN 802.11 and 3G convergence at the service level. This convergence is based on the service delivery infra-structure proposed by IMS (IP Multimedia Subsystem). These strategies are beyond the 3GPP solutions for wireless local area networks (WLAN) access to the IMS. Our main aim from this convergence is to introduce new services such as session mobility/splitting and intelligent call routing between these two access technologies. Our strategies of convergence is independent of network infrastructure and thus fit well with the LTE (Long Term Evolution) infrastructure architecture.

4.1   Introduction

Based on IEEE 802.11 protocol, WiFi technology is mainly developed to provide broadband wireless access to Internet services. Considering the vast deployment, low installation cost, and broadband media of 802.11, providing converged 3G/802.11 services can substantially improve the availability of advanced services for users. In recent years, there have been massive research and proposals for integration of WLAN and 3G infrastructure in network layer [24, 4]. The existing solutions lead to interoperability of these two technologies, however, from the service level point of view these two systems remains disjointed.

In this chapter we introduce our research work to define a service level solution for WLAN 802.11 and 3G convergence. Our solutions are based on the infrastructure proposed in the IMS for service delivery. The convergence of service delivery platforms of different network technologies is the solution to get rid of disjoint service-specific networks [11]. This is known as Service Convergence. Service convergence is a service-level aspect which is not achieved by integration in the network layer alone and but is a concept beyond interoperability of service-specific network technologies. Figure 4.1 depicts the difference between the concepts of traditional inter-operation and convergence. Our work provide the solutions beyond the 3GPP strategies for WLAN access to the IMS. Moreover, since our solutions are in service level, they are network technology independent and fit very well with the proposed infrastructure in LTE (Long Term Evolution) specification.

To get the disjoint systems interoperated, their transport levels should be interconnected, where at signalling level there are the gateways that translate the signalling of one domain in the border of other domain. With interoperation of disjoint systems, users of different domains become reachable for each other. However, in convergence, a service delivery overlay by using a unified signalling protocol employs communication services over all the access technologies. This allows the users of one domain to benefit from the services of the other domain, in addition to the interoperation. Moreover, the combinational Internet/Telecom or Fixed/Mobile services will be feasible in a scalable and standardized approach. The services such as unique fixed/mobile numbering, shared voicemail, spanning buddy list, content sharing between multiple end-devices, session splitting/merging between/from different access technologies,

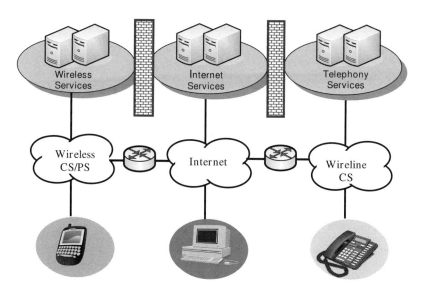

(a) Interoperation of vertical disjoint service domains

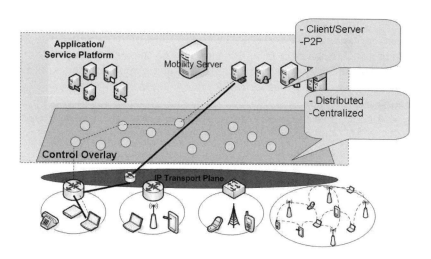

(b) Overlay approach for convergence of multi domains

FIGURE 4.1
From interoperation to convergence.

and plenty of other combined services are the examples of the new paradigm of services in the multitechnology service convergence.

As SIP (Session Initiation Protocol (SIP)) is going to be the dominant protocol for IP-based communication services, and IMS is a service delivery technology which is specified based on SIP, we also consider IMS as the service control overlay technology.

This chapter concentrates on the convergence of 3G/WiFi technology and deployment of communication services based on IMS over WiFi. In Section 4.2, we introduce new win-win strategies for deployment of IMS in WiFi technology that leads to WiFi/3G convergence. By this convergence we aim to deploy the advanced mobility features such as session mobility/splitting between these two access technologies. In this regard, in Section 4.3, we extend the IMS policy-based admission control architecture to be able to handle these kind of session mobility between these two different access technologies. In the current IMS policy-based quality of service (QoS) control system [2], hybrid access technologies should obey a unique policy for their resource allocation. Considering that miscellaneous access technologies have different characteristics in medium access strategy and bandwidth, the unique policy leads to inefficient resource utilization [14]. We improve the architecture of IMS policy-based resource reservation for multiaccess scenarios. Then with the improved architecture, 802.11 WLAN operators are able to define and effectuate their own policies for call admission and resource allocation.

In Section 4.4, by taking advantage of IMS infrastructure, we propose the implementation of a mobility server that can handle the session mobility/splitting between 3G and 802.11 WLAN. This mobility server will be deployed in an application layer and have access to the admission policy of each domain. We show that how by enforcing the proper policies, this kind of session mobility can lead to better utilization of resources in these two network technologies.

4.2 Strategies for IMS Integration in WiFi

Adoption of IMS as the service control overlay leads to the support of advanced telephony services in WiFi domains with a robust and scalable architecture. However, different WiFi operators may choose different strategies to provide IMS services, according to their network configuration, capabilities, and resources.

One strategy, specified by the 3GPP in [1], is to define the proper interfaces to let the clients have access to 3G IMS services via 802.11 WLAN. In such an approach, IMS is not deployed in the WiFi domain. This strategy introduces a Master and Slave architecture. The 3G operator will be considered as the Master and defines all the required functionality for bearer level interconnection as well as the Authentication Authorization and Accounting (AAA) functionality in service control layer.

4.2.1 3GPP Approach for Access to IMS through WiFi

In Release 6, 3GPP specified an architecture with a Master and Slave model to access the 3G services via WLAN [1].

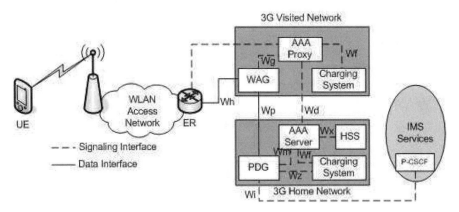

FIGURE 4.2
3GPP approach for access to IMS services via WLAN.

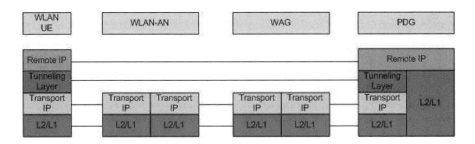

FIGURE 4.3
IP connectivity configuration for 3GPP/WLAN interworking.

The goal of this architecture is to allow the 3G operators to extend their Packet Switched services to the WLAN. These services may include, for example, IMS-based services, location-based services, Multimedia Broadcast/Multicast Services (MBMS), and any service that is built upon the combination of these components. This architecture is depicted in Figure 4.2. As shown in this figure, there are four new functional elements that are defined for the purpose of access to 3G services via WLAN [1].

3GPP AAA Server is located in the 3G home domain. It retrieves authentication information and subscriber's profile from the HLR/HSS for each user, and authenticates the users based on the information received from the HLR/HSS. This information is transferred to WLAN for admission control of the users. In addition, this entity provides the charging information.

3GPP AAA Proxy is used in roaming situations, and it works as a proxy between user equipment (UE) in the WLAN and the 3GPP AAA Server in the 3G home domain.

In addition to these AAA functional elements, two new elements are also defined at bearer level to create secure connection to 3G domain.

WAG – WLAN Access Gateway is the entry point of the mobile network from the WLAN perspective. In the case of roaming, the UE will be connected to the WAG in the visited network. WAG (either in home or in visited network) forwards the data toward the Packet Data Gateway in home domain. WAG also transfers the bearer level charging information to 3GPP AAA Proxy/Server.

PDG – Packet Data Gateway is the edge router in the 3G home domain that establishes and controls the connection of the user to the external network and services (i.e., IMS overlay) located in 3G domain. It allocates remote IP (as depicted in Figure 4.3) to the UE and creates a tunnel between itself and UE in WLAN. Then it maps this tunnel to the external tunnels toward external networks. As shown in Figure 4.3, it can be noted that two IP layers are represented in this protocol stack. In fact, with such configuration, the PDG can be seen as the last hop router for the UE from the external network view, and in addition the 3G Core Network configuration has no impact on the IP connectivity of the UE in WLAN.

Figure 4.4 shows the complete procedure to setup a session for an IMS service. After access point detection and layer 2/layer 1 attachment, UE starts the authentication process with the AAA server. A successful authentication allows UE to get an IP address routable in WLAN domain. Then, the UE retrieves the IP address of the gateway to 3G domain (PDG) from the local DNS (domain name server). In next phase, a secure tunnel between UE and PDG is established. This tunnel enables UE to connect to the DHCP server in 3G domain and obtain (1) a valid IP address in 3G domain and (2) the address of P-CSCF (Proxy Call Session Control Function). After P-CSCF discovering, the normal process of IMS registration and session setup will be followed.

4.2.2 Beyond 3GPP Proposal by Adopting IMS in WiFi

The 3GPP approach for providing WLAN access to the IMS services is satisfactory for certain operators with specific conditions such as (1) the operators that own both of WiFi and 3G access technologies; and (2) WiFi operators who do not desire to invest too much for the deployment of IMS on their own domain. However, for the WiFi operators who have deployed their own core network and provide vast public WLAN access, this architecture may not be acceptable for several reasons as follow.

First, from the business point of view, it is unlikely that such a WiFi oper-

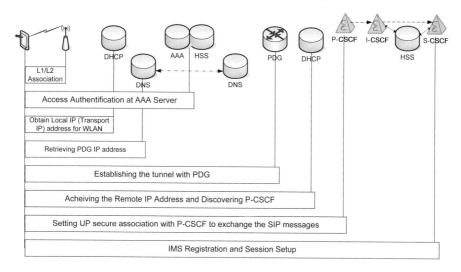

FIGURE 4.4
Different phases of access to IMS services from WLAN.

ator accepts that all the IMS services reside in the 3G domain. Second, in such a Master and Slave interworking architecture, all the policies for interconnection are defined by the 3G operator. Third, the process of setting up the IMS service sessions via WLAN will be so complicated (as depicted in Figure 4.4). Fourth, the remote IP address that allows the user to be connected to the IMS domain is assigned by the 3G domain. Therefore, if 3G has deployed its IMS based on IPv6 (as what is specified in the standard), the end users in WLAN should support a multistack IPv4/IPv6 protocol.

Regarding these issues, WiFi operators who own enough resources and desire to support IMS services may adopt IMS or an IMS-like architecture as the service overlay in their own domain instead of just being connected to and relying on the 3G IMS. Deployment of IMS in the WiFi network domain leads to horizontal convergence of 3G and WiFi technology. In such architecture, a user may be accessible by a unique number via both of UTRAN and 802.11 access technologies. Moreover, the users of each domain can benefit from the services deployed in other domains.

To deploy the IMS overlay, WiFi operators may choose a step-by-step and evolutionary strategy to avoid a sudden huge cost for deploying the IMS services. To this end, we have proposed several phases for adopting IMS in WiFi technology. In the first phase, as depicted in Figure 4.5, the WiFi operators may just implement the functionality of P-CSCF. The P-CSCF will define the security strategy and the compression protocol to reduce the size of SIP messages. For interoperability with IPv6 domain, NAPT (Network Address/Port Translator) and ALG (Application Level Gateway) functionalities are required.

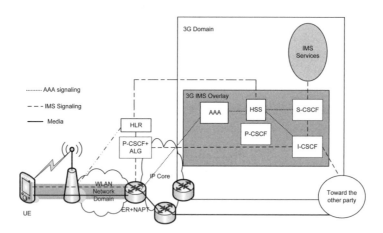

FIGURE 4.5
First phase of deploying IMS in WiFi domain.

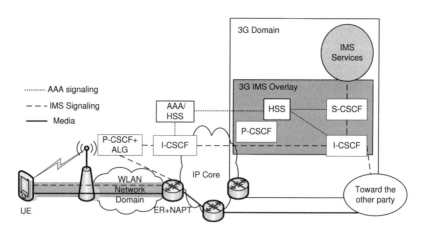

FIGURE 4.6
Second phase of deploying IMS in WiFi domain with AAA functionalities.

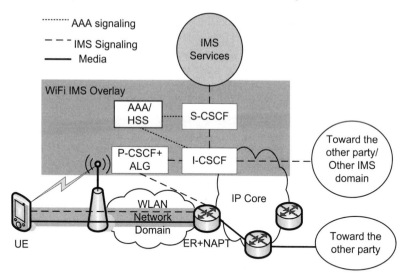

FIGURE 4.7
Last phase of deploying IMS in WiFi domain: complete IMS.

NAPT should be implemented in edge router to replace the IPv4 and IPv6 addresses at the edge of the network in bearer level. ALG functionality, in this phase of IMS deployment, may be added to the P-CSCF. ALG is a STUN [18] or ICE [17] server to resolve the NAT traversal issues in SIP sessions. With such architecture, the end users do not need to support a multi-protocol stack of IPv4/IPv6 in order to access to IMS services. Moreover, in this approach, there is no need of implementation of PDG and WAG. In fact, the WiFi operator connects its network directly to the network backbone without passing through 3G network. However, for session routing, the SIP messages in this phase are still required to be routed via 3G domain.

HLR (Home Location Register) is also implemented in the WiFi domain, and the SSID (service set identifier) will be considered as the attachment point information. SSID is a 32-character unique identifier attached to the header of packets sent over a WLAN that acts as a password when a mobile device tries to connect to the BSS (Basic Service Set). The BSS is an Access Point (AP) that is connected to the wired network and a set of wireless stations. The SSID differentiates one WLAN from another, so all access points and all devices attempting to connect to a specific WLAN must use the same SSID.

In the next phase of IMS deployment (Figure 4.6), WiFi operators deploy also both the HSS (Home Subscriber Server) and I-CSCF (Interrogating Call Session Control Function). In this phase, WiFi operators will be able to carry out AAA functionalities in their own domain. Therefore, the WiFi operator is able to provide the public and private identification for their clients independent of the 3G operator. I-CSCF interrogates HSS to assign a S-CSCF

(Serving Call Session Control Function) to the user according to his service criteria. In this phase, service triggering functions are carried out in 3G domain, then WiFi is able to add some advance and combinational services to the existing services in 3G domain in order to enrich the existing services for the clients.

Finally, in the last phase as depicted in Figure 4.7, the IMS overlay will be deployed completely. S-CSCF in the WiFi domain localizes the clients when they register for IMS services and invokes the services upon their request. SIP-based services will be implemented in this phase in the IMS overlay of the WiFi domain. In this phase at both bearer and signaling level WiFi is independent of 3G and can connect to the rest of the network directly.

The step-by-step IMS deployment provides a smooth migration toward WiFi-with-IMS for the operators. The WiFi operators, according to their strategies and capabilities, may choose one of the proposed architecture.

Finally, it is worth mentioning that there are intrinsic differences in 802.11/WiFi and UTRAN (UMTS Terrestrial Radio Access Network) technologies for channel access and resource allocation. In 802.11 technology, access to the channel is contention-based and then there is no resource reservation system. Considering the absence of dedicated resources in 802.11, in 3GPP solution for WLAN access to IMS services, resource reservation in WLAN access is just ignored. The differences between WLAN 802.11 and 3G technologies for access to IMS in 3GPP solution is summarized in Table 4.1.

In this work we aim to extend the IMS policy-based QoS control system to WiFi. The two main challenges in this way are: (1) QoS architectural limitations in IMS; and (2) random access medium of 802.11. The following two sections address these challenges for this integration.

4.3 Deploying Policy-Based QoS Control System for 802.11-WLAN

Establishing a flexible and scalable end-to-end QoS control mechanism for multimedia services in heterogeneous infrastructure of NGN (next-generation networks) (see Figure 4.1) where access to the services may be achieved via different kinds of wireless-wireline access technologies such as UTRAN, WLAN, xDSL, and cable is a challenging task. In fact, the current QoS mechanism defined in 3G cannot meet the requirement of such heterogeneous networks in NGN. The current QoS control mechanism introduced in 3G does not allow different domains en route of media-flow to reserve the resources homogenously. For the same QoS class, different domains dedicate different amount of resources according to their interpretation of that class. This leads to discordant resource reservation in the data flow route.

In this section, we propose the IMS architectural improvement for QoS

TABLE 4.1 WLAN and UMTS differences for access to IMS

	WLAN	UMTS
Number of actors in roaming case	Three: WLAN AN, VPLMN and HPLMN	Two: VPLMN and HPLMN
G_o, policy control	This is not defined at all in 3GPP for WLAN accesses	This is defined by 3GPP for UTRAN
Notion of "tunnel for signalling"	No notion of "secondary tunnel" is defined for WLAN Interworking, which means that both signaling and data packets will use the same tunnel	The notion of Secondary PDP Context allows differentiating the PDP Context for signalling and the PDP Context for data
Support of IPv4 necessary on the Local IP layer	IPv4 must be supported on the Local IP layer because many WLAN Access Networks will only support IPv4 in early deployments, as IPv6 must be needed for IMS, this means that the WLAN UE must support the dual stack	The use of IPv6 only is possible as long as the SGSN and GGSN can support it; in such a case, the UE can be based on IPv6 only as far as IMS is concerned

control to allow different access technologies (including 802.11) to exchange the QoS policies and limitation of their network dynamically and efficiently.

4.3.1 IMS Policy-Based QoS Control Architecture

Establishment of sessions for multimedia services like voice or video telephony, video streaming, MMS, and video conferencing needs coordination between bearer and session layer for QoS provisioning. After Release 5, with the emergence of IMS, the 3G architecture is a layered architecture with a clean split between transport (e.g., SGSN, GGSN), session (e.g., P-CSCF, I-CSCF, S-CSCF) and service planes [3]. The IMS has introduced a policy-based QoS control system.

Providing QoS is not only a transport level issue; the session layer should be involved, too. This is why 3GPP has chosen the policy-based architecture to provide a high quality transport media with efficient resource utilization. In a policy-based network, policy rules describe behavior of the network in some high-level statements without going to the detail of network element configurations.

Policy rules are a set of conditions and instructions; whenever a request for a service fulfills a condition, the corresponding instruction will be performed.

Figure 4.8 displays the proposed policy based architecture by IETF [26, 25].
The four major functional entities are defined in this architecture:

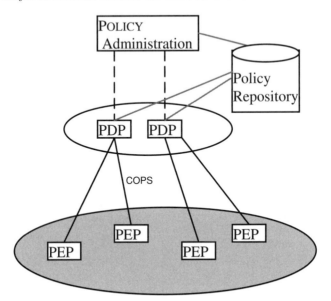

FIGURE 4.8
Application level policy-based QoS control architecture.

    Policy Decision Point (PDP): This is logically a centralized entity that
makes the policy decision according to the policy rules and the dynamic and
static information of the network.

    Policy Repository: All the policy rules reside in this entity. Policy Repos-
itory is usually implemented inside the PDP or separately as a LDAP (lightweight directory access protocol) directory server.

    Policy Administration System: This is the point in which the operator
defines its policies. The Policy Administrator System pushes the defined or
modified policies to the Policy Repository and informs the PDPs about any
modification in policies.

    Policy Enforcement Point (PEP): PEPs enforce the policies in the net-
work. They are network elements (especially edge routers) that will realize the
polices for the resources by using software and hardware features (scheduling,
queuing, classifying, traffic policy, and shaping) in the network.

    In policy management systems, there are two main models for interaction
between PDPs and PEPs: provisioning and outsourcing [25]. In the provisioning model, PDP decides which policy rules should be installed on PEP and
then provisions it for the resource reservation request coming to the PEP.
In contrast, in the outsourcing model, a resource reservation request coming
to the PEP will trigger the process of policy request from PDP. Each model
has some benefits and disadvantages. For example, the outsourcing model in-

creases the signaling load but it is more adaptive to special cases such as link failure or time-dependent polices.

The policy-based architecture defined by IETF is adopted in 3G architecture to establish end-to-end QoS for session based multimedia services. This QoS control system creates coordination between 3G transport level and IMS as the service control overlay.

As the first entry point for a SIP request message (which conveys requested QoS specifications of the service) from a user side, P-CSCF was chosen to host the Policy Decision Function (PDF) in Release 5. PDF is equivalent of PDP and enforces the policy rules on the PEP. In the subsequent releases, PDF was introduced as an independent function, and the G_q interface was introduced between P-CSCF and PDF. With this revision, other non-SIP based servers are also able to express their session QoS requirement to the PDF. GGSN (Gateway GPRS Serving Node) as the gateway of data flow to external network acts as the PEP and translate the policy rules to the IP flow control functions (labeling diff-serv flow and traffic classification, scheduling, traffic policy, and admission control and traffic shaping). To open a gate for a resource reservation request of a data flow, the PEP component of GGSN must verify the request with PDF in the signaling path. The G_o interface makes this coordination feasible. The 3GPP has agreed on the DIAMETER protocol as the communication protocol on the G_o Interface [27].

4.3.2   IMS QoS Control System Architectural Improvement

The limitations of end-to-end (E2E) QoS control of 3GPP can be divided in two categories: (1) QoS Control for multiaccess technologies; and (2) Inter-domain SLA. In this section, we introduce two new architecture for QoS control in a multiaccess paradigm and then in Section 4.3.3 we will introduce our solution for interdomain SLA.

The current IMS QoS Control architecture allows only a unique set of policy rules to be defined for resource reservation in the whole system. In fact, in the case of access to a service via other access technology, the current architecture only considers the QoS policy of the network where the service is provided. However, in a multiaccess paradigm, the QoS signaling and protocol and availability of resources in different technologies may be completely different. For example, in UTRAN (which is the UMTS access technology) the resource reservation is based on PDP, and it is completely different from 802.11 condition. Hence, the policies defined in another network are very likely inefficient in another technology.

In [5] an architecture similar to what is shown in Figure 4.9 is proposed for multiaccess to IMS services. In this architecture, PDF can control the edge routers of different access technologies. Although this architecture improves the 3GPP approach, the solution is limited to the cases where (a) the operators of all access networks are the same or (b) there is a big trust between two operators and the access network operator has agreed that the polices be dictated by the core network operator. Such an architecture is not acceptable for horizontal convergence sought in NGN.

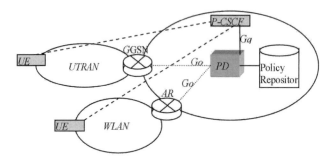

FIGURE 4.9
Modified architecture for multidomain E2E QoS.

To cope with this problem, we have proposed two other architectures which are described in Figures 4.10 and 4.11. In the architecture of Figure 4.10, the Local PDF (LPDF) will exchange the policies with the PDF in the core IMS and control the corresponding edge router. On the other hand, in the architecture in Figure 4.11 we have proposed that Local Policy Repositories of each access network will exchange their policies with a shared PDF that we call that S-PDF. The S-PDF will control the edge routers of a certain access technologies in that proximity.

Each architecture has its benefits and drawbacks, and its adoption depends on the policies and capabilities of the access network operators. In the first architecture, for example, for SIP-based applications P-CSCF as the SIP proxy should be implemented in the access network to transfer the session QoS parameters to the local PDF. This costs more but leads to a more dynamic, scalable and distributed solution. This architecture is more suitable for access networks that have already deployed P-CSCF for their home services.

On the other hand, in the architecture of Figure 4.11, there is no need for deployment of signaling processing elements in the access networks and cost will be decreased. However, the policy exchange can not be as dynamic as the previous architecture and in addition the S-PDF may be the bottleneck of the system. In these two architectures, the policy repositories or PDFs are distributed in the access and core networks of different network technologies/domains. Therefore, all the operators, including access and core operators, express their policies to be considered for the end-to-end resource reservation, and all the domains en route the data flow will reserve their resources considering the policy of other domains as well as theirs. This leads to homogenous end-to-end resource reservation.

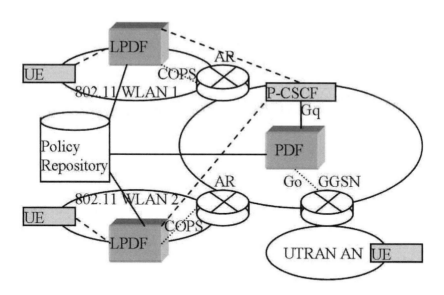

FIGURE 4.10
Modified architecture for multidomain E2E QoS: The access networks own
their Local PDF to control the AR.

4.3.2.1 Advanced Policy Definition

In convergent WiFi/3G networks, according to the different capability of these
access technologies beside the various requirements of different services (e.g.,
bandwidth, delay jitter, end-to-end (E2E) delay, security, etc.), different criteria should be considered for admission policies.

These criteria can be divided into four main categories:

- Access network information: Access parameters are downstream available bandwidth, upstream available bandwidth, link quality condition
  (SNR, PER, retransmission rate), security level, and access cost.

- User preferences: User preferences include expected QoS level for the
  service, preferred access point, the priorities that he/she defines for each
  service, etc.

- Terminal capabilities: User terminal capabilities indicate if it is a multimode terminal or not. If yes, which access technologies it supports. The
  size of the screen, battery life, etc. can be considered as terminal capabilities too.

- Service type: Each service including VoIP, Videophony, Conferencing,
  web, email, File transferring applications, etc., needs different QoS and

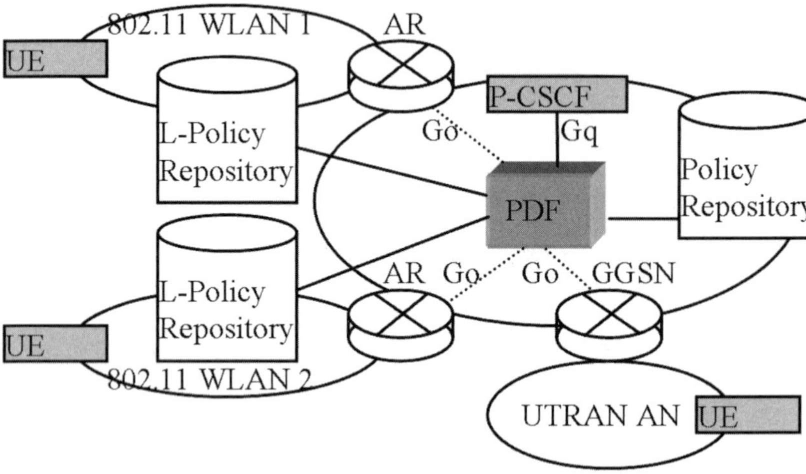

FIGURE 4.11
Modifiedarchitecture for multidomain E2E QoS: the access networks do not have their Local PDF to control the AR but defines their policies themselves.

Security support. Therefore, according to the active sessions of the user, the best candidate access should be selected in Handover time.

4.3.3   SLA Broker: New Actor in Converged Networks

NGN follows the goal of horizontal convergence of hybrid technology in the infrastructure and service level. Each network domain in converged paradigm should be able to exchange SLA (service level agreement) with other network operators and service providers of other domains and technologies. This is very essential because without that, the users of other domains will not benefit from the services of other domains/technologies, and this conflicts with the goal of "horizontal" convergence.

The existing QoS control architecture in 3GPP cannot fulfill all of these requirements because each operator needs to exchange its SLA with all other operators and technologies directly. However, in the NGN paradigm the number of network operators as well as service providers may be high and then it is not feasible for a domain to exchange directly its SLA with all other operators.

In [19], some mechanisms to exchange dynamic SLA between end-user and service networks are introduced, but the solution is not for interdomain and intertechnology architecture. To enable interdomain SLA in a scalable manner, we have proposed the architecture of Figure 4.12 [15]. In our interdomain

SLA model, there is a service provider that enables the service of SLA exchange for all operators in the blended network architecture of NGN. We call this service provider the SLA broker. The SLA broker may be founded by a consortium of major operators of different technologies. With the existence of the SLA broker, each operator defines its policies and SLA for the services it provides and registers them with this SLA broker. Then when the user of one domain requests a service that is hosted in another domain (as in the case of roaming) the SLA broker exchanges the SLA of the involved operators. With such architecture, the operators no longer needs to exchange their SLA with all other operators directly, and this leads to scalable architecture for interoperation of different domains.

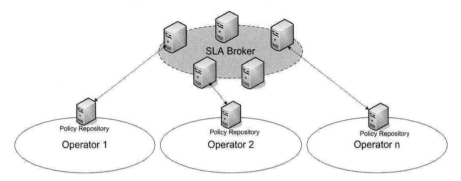

FIGURE 4.12
Service Level Agreement(SLA) broker.

4.4 Advanced Mobility Features Based on IMS over 802.11/3G

Convergence of 3G and WiFi technologies based on IMS brings in several advanced services including advanced mobility features such as session mobility, session splitting, service mobility, and cognitive call routing. In this section we take advantage of IMS service infrastructure and propose the deployment of a mobility server to provide session mobility/splitting over WLAN and 3G access technologies.

The Mobility Server can provide attractive features such as the ability of transferring an active session running on a device to another device connected to another access technology. Session Mobility is one of the aspects of mobility that brings in many advantages including:

- new capabilities for the customers;

- better utilization of resources and load balancing;

- higher user satisfaction level;

- service persistence;

- leverage frequency and duration of service usage.

4.4.1   Session Mobility Components

Session mobility is the process of transferring an active session to another terminal or another interface. There are four main components involved in a session mobility process:

- Triggering Device: The triggering device is the device that triggers the session mobility. In the simplest scenario, the triggering device is one of the two devices that are currently involved in the active session (caller or callee device). However, in an advanced model of session mobility, the triggering device is not necessarily one of the devices involved in the current session. The triggering device should just have authorization for this session transfer request.

- Source Device: The source device, is the device involved in the current session whose session will be transferred to another device.

- Target Device/Interface: The target device/interface is the device/interface that receives the media after session transfer.

- Corresponding Device: The corresponding device is the device of other party of the session that remains unchanged after session mobility.

Session mobility needs the following steps to be completed:

- Device/session discovery

- Authentication of the devices

- Selecting and authorizing the target device for transfer

- Informing the corresponding device about new party

- Negotiating for the new parameters of the session

- Session migration: Media transferring

- Session resume

In the device discovery process, the triggering device discovers the legitimate devices whose capabilities match with the requirements in the query for the target device. The discovered devices all have already been authenticated by the system, and non authenticated devices are not allowed to take part in the discovery process. On the other hand, session discovery is required when the triggering device is not involved in the current session. In this case the triggering device needs to discover different active sessions on the source device to select the one it desires to transfer.

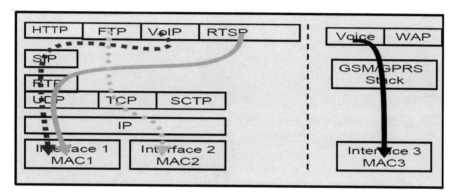

FIGURE 4.13
Different active sessions on a device.

Based on the fact that each session on a terminal uses its proper protocols in each layer of the protocol stack (Figure 4.13), all the sessions on a terminal should be described with a session digest in a manner that they can be distinguished from other sessions. Digest of a session is the minimum piece of information which unify a session.

Session Description Protocol (SDP) [9] describes the sessions with identifying the following information: (1) five tuple information: Source IP address, Destination IP address, Source Port Number, Destination Port Number, and transport protocol name, (2) Session ID, (3) Connection Protocol (Internet, CS, etc.), (4) Media Type (Video, Audio, etc.), (5) Format of the media (G711 voice, H261 Video, MPEG Video, etc). SDP is currently used in SIP-Session Initiation Protocol and RTSP [20]. RTSP is standardized for streaming services and is used more and more in such kinds of applications.

After device discovery, when the target device is chosen, the corresponding device will be informed about the new device parameters and if required, the session renegotiation for new parameters will be accomplished. Then, finally, the media will be established between target device and corresponding device. In some cases, after a while, the source device needs to retrieves the session. We call that session resume or session retrieval. For example, consider that a user transfers a voice session from his cell phone to his laptop when he arrives in the coverage of the WLAN at his home. Then after some minutes he decides to leave his home, so he should be capable of retrieving his session on his cell phone.

In session mobility, there are also some advanced interesting features: session splitting and merging. Session splitting means that a composed session is split over several simple sessions on different devices. For instance, a video-phony, including video and voice, may be split over three sessions as follow: Ingress Video, Egress Video, and Voice. In contrast, session merging is required to merge a sessions that is split over several devices to a single session on a unique device.

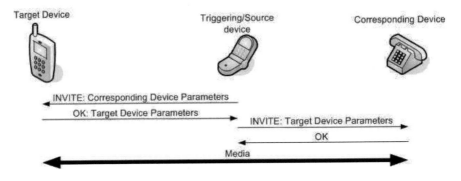

FIGURE 4.14
SIP-based third party control approach for session mobility.

Session Mobility solutions are proposed in different layers: In network layer, Mobile IPv6 (MIPv6) with some modification will be able to support session mobility [12]. MIPv6 makes a MN (Mobile Node) accessible when it changes its IP address by providing a binding between Care-of Address (New Address) and Home Address of MN. If MIPv6 creates a binding between two Home IP Addresses of two different devices, it will be able to provide session mobility. The main limitation of MIPv6 solution for session mobility is that all the active sessions of a device will be transferred together. There are also some solutions for session mobility in transport layer such as TCP-Migrate [22] that supports session mobility for a certain transport protocol (TCP). However, we focus on the solutions based on SIP for two main reasons: First, IMS is based on SIP. Second, session mobility solutions based on SIP in comparison to the other approaches are much more flexible in supporting advanced features like session splitting and merging.

4.4.2 SIP-Based Session Mobility Approaches

Two approaches for SIP-based session mobility exists: Third-party control and the REFER method. To simplify the scenarios of our examples, we ignore the device discovery phase and we assume that the target device is already selected. In addition, we consider that triggering and source device are the same.

- **Third-party control** [21]: The signaling flow of this method is depicted in Figure 4.14. In this figure, D1 is the triggering/source device and D2 is considered as target device. In addition, the corresponding device is indicated by CD. To trigger the session mobility, D1 sends an INVITE request to D2 conveying the SDP parameters of the corresponding device. Then D2 responds with an OK and indicates its preferred SDP parameters to D1. Then D1 use these parameters and sends an INVITE to CD including the SDP parameters of D2. If CD agrees with these parameters, it sends an OK to D1, and then the media will be transferred to the D2. In this method there is no direct session negotiation between CD and D2. In fact, all the SIP messages will be passed through D1; this is why this method is called third-party controlled. However, the media may be end-to-end or again being proxied by D1.

  This method easily supports session splitting [21]. But there are some essential limitations: First, the session transfer will be triggered always by the triggering device. Second, even after transfer, the triggering device should remain active for session renegotiation during the session; therefore, if the triggering device gets a battery failure, or leaves the area, the session will be lost.

- **REFER METHOD** [21]: The REFER method is defined by IETF Sipping working group in a separate RFC [23] as an extension to the SIP methods defined in RFC 3261.

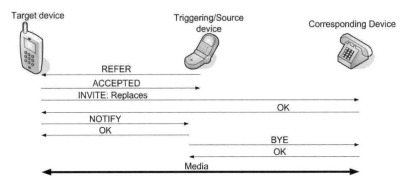

FIGURE 4.15
Session mobility based on the SIP REFER method.

Figure 4.15 describes the session mobility signalling flow by using the REFER method. D1 sends REFER to the D2 asking to accept its active session. If D2 accepts, it will send an ACCEPTED to D2. And then it creates an INVITE targeting to CD. This INVITE contains a Replace header, explaining the modification of the session parameters. CD informs D2 its acceptance with an OK. Then D2 sends a NOTIFY to D1 to inform it about the success of the negotiations. Then, in the last phase, D1 sends a BYE to CD and quits the session. In this time, the media will be transferred to D2. This method copes with the main draw back of the third-party control mode. It means that, the triggering device will be no more in the session signalling. However, with this method the implementation of the session splitting is more complicated [21]. In fact, if triggering device desires to split its session over D1 and D2, it should send two separate REFERs to each device; in consequence, each of these two devices will contact the CD separately by sending an IN-VITE. There is no guaranty that these two INVITEs will arrive in the same time, hence there will be the problem of session synchronization.

Because IMS uses SIP, SIP-based session mobility is easy to be adopted in IMS. However the implementation of these SIP-based approaches is not that straightforward, and there are two important issues, as follows: (1) third-party controlled method is not accepted in IMS because of security problems. In fact, it is not acceptable that a transferred session will be controlled by a terminal (triggering device) that is not involved anymore in the session. (2) using the REFER method for session mobility as proposed by IETF, push a huge amount of signalling load in the radio links. Figure 4.16 shows that in REFER method (without modification) a session transfer requires the exchange of 18 signaling messages, which all pass through the access networks. These issues are our main motivation to develop a mobility server that connects to IMS via the ISC interface, similarly to other application services. The mobility server handles the mobility features that the mobility management mechanism of the network (like mobile IP) cannot manage. These features include session mobility, session splitting/merging, personal mobility, service mobility and service adaptation to the user context.

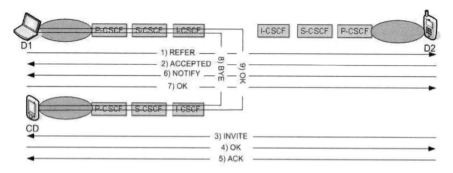

FIGURE 4.16
REFER method session mobility without mobility server in IMS.

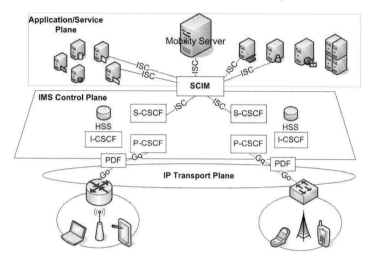

FIGURE 4.17
Position of mobility server in IMS architecture.

Figure 4.17 shows the overall architecture and the situation of the mobility server. Numerous advantages may be achieved by deploying such a mobility server. As a matter of fact, because it is deployed as a service and is connected to IMS via the same interface as other application servers, it is able to cooperate with other services to provide enhanced features in mobility management. This cooperation can be provided by SCIM (Service Capability Interaction Manager). The SCIM is a special kind of SIP application server that performs feature interaction management, and connects to application servers using the same ISC interface. For instance, the collaboration of the Location Server and this server can bring in location aware handover. The location information provided by GPS and analyzed by the location server can be sent to this mobility server. Then, according to the location, movement direction of the user and the session QoS parameters, the best access technology in that proximity will be chosen in handover process. The focus of this paper is only on the session mobility feature that this server can provide by using the service convergence that IMS brings for heterogeneous access technologies.

4.4.3 Session Mobility Managed in Application Level

As we mentioned before, the two SIP-based session mobility approaches proposed by IETF can not be used directly in IMS. By deploying the mobility server, we propose a combination of these two approaches to benefit the strong points of each approach, as well as cope with their drawbacks. Figure 4.18 shows the session mobility signaling flow with the existence of a mobility server in IMS. The mobility server receives the REFER method from the trig-

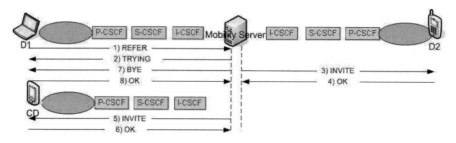

FIGURE 4.18
Session mobility based on mobility server in IMS.

gering device. Then it starts an approach like the third-party control method. By using this hybrid method, the number of SIP messages that pass through the access network will be decreased to half. In addition, there is no more the problem of the third-party control method requiring that even after session transfer, the session should be controlled by the previous device. In our method, the mobility server receives the REFER from D1 requesting transfer of the session to D2. Then, the mobility server sends an INVITE with CD parameters to the D2. If D2 accepts, it will send an OK with its parameters. Then the mobility server provides another INVITE destined to CD, conveying D2 parameters. After acceptance of CD, the mobility server sends BYE to D1, and the session will be established between D1 and CD.

Advanced features can be considered in this architecture for session mobility:

Device Discovery: Each domain can deploy local device/service discovery mechanism. Device discovery enhance the session mobility features. With contact to Service Discovery Directory Agent, IMS clients will be able to discover the devices with different capabilities in that proximity. Each subscriber, according to its profile is authorized to use a certain number of devices.

Combination of Device Discovery and Session Mobility Process: The process of device discovery and session mobility may be combined. In this approach, instead of indicating the exact contact information of the new device, the originating device, just sends the mobility server a description of the capabilities of the target device that the user desires. Then the mobility server sends a query to the service discovery directory agent (DA). The DA replies back to this request with a list of available devices compatible to the request. The signalling flow is depicted in Figure 4.19.

Session Splitting: Figure 4.20 shows our session-splitting approach. The triggering device sends REFER to the mobility server. In this REFER, the triggering device, indicates the desired target devices for each session component. The mobility server, according to the received REFER, creates separate INVITEs with the relevant SDP parameters for each of the target devices. Then each of the target devices receives only the SDP parameters of a part of the original combinational session. Each target device replies with an OK indicating its desired parameters for their relevant session. Then, the triggering device concatenates all the SDP parameters of each target device in a single INVITE and sends it to the corresponding device.

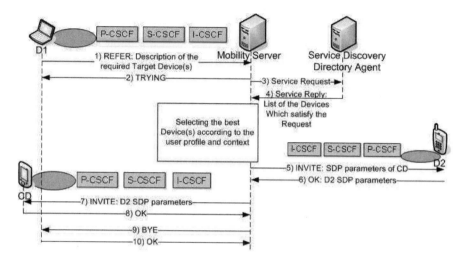

FIGURE 4.19
Combination of device discovery and session mobility signaling.

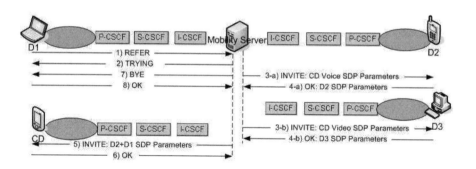

FIGURE 4.20
Session splitting with mobility server in IMS.

4.4.4 Session Mobility Server Advantages

The advantage of a mobility server similar to what we have proposed is that it can be connected to the policy decision function (PDF) and defines the session mobility strategies, considering the policy and available resources of each 3G and 802.11 access technologies. By defining advanced strategies, this mobility server enables (1) selecting the best access compatible to user constraints (i.e., cost constraints) and required resources for the active services; (2) load balancing among different network technologies; and (3) having the capability of splitting different ongoing sessions belonging to one user over different accesses (if user is accessible simultaneously via different connections). The advanced policies for session mobility can be defined based on:

1. **Better QoS**: The user may trigger the session mobility process to receive better QoS via other access technology.

2. **Service Request**: If the requested service can not be supported in the current access technology (due to policy rules, etc.), the user may trigger the session transfer.

3. **Availability of better price**: The active session of a user may cost cheaper via another access technology. In this case, the user may transfer its session to achieve better price.

4. **Load Balancing**: The network operator may transfer some active sessions from one access technology to another to achieve a better resource utilization. This decision should be based on the available resources in each access technology

4.4.4.1   Mobility Server and Load Balancing

In the overlapping area, two tiers of access technologies with hybrid characteristics and different QoS provisioning systems are available. Then comes the problem of optimized distribution of multiclass traffic over the hybrid shared resources of these two access technologies in order to reach the best resource utilization and user satisfaction. Admission policies for the new user service requests in the overlapping area influence the load balancing over the hybrid system. However, in addition to the new requests, there are also active sessions that may arrive in the overlapping areas. For instance, consider a user with a video-phony session via a 3G system who arrives in the WLAN/3G overlapping area. The transfer of such sessions from one access technology to another based on the availability of the resources can improve significantly the load balancing in the system. The mobility server is able to provide such a feature because it has access to the PDF and is aware of the available resources in each technology.

For load balancing our goal is to select an access technology for a client according to its active sessions in a manner that the set of overlapping accesses can admit the most number of clients of each service. Access networks should be classified for different applications regarding their performance and capabilities. For example, one efficient strategy may be allocating access technologies with limited bandwidth but dedicated resources to the voice, and on the other hand considering accesses with higher bandwidth and common resources to the bandwidth consuming applications.

The resources in 3G domain is organized and deterministic. 3G guaranties a certain amount of resources for each service class. On the other hand, 802.11 WLAN proposes a random access media with shared resources. Therefore, in 802.11 the allowed number of each service class in the system can be very flexible.

In order to specify the limit for the number of each service class, a performance metric should be defined in WLAN. In 802.11 WLAN, collision is the major phenomena that affects the performance of the system. To control the collision rate in a WLAN, admission of users should be limited. On the other hand, the admission rules should not be so strict that resources remain unused and idle. Taking these two issues in hand, we consider $1 - (P[idle] + P[collision])$ as the performance metric in WLAN. Figure 4.21 illustrates the general pattern of $1 - (P[idle] + P[collision])$ in WLAN based on the number of two service classes (voice and data) in the system. $P[idle]$ is the probability that a transmission time slot in 802.11 medium access remains idle, and $P[collision]$ is the probability of collision in a transmission time slot.

With the increase (decrease) in the total number of admitted stations the $P[idle]$ merges to zero (one), however, the $P[collision]$ limits to one (zero). Therefore, $P[idle] + P[collision]$ is always greater than a certain minimum level. Considering two kinds of services, voice and data, with n_v/n_d as the number of the voice/data users, this minimum level is obtained when $(n_v, n_d) = (n_{v,opt}, n_{d,opt})$. The minimum level indicates the most efficient resource utilization in WLAN where the resources are neither remain idle nor is the network overcharged with a high collision rate.

However, this optimum point in WLAN does not necessarily optimize the performance of the whole hybrid system. Therefore, some conditions such as $\Gamma_{d,Limit}$ as the throughput limit (bits/sec) of data services in the whole overlay system and B_{limit} for the limit of blocking rate of voice services are also enforced. Consequently, an optimization problem is defined as the following:

$$\min \ (n_v, n_d)|(P[idle] + P[collision]) \tag{4.1}$$

subject to

$$
\begin{aligned}
\rho_v &< 1 \\
\Gamma_d &\geq \Gamma_{d,limit} \\
Call - Block &\geq CB_{Limit}
\end{aligned}
$$

$$\tag{4.2}$$

ρ_v is the saturation factor for voice traffic in the access point (AP). As explained in Section 4.4.1, when there are n_v voice sessions with bit rate of λ in a WLAN, the AP's traffic rate will be $n_v \times \lambda$. Now, if we consider that AP manages to access the channel with the rate μ to transmit its traffic, $\rho_v = \dfrac{n_v \times \lambda}{\mu}$ indicates the voice traffic queue size in the AP and represents the saturation status of the AP.

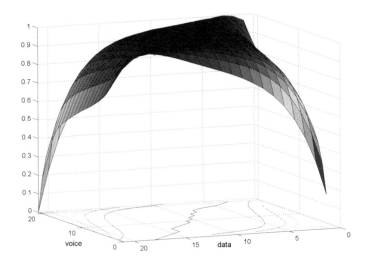

FIGURE 4.21
Performance metric for admission control: 1-(P[collision]+P[idle]).

Γ_d is the throughput (bits/sec) of data services in the whole overlay system. This throughput should be guaranteed to remain higher at a defined limit in the whole system. Considering $\Gamma_d \geq \Gamma_{d,limit}$, the decision for a session transfer/admission in the WLAN will be taken based on the whole cellular/WLAN system data throughput.

Based on this admission strategy, in our system the mobility server decides to transfer a session from one access technology to another. Moreover, we also consider session splitting as another feature of the mobility server to improve the load balancing. With session splitting, the elements of a complex session can be distributed over 802.11 and 3G interfaces based on their required resources. For instance for a video-phony, the data and video part can be sent/received from 802.11, and the voice can be kept on 3G interface. This can help make a better utilization of resources. In the next section we show the effects of session mobility/splitting on the resource utilization efficiency.

4.4.5 Simulation Model and Results

Our simulation is carried out in a two overlapping 3G/WLAN testbed by using NS2.28. WLAN is simulated based on IEEE 802.11b MAC technology with 11 Mbps bandwidth and the parameters indicated in the Table 4.2. Moreover, the bandwidth of 3G technology is considered to be 2 Mbps. Two kinds of traffic classes are considered for voice and data. The data traffics are considered to have the maximum bit rate of 100 Kbps and packet size of 500 bytes. For the

TABLE 4.2 Parameter values of 802.11b and 802.11g

	802.11b	**802.11g**
DIFS	50 μs	28 μs
SIFS	10 μs	10 μs
Slot time	20 μs	9 μs
CWmin	32	16
CWmax	1024	1024
Retry Limit	7	7
Supported data rates	1, 2, 5.5 and 11 Mbps	6, 9, 12, 18, 24, 36, 48 and 54 Mbps
ACK frame	10.2 μs	2.1 μs
PLCP and preamble	192 μs	24 μs
MAC header + FCS	24.7 μs	5 μs

voice traffic we consider the iSAC codec with a bit rate of 10 Kbps in WLAN. However, in the 3G we consider the bit rate of 32 Kbps for each voice session.

In the 3G, 1 Mb of the bandwidth is shared for all the data traffic. However, the rest of the 1 Mb is divided by 32 Kbps and creates 32 places for voice sessions. Moreover, if the data resources are not completely used, voice sessions can also occupy the available resources in this part. According to the organized resource sharing system in the 3G, the available resources are calculated simply based on the number of admitted sessions of each traffic class. However, for WLAN it is the saturation factor of AP (ρ) that specifies the available resources.

The new sessions are admitted in the system one by one. With the probability of 0.5, each new session is either voice or data. In each simulation round the new sessions are admitted to the system until the sum of the requested resources are equal to the maximum capacity of the system. In each simulation round, we limit the maximum number of the requests that are in the overlapping area (in the coverage of WLAN). For instance, in a simulation round we consider that only 10% of the users are in this overlapping area. Then, we vary this percentage from 5 to 35% and repeat the simulation in a separate round for each value. Subsequently, we calculate the resource utilization factor (RUF) in each round. RUF is equal to the allocated resources over the requested resources.

In the Figure 4.22, the RUF is demonstrated versus the network requested load and the overlapping area population limit. Moreover, we have calculated the call blocking ratio in the system to accommodate overlapping population limit is equal to 25%. Figure 4.23 shows the results for the voice call blocking ratio.

We have carried out the simulation for three different cases as follows, (1) we consider only the admission policies we have defined in the equation (4.1) with no session mobility; (2) we add the session mobility feature to improve the load balancing by transferring the sessions from one technology to the other

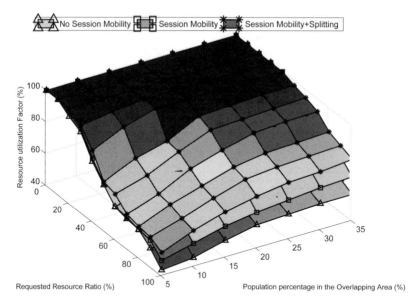

FIGURE 4.22
Resource utilization factor (RUF).

according to their characteristics and available resources; (3) finally we add the session splitting feature to evaluate how it can improve the load balancing.

As displayed in Figure 4.22 when considering session mobility+splitting, the RUF remains always higher than 75%. In addition if the WLAN hot-spot is located in the crowded area in a manner that more than 35% of the users are in the overlapping area, even with a high resource request, the RUF remains more than 85%. It means that the resources are allocated efficiently. However, in case admission control only distributes the load for newly arriving requests, RUF can fall down to 40% in high resource request situations. Moreover, the voice call blocking ratio remains only less than 0.002 with session mobility+splitting.

4.5 Related Work

To reach a scalable and reliable architecture for horizontal Fixed/Mobile Convergence, CableLabs has decided to adopt IMS as the service control overlay in the PacketCable 2.0 project [6]. In this frame, CableLabs is collaborating with 3GPP in order to modify and reproduce some IMS aspects in order to extend IMS services to broadband cable technologies. In [16], a step-by-step strategy to move toward cable technology with IMS is defined. Moreover, all of

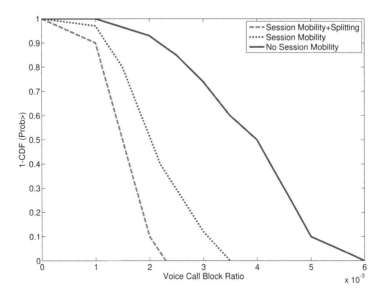

FIGURE 4.23
Voice call block ratio.

the important challenges including SIP protocol compatibility, resource reservation, and unified authentication system is discussed in this paper. In [13], the details of the required architecture and signaling flow for a non-SIP based device access to PacketCable IMS are introduced.

The work on getting IMS integrated in broadband fixed technologies is also being pursued by a standardization body in ETSI (European Telecommunications Standards Institute), called Telecommunications and Internet converged Services and Protocols for Advanced Networking (TISPAN) [8] working group in ETSI. TISPAN is considering some part of the work on IMS done by 3GPP as their Service Layer Model [7]. The TISPAN specifications are supposed to stretch IMS services to broadband xDSL subscribers. The specified architecture is structured according to a service layer and an IP-based transport layer. The service layer includes four major subsystems as follows.

- The core IP Multimedia Subsystem (IMS)

- The PSTN/ISDN Emulation Subsystem (PES)

- Other multimedia subsystems (e.g., IPTV Dedicated Subsystem) and applications

- Common components (i.e., used by several subsystems) such as those required for accessing applications, charging functions, user profile management, security management, routing data bases (e.g., ENUM), etc.

The "Core IMS" is a subset of the 3GPP IMS, which is restricted to the session control functionalities. Application Servers (AS) and transport/media related functions such as the Multimedia Resource Function Processors (MRFP) and the IMS Media Gateway function (IMS-MGW) are considered to be outside the "Core IMS". The transport layer provides IP-connectivity to user equipment under the control of the network attachment subsystem (NASS) and the resource and admission control subsystem (RACS). These subsystems hide the transport technology used in access and core networks below the IP layer. NASS provides the required functionalities for IP address provisioning, IP layer authentication, authorization of network access according to user profile, etc. RACS is the TISPAN NGN subsystem responsible for the implementation of procedures and mechanisms handling policy-based resource reservation and admission control for both unicast and multicast traffic in access networks and core networks. TISPAN architecture is proposed for broadband access technologies like xDSL.

Moreover, in [1], 3GPP has proposed an architecture to allow WLAN access to 3G services. This is a general architecture to provide access to all services in 3G domain including IMS services. With the proposed architecture the roaming of 3G clients in WLAN domain is feasible. This is a Master-Slave architecture where 3G controls the access to services and defines the corresponding policies. Finally, we should mention that there are also some researches for IMS integration in WiMAX. In [10] different levels of service integration between 3G and WiMAX based on IMS is analyzed.

4.6 Conclusion

In this paper different strategies for adopting IMS in WiFi technology were proposed. IMS, as the most complete service control overlay, can bring in horizontal convergence of different access technologies which consequently leads to advance services such as intertechnology session mobility. We presented the challenges and limitations of the proposed approach by the 3GPP for 3G-WLAN interconnection. Then, to cope with these limitations, we proposed a three phase-by-phase strategies of IMS deployment in WiFi domains. The defined strategies allow the WiFi operators to interconnect their networks to 3G domain in a win and win condition. The defined strategies get over the existing problems in 3GPP approach such as Master and Slave interworking architecture, complicated session set up process, and the requirement for multistack support in end-user device.

Moreover, we focused on the extension of an IMS-based QoS control system to be applicable to a multiaccess and hybrid technology domain. In this regard, we proposed two different solutions that allow all the access technologies to express their resource allocation policies in a multiaccess IMS domain. Subsequently, we took the advantage of IMS architecture and introduced a mobility server over hybrid 802.11/3G. The mobility server provides session mobility between the devices in different access technologies. SIP-based session mobility approaches are modified to reduce the required signalling for session transfer. One of the distinguished advantages of this mobility server is that it is defined at service level and is accessible via the same interface as other application services. Therefore, with the collaboration of mobility servers and other servers such as directory agent service discovery, and new advanced features such as combined device discovery and session mobility can be provided. We have also shown how session mobility can lead to better utilization of resources in a hybrid 802.11/3G domain. According to our simulation results, defining the efficient policies and transfer of the sessions from one technology to another based on their QoS requirements lead to improvement in resource utilization up to 30%.

Bibliography

[1] 3GPP. 3GPP System to Wirless Local Area Network (WLAN) Interworking; System Description (Release 7). 3GPP TS 23.234, v7.6.0, 3rd Generation Partnership Project.

[2] 3GPP. Ent-to-End QoS Signaling Flows – Release 6. 3GPP TS 29.208, v6.7.0, 3rd Generation Partnership Project, June 2007.

[3] 3GPP. IP Multimedia Subsystem (IMS) – Release 8. 3GPP TS 23.228, v8.2.0, 3rd Generation Partnership Project, Sept. 2007.

[4] O. Andrisano, A. Bazzi, M. Diolaiti, C. Gambetti, and G. Pasolini. UMTS and WLAN Integration: Architectural Solution and Performance. *Proc. of IEEE PIMRC*, pages 1769–1775, Sept. 2005.

[5] W. Böhm and P. Braun. Policy Base Architecture for the UMTS Multimedia Domain. *Proc. of IEEE NCA*, pages 275–285, April 2003.

[6] CableLabs. PacketCable 2.0 Specifications, `http://www.packetcable.com/specifications/specifications20.html`.

[7] ETSI. TISPAN IP Multimedia Subsystem Core Component. Technical Report ES 282 007, Version 1.2.6, 2007.

[8] ETSI. TISPAN NGN Functional Architecture. Technical Report ES 282 001, 2007.

[9] M. Handley, V. Jacobson, and C. Perkins. SDP: Session Description Protocol. IETF RFC 4566, July 2006.

[10] A. Hasswa, A. Taha, and H. Hassanein. Interworking of WiMAX and 3GPP networks based on IMS (IP Multimedia Systems) infrastructure and services. *Proc. of IEEE Conference on Local Computer Networks (LCN)*, pages 931–938, Oct. 2007.

[11] C. Low. Integrating communication services. *IEEE Communications Magazine*, pages 164–169, June 1997.

[12] W. Lu, A. Lo, and I. Niemegeers. Session mobility support for personal networks using Mobile IPv6 and VNAT. *Proc. of the ASWN Workshop*, June 2005.

[13] M. Mani and N. Crespi. Access to IP multimedia subsystem of UMTS via packetcable networks. *Proc. of IEEE WCNC*, pages 2459–2465, March 2005.

[14] M. Mani and N. Crespi. New QoS control mechanism based on extension to SIP for access to UMTS core network via different kinds of access networks. *Proc. of IEEE WiMob*, 2, 2005.

[15] M. Mani and N. Crespi. Inter-domain QoS control mechanism in IMS based horizontal converged networks. *Proc. of IEEE ICNS*, 2007.

[16] M. Mani and N. Crespi. How IMS can enable converged services for cable and 3G technologies: a survey. *EURASIP Journal on Wireless Communication and Networking*, 2008.

[17] J. Rosenberg. Interactive Connectivity Establishment (ICE): A Methodology for Network Address Translator (NAT) Traversal for Offer/Answer Protocols. IETF Internet Draft draft-ietfmmusic-ice-19.txt, 2007. work in progress.

[18] J. Rosenberg, J. Weinberger, C. Huitema, and R. Mahy. STUN – Simple Traversal of User Datagram Protocol (UDP) through Network Address Translators (nats). IETF RFC 3489, March 2003.

[19] S. Salsano and L. Veltri. QoS control by means of COPS to support SIP-based applications. *IEEE Network*, pages 27–33, March 2002.

[20] H. Schulzrinne, A. Rao, and R. Lanphier. Real Time Streaming Protocol. IETF RFC 2326, 1998.

[21] R. Shacham, H. Schulzrinne, S. Thakolsri, and W. Kellerer. Session Initiation Protocol (SIP) Session Mobility. IETF Internet Draft draft-shacham-sipping-session-mobility-05.txt, 2007. work in progress.

[22] A. C. Snoeren. *A session-based approach to Internet mobility.* PhD thesis, Department of Electrical Engineering and Computer Science, MIT, 2002.

[23] R. Sparks. The Session Initiation Protocol REFER Method. IETF RFC 3515, April 2003.

[24] V.K. Varma, S. Ramesh, K.D. Wong, M. Barton, G. Hayward, and J.A Friedhoffer. Mobility management in integrated UMTS/WLAN networks. *Proc. of IEEE International Conference on Communications (ICC),* pages 1048–1053, May 2003.

[25] A. Westerinen. Terminology for Policy Based Management. IETF RFC 3198, Nov. 2001.

[26] R. Yavatkar, D. Pendarakis, and R. Guerin. A Framework for Policy-based Admission Control. IETF RFC 2753, Jan. 2000.

[27] W. Zhuang. Policy-based QoS architecture in the IP multimedia subsystem of UMTS. *IEEE Network,* pages 51–57, May/June 2003.

Part II

Cross-Layer Design and Session Continuity

5

Mobile Virtual Private Networks Architectures: Issues and Challenges

Christian Makaya

Telcordia Technologies, Piscataway, NJ
Email: chrismak@ieee.org

CONTENTS

A Virtual Private Network (VPN) is a concept used to build a secure and private communication path on top of a public communication network such as the Internet. In other words, VPN is an overlay network that uses the public network to carry data traffic between corporate sites and users, maintaining privacy through the use of tunneling protocols and security procedures. VPNs were designed to support secure remote access for static connections and are based on the assumption that connection information (e.g., IP address) would not change. Mobile users, however, access information through wireless networks, and their underlying access networks may change over time. Due to their dependency on static, persistent connection information, traditional VPNs are not suitable for mobile users who wish to access secure information through wireless networks. Furthermore, using wireless networks to carry sensitive data introduces several security issues. Mobile VPN technology attempts to extend the VPN concept to support secure data access over public, unsecured wireless networks. This chapter presents several challenges related to Mobile IPsec (Internet Protocol Security) VPNs architectures and discusses some widely used solutions and implementations.

5.1   Introduction

Secure remote access to sensitive information is crucial in today's corporate environments. With the increasing benefits and efficiencies of a mobile workforce, corporations need to provide their mobile employees with seamless and secure access to the corporate Intranet and to corporate applications while they are on the move. E-mail is perhaps the most common and widely used corporate application, however, other applications used by sales agents, data collectors, field operators, etc., will need to be available over public wireless networks. Such users are seeking high-speed secure connections, good coverage, mobility, and ease of use.

Network access security model is usually subdivided into three steps: network access authentication, secure association, and access control and ciphering. During network access authentication, the mobile node (MN) and authentication server (e.g., AAA server) establish security association through Extensible Authentication Protocol (EAP) [1]. Secure association is a link-layer specific procedure to attach in a secure manner to a point-of attachment (PoA). On the other hand, the access control enforces link-layer data frames to be exchanged between MN and PoA only after a successful execution of network access authentication and secure association [16].

The implementation of VPNs can be done at the link layer by using L2 Tunneling Protocol (L2TP) [14], at the network layer based on IPsec [13], or at the transport layer by using the Transport Layer Security/Secure Socket Layer (TLS/SSL) [4]. Remote secure access to enterprise resources is usually based on VPNs techniques [4, 13, 14]. VPN in this chapter refers to IPsec-based virtual private networks implemented at the network layer as designed in [13]. The negotiation of secure communication between two IPsec-enabled devices is done by using Internet Key Exchange (IKE) [5] and IPsec protocols as specified in [13]. VPNs are logical networks that connect physical networks or devices by using tunneling protocols that effectively create private network over public networks paths without compromising security.

These conventional or legacy VPNs enable secure remote access from a stationary laptop. However, they are not appropriate for mobile users who require access from handset devices (e.g., PDAs, smartphones, notebooks). In fact, due to heavy authentication procedure, session response time is very low, and mobile users are often disappointed. Mobile workforces, for example, this induces productivity losses for the enterprise and eventually for the client. Example of corporate users that may realize significant productivity gains when appropriate VPN is deployed include mobile executives, field technical or operations employees, and sales agents. Such users are seeking high-speed secure connections, good coverage, mobility, and ease of use. This example shows the interest and necessity of efficient Mobile VPN (MVPN) solutions. Mobile VPNs are defined as extensions or the next-generation of legacy VPNs designed specifically for enterprises with workers who are mobile inside and outside the office. MVPN architecture enables mobile users to establish secure and transparent connections to secured services and applications such as corporate Intranet.

Mobile VPNs are defined as extensions of legacy VPNs and are designed specifically for enterprises with workers who are on the move, both inside and outside the office. Current business demands are making it less acceptable to disconnected from critical data and applications while traveling. Anytime, anywhere connectivity is becoming the norm [8]. Mobile VPN solutions are designed to meet these requirements and provide a seamless user experience without sacrificing the security. Mobile VPNs combine proven security features, traditionally found in conventional VPNs, with mobility and simplicity features typically associated with wireless networks. Mobile VPN solutions must allow the following benefits to corporate, regardless of the network connection type: enforced enterprise security requirements, reduced connectivity cost, increased employee productivity, enforced device protection, enabled optimization of wireless networks resources, etc. There is a tremendous market opportunity for operators or service providers who can meet the needs of these mobile workers or users and their respective corporations.

This chapter presents a comprehensive analysis of Mobile IPsec VPN architectures and discusses some widely used and standardized solutions. The remainder of this chapter is organized as follows. Section 5.2 presents challenges facing Mobile VPNs while Section 5.3 describes the main capabilities of MVPN solutions. In Section 5.4, existing MVPN architectures are presented following by Section 5.5 where requirements of MVPN solutions for next-generation wireless networks are described. Finally, Section 5.6 concludes the chapter.

5.2   Challenges and Issues

With the increasing number of mobile workers, enterprises need to extended the security benefits of legacy VPNs without compromising their existing security policies and rules. In fact, enterprises need to provide mobile employees secure and seamless VPN access to corporate Intranet data and applications, anytime and anywhere. In the mobile context, provisioning these services must guarantee the following features already supported in legacy VPNs: message integrity, privacy, authentication, encryption, and replay protection, to cite just few.

MVPN technology adds some capabilities to solve issues inherent to wireless networks such as handoff delay and session disruption while using encrypted and authenticated tunnels similar to conventional VPNs remote access. During the past few years, mobile VPN adoption was minimal in comparison to conventional remote access VPNs. In fact, wireless networks adoption has been hindered in corporate environment due to several critical problems such as application persistence security, end-users complexity, and roaming across heterogeneous networks. The adoption rate of wireless networks has been increased very fast with the requirement of "always best connected and protected" [8] by mobile users/workers and the exponential usage of multimedia services. Moreover, wireless networks and mobile VPNs are frequently used to improve the usability and performance of services and applications.

It is well known that there are many threats and risks inherent in wireless networks: authorization, data integrity and security, and network and device protection. Several efforts have been deployed in order to solve these issues, and various techniques may be used for network security and privacy:

- **Authentication** helps to provide a reasonable assurance that the user and device attempting to connect are, in fact, who they claim to be.

- **Encryption** can be used to assure that the data in transmission has not been altered and that no one can eavesdrop on the transmission.

- **Policy** can be used to further protect both the network and the device, providing controlled access to each.

The number of security threats that exist for communications across the Internet are, of course, legion. Placing a firewall at the perimeter of the enterprise network offers some protection because it helps to secure the type of traffic allowed in or out of the enterprise. But this approach is insufficient with regard to validating the integrity of the data and the identity of the mobile user or device – both essential aspects of mobile computing security. Moreover, it is well documented that the lack of strong authentication and encryption of public hotspots (WiFi, for example) available in airports, coffee shops, and other public areas induces security risks for enterprise network resources when mobile workers connect via these hotspots and try to get access to their corporate intranet.

On the other hand, one of the most challenging issues in next-generation wireless networks is the tradeoff between security, scalability, and efficiency. This tradeoff is significant when the network connectivity provider is different from the services provider. The level of scalability of a given scheme or mechanism shows how it could adapt when the number of subscribers and the size of networks increase. The efficiency of a scheme is usually related to performance and quality of service (QoS) metrics such as delay, signaling overhead, and service continuity. For security support, several criteria like mutual authentication, privacy, non repudiation and preestablished trust relationships are important.

A key element of a secure corporate network is the appropriate design and configuration of VPNs, including authentication, confidentiality, and integrity. MVPN appears as an always-on service and in order to succeed, it must improve mobile user experience while delivering many of the same security capabilities offered by conventional VPNs (e.g., IPsec, TLS, L2TP). In other words, MVPN solutions should provide secure, persistent, and transparent access to corporate network resources and applications.

Conventional VPNs are dependent on a stable IP address. In fact, an IPsec [13] VPN tunnel is created between two IPsec gateways or between an IPsec gateway and a remote user device who has an IPsec VPN client installed. This tunnel is tightly coupled with the IP address of each endpoint. However, in the wireless environment, when mobile users move from one subnet to another, their IP address is likely to change. Hence, the associations binding VPN tunnels break, and the VPN connection cannot survive. This leads to application failures and session disconnects. The end-user must then restart a new session from scratch.

Unlike traditional IPsec VPNs, MVPN tunnels are not tied to the permanent IP addresses. Each tunnel is bound to a temporary (*logical*) IP address. The mobile device maintains its association with the logical IP address no matter where it may roam. MVPN clients use different permanent IP addresses bound to the attached access network while retaining one logical IP address. The logical IP address is used as an endpoint for any communication to/from the mobile device. Hence, applications running on the mobile device and inside the corporate network are unaware of the user's physical or network transition.

Although, MVPNs are conceptually similar to legacy VPNs, specific characteristics of wireless networks and mobile devices introduce challenges to the design of MVPN solutions. In fact, designing and implementing Mobile VPN face the following constraints:

1. With limited processing power, battery power, resources, and memory of mobile devices, heavy security operations such as key generation and encryption may take more time.

2. Although MVPNs allow secure exchange of data, security of the device itself is also important. In fact, some data may be stored in the device and need to be protected.

3. Speed and delay over wireless links may induce timeout issues for delay sensitive applications such as authentication.

4. The diversity of the underlying network infrastructures and technologies introduces interoperability issues.

5. With the growing number of mobile workers and escalated amount of mobile data traffic, the loads associated with concurrent connections and simultaneous authentications must be estimated to determine if the IPsec gateway can handle the required traffic demands.

6. Increasingly, mobile devices are now equipped with multiinterfaces enabling connections to different access networks and technologies. While

roaming between different networks or technologies, MVPNs must integrate roaming capabilities across different mobility management protocols into the security solution.

5.3 Capabilities of Mobile VPN Protocols

In order to succeed, MVPNs must improve the mobile user experience while delivering many of the same security capabilities offered by conventional VPNs. In other words, MVPN solutions should provide secure, persistent, and transparent access to corporate network resources and applications. Mobile VPNs inherit some concepts of conventional VPNs. However, there are several capabilities that differentiate both techniques [17]:

1. *Network independence*: mobile VPNs must operate over any kind of network, public or private, wired or wireless including Ethernet, residential broadband connections (e.g., cable, DSL), WiFi hotspots, wireless MAN (e.g., GPRS, EV-DO, UMTS, LTE, WiMAX).

2. *Application persistence*: application session persistence means that standard network applications remain connected to their peers, preventing the loss of valuable user time and data. When mobile VPN devices lose connectivity, the mobile VPN server queues undeliverable data, resending them when the device later regains connectivity. Some applications still may timeout when using a mobile VPN, but many can survive moderate disruptions.

3. *Network persistence*: network session persistence means that users do not have to repeat the login process when they move from one subnet to another, or when they go out of coverage area and return, or suspend and resume sessions. In legacy VPN, users loss their IP address and connection each time they roam into a different network. To avoid this, the virtual or logical IP address of each mobile VPN client remains the same when the access network-based IP address changes. By communicating through this persistent IP address, clients can avoid network connection reset and reauthentication.

4. *Transparent suspend/resume*: when mobile devices go to idle mode to conserve battery power, the mobile VPN server maintains enough information about the devices and clients so that they can "awaken" and resume their communication without disrupting the VPN tunnel or associated state.

5. *Network transparency*: the mobile VPN solution delivers the same degree of security, application interface, and user experience, regardless of the underlying access network.

6. *Wireless optimization*: wireless link optimization means that data are transmitted as efficiently as possible. Some mobile VPNs optimize network performance by adjusting to each links' characteristics, avoiding, for example, unnecessary but costly fragmentation and retransmission.

7. *Network aware rules*: like legacy VPNs, mobile VPNs enforce security policies. However, mobile VPN rules also can incorporate network characteristics, for example, disabling encryption when connected to a trusted Ethernet, blocking file transfer over low-bandwidth of high-tariffs links, or automating Wi-Fi hotspot login before VPN tunnel establishment (i.e., optimistic access).

8. *Policy-based roaming among multiple networks*: mobile VPNs manage the use of multiple links with roaming policies. For example, policies can use signal strength threshold to trigger roaming (handover) or use speed, pricing, and preferences to select and log the device into the best available network. The ability to manage multiple links is critical, as the trend towards multiinterfaces and multihoming devices accelerates.

Persistence is the primary reason that mobile users choose a mobile VPN. However, mobile VPNs must also improve application performance. In fact, for deployment success, mobile VPNs must improve user experience while delivering many of the same security capabilities offered by conventional IPsec and TLS VPNs. When using IPsec for mobile deployments, the performance may degrade significantly, up to 40%. Such performance degradations must be managed and accounted when considering applications appropriate for mobile VPNs. While mobile VPNs support a broad mix of IP-based applications, session persistence cannot make lengthy outages invisible to real-time applications like streaming media or ongoing voice over IP (VoIP) calls.

5.4 Mobile VPN Architectures

Several mobile VPNs architectures have been proposed in the literature, each with pros and cons. In this section, we describe the most representative architectures considered in the standards bodies.

5.4.1 Mobile VPN Models

VPN can be seen as an overlay network that uses the public network to carry data traffic between corporate sites and users, maintaining privacy through the use of tunneling protocols and security procedures. Moreover, VPN can be categorized into two models: the *end-to-end* model and *network-based* model. In the *end-to-end* model, the user or site connects to the enterprise resources over a secure tunnel using the underlying network as a simple data conduit. In the *network-based* model, the network service provider (NSP) implements VPN-aware router within the network. This router usually setup two secure tunnels, one from the user or site to the router itself and the other from the router to the enterprise. The network-based model needs trust in the NSP to maintain the security associations with the endpoints at the VPN router.

Although the end-to-end model makes VPN service access facilities independent of the service provider, it cannot support scalable growth and requires large investments by the enterprise. The network-based model, on the other hand, can perform traffic aggregation at the tunnel concatenation points for better scalability and network resource usage, and therefore, can cost-effectively offer VPN services. Thus, a network-based VPN is a preferred solution for remote access VPNs to support stationary users as VPN usage grows [3]. But with the growth in the number of mobile users, an important issue to explore is whether the existing network-based VPN architecture is suited for mobile users. With the emergence of high-speed wireless data services in 3G and Beyond 3G (i.e., LTE) wireless technologies, VPN usage from mobile nodes (that is, mobile VPN services) will grow exponentially.

5.4.2 IPsec over Wireless Links

IPsec [13] provides network layer security by authenticating users and encrypting their data, thus preventing eavesdropping, identity spoofing, and traffic sniffing. IPsec also provides message integrity, privacy, authentication, and replay protection. Additionally, IPsec establishes a mechanism to negotiate security algorithms and keys required to establish point-to-point security.

However, IPsec on its own is not well suited for wireless networks. IPsec protects the source IP address, which must remain static for the duration of the secure tunnel to validate the integrity of the sender. While this provides effective protection against spoofing, it also means that an IPsec VPN connection cannot survive wireless coverage gaps, loss of connectivity, or networks transitions where the source address may change or be released. It is well known that IPsec has poor wireless performance and no application level control. Combining Mobile IP (MIPv4) [6] and IPsec is the basis of the MVPN architecture. Mobile IP enables mobile node to remain reachable while moving from one IP network to another. For this reason, MIPv4 is sometimes used to augment IPsec for the purpose of hiding IP address changes while roaming to support network persistence.

Several layers of encapsulation and tunneling are required for this combination. This can include an IPsec encapsulation for protecting the endpoint data, a mobile IP encapsulation to hide the address changes, and a second IPsec encapsulation for Home Agent (HA) and Foreign Agent (FA) security.

This excessive use of encapsulation and tunneling and the associated overhead make this approach inappropriate for most wireless networks. On the other hand, there are some difficulties in combining MIPv4 with firewall and QoS management. For example, the packet-filtering specification of firewall should be dynamically updated with the new care-of address (CoA) in accordance with route optimization concept. Similarly, QoS policies of routers should be updated with the new CoA. However, there is no standard method for these controls. Furthermore, combining IPsec with MIPv4 has at least two mains shortcomings:

1. Using pre-shared keys with dynamic IP addresses (i.e., CoA) require the IKE Aggressive Mode [9] which exposes the users identities (IDs) during the protocol exchange since it is sent in the clear.

2. IPsec initiators with dynamic addresses require the responder to accept all IP addresses.

Resolving MVPNs weaknesses in wireless environment is one of the missions of MOBIKE protocol [7]. MOBIKE [7] provides a method to keep the connection with the VPN gateway active for a host with multiple IP addresses and/or where IP addresses may change over time (for example, due to mobility). However, MOBIKE does not address application or session persistence.

5.4.3   IETF Mobile VPN Solution

Mobile IPv4 (MIPv4) [6] has been proposed for mobility management at the IP layer and allows a mobile device to remain reachable despite its movement. By combining MIPv4 and IPsec, the registration procedure will fail when the mobile node (MN) uses the foreign agent (FA) care-of address (CoA) in the visited network. In fact, the foreign agent cannot recognize an encrypted MIP registration message. To solve this issue, the IETF has proposed a solution [19] for coexistence of MIPv4 and IPsec for mobile users. In this solution, MIPv4 is used when the MN is inside the home network and the VPN tunnel endpoint address is used for the MIPv4 registration when the MN is outside.

The IETF MVPN solution places an external home agent (x-HA) outside of the home network that supports mobility for VPN users who travel outside of the home network. When an MN moves out of the home network, it must register to both the x-HA and the i-HA (internal HA located in the home network). The MN does not encrypt the MIP registration message with IPsec ESP (Encapsulating Security Payload) when it registers to the x-HA because the registration message is sent directly to the x-HA without passing the VPN gateway. Another layer of MIPv4 is used underneath IPsec in order to overcome renegotiation of IPsec-VPN tunnels after each movement of the MN. This makes IPsec unaware of movement and the MN can freely traverse the external network without disrupting the VPN connection.

Figure 5.1 shows an overview of connection setup when MN is in foreign

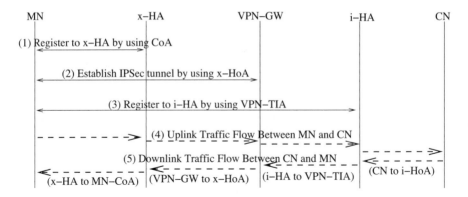

FIGURE 5.1
IETF Mobile VPN messages exchange procedure.

network, VPN-TIA means VPN Tunnel Inner Address [19]. After completion of the first three steps, the triple tunnels (i-MIP, IPsec and x-MIP) are constructed to enable session continuity, secure communication, and MN reachability from Intranet.

The IETF MVPN [19] leads to some issues, including where the x-HA should be located, what trust mechanisms are required with the x-HA and the overhead of three tunnels required by the scheme. The placement of x-HA will impact the handoff latency and end-to-end data delivery delay. According to [19], only one x-HA is placed through the Internet. This choice would not only degrade the performance of x-HA due to heavy overload but also, if the MN is getting far from the x-HA, the handoff delay and data delivery time will increase significantly. In addition, the x-HA is outside of the demilitarized zone (DMZ) and might not be under the control of the VPN administrator.

Moreover, there should be a trusted mechanism to assign the x-HA when the MN roams outside of its home network. In fact, the x-HA must be authenticated and authorized before it is assigned to an MN. In the IETF MVPN solution, a packet will be encapsulated by three extra headers. Figure 5.2 shows three times packet encapsulation for downlink traffic. This may degrade the performance of real-time applications because the payload is often short in real-time applications and would lead to excessive fragmentation. Furthermore, the encapsulation and decapsulation of three headers would induce extra signaling overhead. Another critical problem is caused by triangular routing when two mobile VPN users are located in foreign networks and want to communicate. In this case, the traffic is sent to the i-HA via the x-HA and a VPN gateway, for uplink (from MN to CN) communication. For downlink (from CN to MN) communication, traffic is relayed from the i-HA via the VPN gateway and the x-HA based on binding information in the i-HA. This procedure is known as triangular routing and is illustrated by steps (4) and (5) in Figure 5.1.

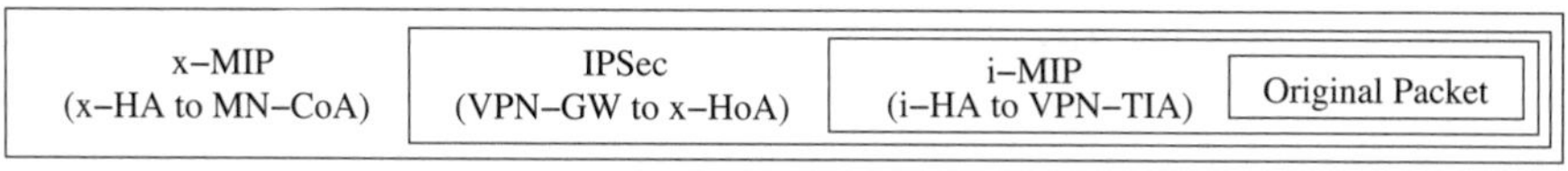

x–MIP (x–HA to MN–CoA)	IPSec (VPN–GW to x–HoA)	i–MIP (i–HA to VPN–TIA)	Original Packet

FIGURE 5.2
Downlink packet encapsulation with IETF MVPN.

Mobile IPv6 (MIPv6) [12] has been specified to allow nodes to remain reachable while moving in the IPv6 Internet and also to solve shortcomings of MIPv4. In MIPv6, VPN peers continue to maintain security associations without being affected by changing of IP addresses [2]. Security mechanisms for the control traffic between MN and HA are discussed in [2]. If this control traffic is not well protected, MNs and correspondent nodes (CNs) are vulnerable to several threats, such as man-in-the-middle, passive wiretapping, and denial-of-service (DoS). However, we should note that the choice of IPsec for securing MIPv6 signaling was historically based on the prevailing thinking of the IPv6 community. The current discussion within IETF is about the design of an alternative security architecture for MIPv6.

In order to enhance performance of IETF MVPN, some proposal are available in the literature. For example, authors in [15] suggested a dynamic allocation of x-HA depending on the current location of the MN. This proposal solves the registration failure issue and reduces the handoff delay compared to the scheme proposed in [19]. However, the triangular routing issue remains.

5.4.4 Secure Handover Protocols

Handoff and roaming management is very important for heterogeneous radio access networks environment. Usually, the main goal of handoff optimization schemes is the minimization of handoff latency, packet loss, and signaling overhead. However, due to different existing security mechanisms across heterogeneous access networks, support of secure handoff across various access technologies (e.g., WiFi, EV-DO, UMTS, WiMAX, and LTE) is now very important. The integration of these access technologies is not only limited at the architecture point of view but also in terms of security. The security association (SA) procedures may differ across access network technologies, and it is necessary to secure handoff signaling and data across these networks technologies to avoid attacks and threats and to guarantee better quality of service (QoS).

When roaming between two access networks, security associations (SAs) for secure handoff operation need to be established. SA is based on the security parameter index (SPI), destination IP address, and protocol type such as Authentication Header (AH) and Encapsulating Security Payload (ESP). During roaming between two access networks, the mobile node needs to be able to handoff toward a visited network with which it does not have a statically established SA. In other words, dynamic SA establishment should be supported. In case of heterogeneous access networks environment, usage of one SA for securing different access technologies could allow optimization of SA establishment procedure. This means that the previously established SA is reused for reauthentication during handover.

Usage of AAA protocols (e.g., RADIUS or DIAMETER) has been mandated to support secure operations. However, these protocols increase overall handoff delay, especially when the home network is far away from the visited network. In order to optimize handoff procedure, it is important to minimize the delay introduced by the AAA protocols during handoff. Several schemes try to provide secure and seamless handoff across heterogeneous networks by using AAA context transfer concept. The concept of context transfers has been introduced to reduce delays associated with AAA operations. However, context transfer techniques assume that the target network can support the service provided in the serving network. This might not always be the case and the home network must be contacted for the renegotiation of offered services, which leads to more delays and performance degradation. Low-latency handover is possible within the same AAA domain. However, if there is no trust relationship between the serving and target networks, execution of authentication procedures is required for roaming between the two AAA domains and full authentication cannot be avoided.

Performance evaluation of secure handoff protocols could be based on the following criteria: security signaling latency, cryptographic processing time, transmission delay, handoff latency, and packet loss. The security signaling latency is defined as the time elapsed between the sending of the first authentication message until the reception of the acknowledgment for the last authentication message. The cryptographic processing delay is the time spent by the mobile node to perform different cryptographic operations during the authentication. The transmission or end-to-end delay is the time it takes a packet to reach its destination. The handoff latency usually represents the time elapsed when a decision to handover is executed until the traffic is redirected to the new interface.

The Media-Independent Pre-Authentication (MPA) framework [18] has been defined to allow service continuity across different link-layer technologies and AAA domains using existing media-specific secure associations procedures. This framework has been adopted by the IRTF MOBOPTS Working Group [11] and has been introduced into the IEEE 802.21 – Media Independent Handover Working Group [10] within the Security Task Group (802.21a). MPA allows MNs to perform proactive authentication, meaning that MNs can authenticate themselves to candidate target networks at the IP-layer before attaching to those networks at the link-layer. MPA takes the authentication process out of the critical phase, reducing handover delay to just the link-layer switching latency. This significantly reduces the time the MN is in a disconnected state unable to send or receive information. The MPA platform can be integrated with existing mobility management protocols for a secure and low-latency mobility solution. In order to allow service continuity across different link-layer technologies, AAA domains and to use existing media-specific secure association schemes, the IEEE 802.21 working group is working on proactive authentication approach.

5.5   Requirements for MVPN Architectures

The design of mobile VPN solutions should take into account several requirements including protection of VPN gateways from malicious traffic, handoff of VPN sessions as well as scalability and usability issues. The reservation and guarantee of QoS is also crucial. In fact, an MN should be able to reserve QoS required for an application when a VPN session is setup for a bandwidth consuming application, such as media-streaming. Furthermore, handoff of a VPN session should be supported. In terms of usability, the mobile user should be able to continue an ongoing application even when his device switches between different access networks technologies.

Security mechanisms integrated into the VPN technology cause a considerable data size expansion, due to IPsec, which is especially in wireless networks undesirable. Even worse is the fact, that mechanisms like header compression techniques cannot be applied anymore, since the relevant headers are encrypted. This leads to much more bits per IP packets that have to be transmitted over the lossy and unreliable wireless channel. Hence, it is important to have a mechanism that enables the application of header compression techniques in IPsec-secured VPNs, which makes it possible to reduce data size expansion and consequently the packet-loss rate and transmissions costs.

With mutual authentication between the mobile node and the VPN gateways, a malicious node and/or entity cannot mount a replay attack by stealing the authentication parameters of the user and impersonate a valid network, since it doesn't posses a valid certificate. Security mechanisms integrated into the IPsec VPN technology cause a considerable data size expansion, which is undesirable in wireless networks. Even worse is the fact that mechanisms like header compression cannot be applied since the relevant headers are encrypted. This leads to many more bits per IP packet that must be transmitted over lossy and unreliable wireless links. This processing capability can have a significant impact on scalability. Hence, it is important to adopt mechanisms that enable header compression in IPsec-secured VPNs in order to reduce data size expansion and increase transmission efficiencies.

A key element of secure networks is the proper design and configuration of VPNs, including authentication, confidentiality and integrity. Since Mobile VPNs combine proven security features, found in conventional VPNs, with mobility typically associated with wireless networks, the main objectives in designing Mobile VPNs architectures can be summarized as follows:

- *Interoperability* that ensures the security solutions can avoid interoperability problems, for example, by using generic solutions applicable to the most networks applications and service scenarios;

- *Usability* that makes it easy for the end-users to use the security-enabled services;

- *Availability* that enforces networks and services not to be disrupted or interrupted by, for example, malicious attacks;

- *Cost-effectiveness* that minimizes the additional costs of security and makes it lower than the cost of risks;

- *QoS guarantee* that requires security solutions like cryptographic algorithms to meet QoS constraints of multimedia applications.

Since encryption and IPsec session establishment require a significant processing, scalability aspects should be considered when designing a Mobile VPN solution. A MVPN solution should be designed to deal with wireless security, roaming, coverage gaps, performance, and handset device support. Moreover, security should not ignore QoS aspects. In fact, without QoS requirements support, all VPN traffic will be treated equally and sensitive applications are vulnerable to delay. Packet loss recovery improves the quality and performance of real-time application transmissions such as VoIP, video streaming, and instant messaging by automatically replacing lost or missing packets without having to retransmit them. Otherwise, occurrences of jitter continue and sometimes could be higher than the tolerance levels for example for real-time video.

Security is an imperative goal, but it must be implemented without substantial impact to productivity or usability. Since encryption and IPsec session establishment require significant processing, scalability issues must be considered when designing a Mobile VPN solution. A MVPN solution should be designed to deal with characteristics common to wireless networks including security, roaming, coverage gaps, latency, speed, performance, and handset device support. Mobile users should be able to continue ongoing applications even when their device switches between different access technologies and networks. Support of packet loss recovery allows MVPN solution to enable session persistence and to handle coverage gaps.

Furthermore, the design of Mobile VPN solution must take into account following requirements:

- Maintain secure connection while mobile user (device) moves from one network to another, IP address changes, and encountering coverage gaps;

- Make the network appear fixed and unchanging to the end user regardless access networks information (IP address, coverage gaps) change;

- Use standards-based authentication and encryption technologies;

- Handle coverage gaps.

Multiple access networks technologies is a reality now. This heterogeneity mandates that whatever security choices are made should be applicable across all networks in use or readily foreseen. With increasing numbers of heterogeneous access technologies, MVPN solutions need to be network agnostic and operate across any network type (e.g., private or public wireline or wireless networks). In fact, Mobile VPN technology must be able to adapt to access network changes and maintain secure connection even if the underlying network changes.

5.6 Conclusion

Mobile VPNs offer the promise of secure, remote access to sensitive corporate data and applications while on the move without compromising security. As more of the corporate workforce becomes mobile, Mobile VPN support is crucial to securing sensitive data and driving greater worker efficiency. Mobile VPNs extend the concept of VPNs to the mobile environment, allowing mobile workers to establish IPsec VPN tunnel from their mobile handset devices to an IPsec gateway over existing Internet connections. The design and implementation of Mobile VPNs raise several challenges. Hence, several requirements and constraints should be considered in their design, including application and network persistence, wireless resource optimization, enterprise security enforcement, connectivity cost reduction, and employee productivity improvements.

In this chapter, issues and challenges related to mobile VPNs have been presented as well as existing standardized solutions and architectures. The presented solutions show that several open issues remain and additional work is needed. The optimized Mobile VPN architectures must provide an end-to-end security solution using industry standard protocols to support interoperability. The ultimate success for deploying Mobile VPN depends on the ability to effectively and efficiently manage complex configurations and provide mobile users with seamless and secure communication experiences.

Bibliography

[1] B. Aboba, L. Blunk, J. Vollbrecht, J. Carlson, and H. Levkowetz. Extensible Authentication Protocol. IETF RFC 3748, June 2004.

[2] J. Arkko, V. Devarapalli, and F. Dupont. Using IPSec to Protect Mobile IPv6 Signaling Between Mobile Nodes and Home Agents. IETF RFC 3776, June 2004.

[3] R. Cohen and G. Kaempfer. On the Cost of Virtual Private Networks. *IEEE/ACM Transactions on Networking*, 8(6):775–784, Dec. 2000.

[4] T. Dierks and E. Rescorla. The Transport Layer Security (TLS) Protocol - version 1.2. IETF RFC 5246, Aug. 2008.

[5] C. Kaufman (Ed.). Internet Key Exchange (IKEv2) Protocol. IETF RFC 4306, Dec. 2005.

[6] C. Perkins (Ed.). IP Mobility Support for IPv4. IETF RFC 3344, Aug. 2002.

[7] P. Eronen (Ed.). IKEv2 Mobility and Multihoming Protocol (MOBIKE). IETF RFC 4555, June 2006.

[8] E. Gustafsson and A. Jonsson. Always best connected. *IEEE Wireless Communications*, 10(1):49–55, Feb. 2003.

[9] D. Harkins and D. Carrel. The Internet Key Exchange (IKE). IETF RFC 2409, Nov. 1998.

[10] IEEE 802.21. http://www.ieee802.org/21/.

[11] Internet Research Task Force (IRTF). http://www.irtf.org/.

[12] D. B. Johnson, C. E. Perkins, and J. Arkko. Mobility Support in IPv6. IETF RFC 3775, June 2004.

[13] S. Kent and K. Seo. Security Architecture for the Internet Protocol. IETF RFC 4301, Dec. 2005.

[14] J. Lau, M. Townsley, and I. Goyret. Layer Two Tunneling Protocol - version 3 (L2TPv3). IETF RFC 3931, Mar. 2005.

[15] Y.-W. Liu, J.-C. Chen, and L.-W. Lin. Dynamic external home agent assignment in mobile VPN. In *Proc. of IEEE Vehicular Technology Conference (VTC'04)*, volume 5, pages 3281–3285, Los Angeles, CA, Sept. 2004.

[16] Y. Ohba, M. Meylemans, and S. Das. Media Independent Handover Security Tutorial. IEEE 802.21/Media Independent Handover Working Group, Mar. 2008.

[17] L. Phifer. Mobile VPNs: Enabling on-the-go workforces. *Business Communications Reviews*, pages 36–40, Oct. 2006.

[18] A. Dutta *et al.* A Framework of Media-Independent Pre-Authentication (MPA) for Inter-domain Handover Optimization. IETF Internet Draft draft-irtf-mobopts-mpa-framework-05, Feb. 2009. (work in progress).

[19] S. Vaarala and E. Klovning. Mobile IPv4 Traversal accross IPSec-based VPN Gateways. IETF RFC 5265, June 2008.

6

Cross-Layer Handover for Mobile WiMAX
Networks

Melody Moh

Department of Computer Science, San Jose State University, USA
Email: moh@cs.sjsu.edu

Pat Jangyodsuk

Department of Computer Science, San Jose State University, USA

Phuong Huynh

Department of Computer Science, San Jose State University, USA

CONTENTS

IEEE 802.16, also widely known as WiMAX (Worldwide Interoperability for Microwave Access), is an emerging broadband wireless standard. The amendment 802.16e has been defined to support mobility and other extensions. Their seamless support over an efficient handover (HO) mechanism is vital to the success of WiMAX networks. This chapter first presents an overview of the IEEE 802.16 standard, followed by a survey of major related work proposed to enhance handover in 802.16. Then, we describe a promising cross-layer design of WIMAX HO, the FMIPv6 (Fast Mobile IPv6 Handoff Protocol), and our newly proposed improved scheme. The enhancement is achieved by adopting two features: reducing control messages and eliminating duplicated care-of-addresses. The resulting HO delay is therefore decreased, which also improves other QoS parameters. Next, we describe the evaluation of supporting video and voice-over-IP traffic over four cross-layer HO schemes: FMIPv6, two existing enhancements to FMIPv6, and the newly proposed scheme. We evaluate the HO delay, HO packet loss, network throughput; PSNR (*Peak Signal to Noise Ratio*) and MOS (*Mean Opinion Score*) for video quality; and R-factor and MOS for voice quality. We found that the proposed enhancement has achieved considerably improvement over the existing schemes, especially under heavy network conditions and fast mobility speed.

6.1   Introduction

In the world of rapidly changing wireless technologies, WiMAX (Worldwide Interoperability for Microwave Access) has emerged to becoming one of the most promising wireless metropolitan area network technologies [11]. The standard is built on the key premises such as open architecture IP (Internet Protocol)-based network, high bandwidth equivalent of wire-line xDSL (Extended Digital Subscriber Line) technology, mobility support, QoS (Quality of Service) support, and low cost of deployment and maintenance. Supporting seamless mobility with QoS is one of the key challenges in offering real-time services. During handover (HO), when a mobile node moves from one place to another, it is essential to provide a satisfactory level of video or voice quality; disruptions or even intermittent disconnections of transmission due to long HO processing time would severely degrade the level of services.

While the WiMAX technology has enjoyed many advantages such as high data rate, wide transmission range, and flexible scheduling, it, however, suffers from long-delay HO. This drawback is mainly due to network-layer HO latency. Specifically, the binding update in MIPv6 (Mobile Internet Protocol version 6) could take a long time and thus becomes a bottleneck. Acknowledging the problem, several major proposals have been presented at the Internet Engineering Task Force (IETF). One of them is the Fast MIPv6 (FMIPv6) [18], which took a cross-layer approach while introducing packet tunneling between current and target base stations to eliminate packet loss [12]. It has improved both delay and loss performance during a HO process [16]. FMIPv6, however, requires a sequence of preparation messages. Only when all these messages have been received in time the predictive mode may proceed. If, however, some messages are delayed or lost, the reactive mode will be triggered, and will cause a long delay in FMIPv6.

In this chapter, we first describe the FMIPv6 cross-layer HO scheme. Next, we present a new algorithm to solve the above-mentioned issue. The first idea is to merge the co-functioning data-link layer and network layer messages [4]. This will reduce the number of messages required and avoid message lost during heavy network conditions. Furthermore, we adopt the Temporary Care-of-Address (tCoA) generation method [20]. This eliminates the duplicated CoA problem and therefore further reduces HO delay. Finally, by combining the two features, we can further eliminate some other HO messages since they are no longer needed.

The chapter is organized as follows. Section 6.2 presents the background information. Related study is explained in Section 6.3. Section 6.4 describes two cross-layer HO approaches, the FMIPv6 and the proposed enhanced scheme. Performance evaluation is illustrated in Section 6.5. Finally, Section 6.6 concludes the chapter.

6.2 Background Concepts

In this section, we first give a brief overview of the IEEE 802.16 networks. This is followed by a detailed description of the IEEE 802.16e HO process [11], and a concise explanation of MIPv4 [24] and MIPv6 [14] protocols.

6.2.1 IEEE 802.16 Networks

Based on the IEEE 802.16 standard, WiMAX is an emerging wireless networking technology that brings many advantages to end-users. Besides the benefits of lower cost and higher bandwidth compared to wired networks, WiMAX supports non-line-of-sight signals that help increasing network coverage and signal reliability. The high bandwidth and wide range of WiMAX make it a promising wireless alternative to cable and DSL (Digital Subscriber Line) for the "last mile" connectivity in broadband access that provides data and telecommunications services [8].

WiMAX, however, has some weaknesses when implemented in practical networks. Theoretically, WiMAX can transmit at the rate of 70 Mbps within 30 to 50 Km. In reality, however, it can only happen under ideal condition within a completely clear area with no obstacle and with line-of-sight (LOS) links. The truth is, within a normal environment with non-line-of-sight links, WiMAX can transmit well only within the range of 10 Km. Besides that, the existence of some environment factors such as obstacles (buildings and walls), noise, and raining can interrupt the connection. Furthermore, interferences caused by nearby wireless devices have negative effects to the QoS during the transmission, and cause difficulties for mobile users, especially when there is the need to handover between networks.

6.2.2 IEEE 802.16e Handover

IEEE 802.16e [11], also known as Mobile WiMAX, is an amendment to the IEEE 802.16 standard. It was proposed in 2005 to give more supports for mobility in the network. The most attractive researched direction in 802.16e technology is how to handle HO process between WiMAX networks or between WiMAX and other networks. Formally, HO is the process where a mobile station (MS) leaves the cell-range controlled by a base station (BS), and enters the cell-range controlled by another BS. Basically, the procedure is divided into three main stages: network topology acquisition, HO decision and initiation, and network reentry. The first one takes place when the mobile node (MN) acquires information about its network such as the list of neighbor base stations or information of uplink and downlink for synchronization. HO Decision and Initiation stage is used when the signal between the MN and current BS is getting weaker and the MN needs to perform HO to a new BS with better signal strength. The last stage, network re-entry, happens after the MN finishes synchronizing with the new BS, and starts to establish a new connection with that BS in the new network. The detailed description of those stages are included in following sections.

6.2.2.1 Network Topology Acquisition

The network topology acquisition procedure is shown in Figure 6.1 and described below. The mobile station (MS) and the serving BS (SBS) participate in this phase with the help of the backbone network. The BS periodically broadcasts Mobile Neighbor Advertisement (MOB_NBR-ADV) messages that contain the channel information of neighboring BS in both physical and link layers. Based on these messages, the MS can synchronize with neighboring BSs without listening to Downlink Channel Descriptor (DCD) and Uplink Channel Descriptor (UCD) broadcast messages. The MS will scan through the neighboring BS to choose candidates for HO. Then, the MS sends the Mobile Scanning Request (MOB_SCN-REQ), which gives a list of potential target BS (TBS) to the SBS to request downlink synchronizing with the neighboring BS; the MS also waits for MOB_SCN-RSP (Response) from SBS to allocate

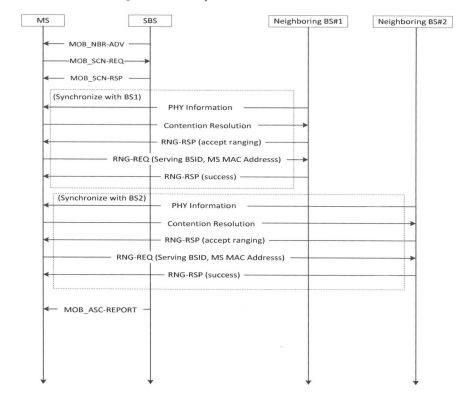

FIGURE 6.1
Network topology acquisition in IEEE 802.16.

a scanning duration. The SBS, upon the receiving of TBS list, will negotiate with those BS to obtain a unicast ranging opportunity for the MS. After this, the next step is ranging. Thus, the ranging will be non contention-based if the negotiation is successful; otherwise, it will be a contention-based CDMA (Code Division Multiple Access) ranging procedure.

The ranging process starts when the MS tries to synchronize and adjust uplink parameters. It may have multiple Ranging Request (RNG-REQ) and Ranging Response (RNG-RSP) messages exchanged between MS and its neighboring BS, and end when the MS finishes synchronizing with all the neighboring BS. During the scanning process, all the incoming data destined to MS will be buffered by the SBS and sent to MS later.

6.2.2.2 Handover Decision and Initiation

HO decision and initiation can be triggered either by the MS using Mobile MS HO Request (MOB_MSHO-REQ) or by the BS using mobile base station HO Request (MOB_BSHO-REQ) messages. For simplicity, the description as-

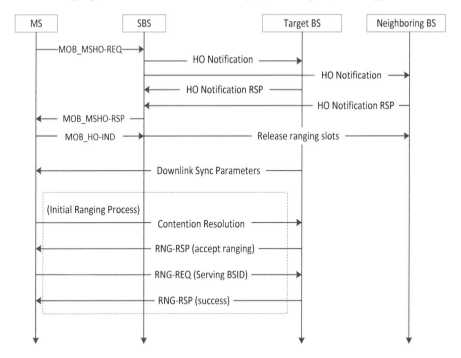

FIGURE 6.2
HO decision and initiation in IEEE 802.16.

sumes that the MS originates this step. Referring to Figure 6.2, by the end of the previous phase, the MS makes a decision about its possible TBS and sends MOB_MSHO-REQ to inform the SBS. The SBS will then send HO notification messages to all the TBS and receive their responses. After the MS receives MOB_HO-RSP (Mobile Handover Response) from the SBS, it will choose its final TBS and send the Mobile Handoff Indication (MOB_HO-IND) message to inform SBS about its decision to perform a HO. The SBS can negotiate with the TBS the allocation of a ranging opportunity. The MS uses this opportunity to synchronize the downlink of TBS and obtains downlink and uplink parameters through DCD/UCD messages. That information is conveyed by exchanging RNG-REQ/RSP between MS and TBS.

6.2.2.3   Network Reentry Procedure

After finishes synchronizing all the physical parameters with the TBS, the MS starts the network reentry process to establish connectivity with the new BS. As shown in Figure 6.3, this process includes capability negotiation, MS authorization and new BS registration. The TBS asks for the MS's information from SBS through backbone network and uses it to authorize the MS. After authorization and registration are successfully, the MS sends MOB_HO-IND

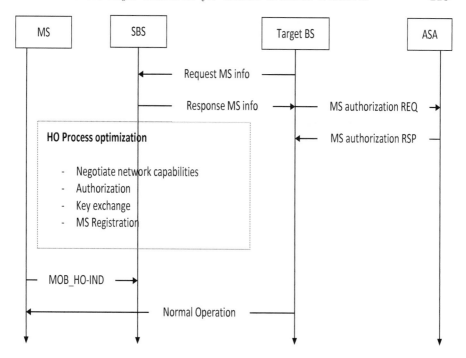

FIGURE 6.3
Authorization and registration procedure in IEEE 802.16.

to SBS to notify the completion of HO and releases the connection with SBS. Since the MS has to wait until the reentry procedure is finished to start its normal communication with the new BS, the duration of this phase is included in the total HO latency.

6.2.2.4 Advantages and Current Issues

The IEEE 802.16e HO procedure described above has the advantage of simplicity. Also, its low cost both in term of hardware and software makes this mechanism relatively easier to be implemented today. However, the disruption time in HO is still high and is intolerant for some applications that are sensitive to such as streaming services or real-time conferences. To overcome this obstacle, many improved algorithms have been proposed involving both link layer and IP layer HO; some major ones will be described in Section 6.3.

6.2.3 Mobile IPv6 Handover

MIP (Mobile Internet Protocol) is a protocol designed to sustain an IP connection while a MN moves across an IP domain. When a MN moves from one network to another, its physical location changed. Yet, its IP address which is used to identify its location does not change. When another node sends packets to the MN using its IP address, the packets will not reach the MN's new location. MIPv4 [24] solves the issue by introducing two new servers called Home Agent (HA) and Foreign Agent (FA). The FA records the corresponding home IP address and foreign IP address (CoA or Care-of-Address) for a registered MN.

MIPv6 [14] surpasses its former MIPv4 by enabling route optimization. In MIPv6, packets do not need to be routed back to the HA but instead go directly to a Corresponding Node (CN) which will then forward them to MN, resulting in a shorter route. Nevertheless, if CN does not know the current CoA of MN, it has to send packets to HA and has them encapsulated and sent to the MN. Therefore, the MN needs to keep CN and HA informed about its updated CoA by periodically sending the Binding Update (BU) messages regardless of location changes.

Typically, MIPv6 HO is slow when combining naively with a layer 2 HO. As shown in Figure 6.4, after the PHY (Physical) and MAC (Media Access Control) layers are ready, MN will have to wait for the Router Advertisement message to acknowledge that it has successfully changed to a new network. From the Router Advertisement message, the MN quickly learns the MAC address of the access router (AR) that it is attached to, and can send the BU without any delay after link layer HO. Then, it will send the BU message to HA and wait for the confirmation response BAck (Binding Acknowledgment). However, this is a one-way-learning process since the AR cannot obtain MN's address at the same time as MN does. The AR has to perform address resolution to obtain the MN's MAC address before sending any packet to the MN. This causes degradation in network performance, especially when the MN moves quickly and frequently between many point of attachments. Both layer 2 and layer 3 delays contribute significantly to the overall delay, which is relatively long and unbearable for some real-time applications such as voice-over IP (VoIP) and video streaming.

6.3 Related Studies

There have been many attempts in this field to improve HO performance for supporting seamless mobility. The basic IEEE 802.16e HO procedure discussed in the previous section is the simplest, yet it lacks the support of layer 3 (IP layer) HO and strong QoS. Along with the rapid development of mobile devices and mobile network technologies, there is an urge for better HO mechanisms that achieves very small delay and loss while supporting on-the-go users in a wider network area. In this chapter, we will survey some recent studies in this area.

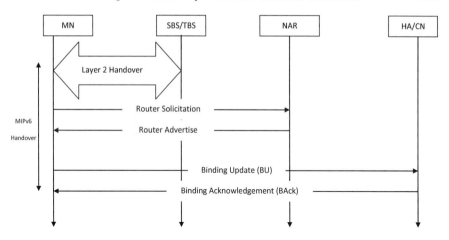

FIGURE 6.4
MIPv6 handover management procedure.

6.3.1 Enhancements of IEEE 802.16e Handover

In 2006, Cho *et al.* [5] proposed an improved scheme for layer 2 hard HO by exploiting the uplink and downlink signals. In the proposed scheme, the current BS periodically monitors uplink and downlink signal strength based on Uplink/Downlink (UL/DL) packets received from the MN, and initializes the HO process early if that signal strength is less than a certain threshold. This scheme is most suitable for the MN that has both uplink and downlink traffic traveling over the air interface. It helps to reduce the number of ping-pong HO and the outage probability. However, there is much workload put on the current BS since it has to continuously repeat the measurement.

Unlike [5] who focused on preventing unnecessary handovers, Chen *et al.* [3] introduced the idea of using a pre-coordinate mechanism to reduce the disruption time during the HO process. In their mechanism, the SBS identifies the location of a MN every 10 seconds, predicts a TBS based on the movement direction of the MN, and starts to precoordinate with the TBS. When the MN approaches the boundary, the SBS will broadcast the responded agreement from the TBS to the MN, so HO process can be triggered immediately.

6.3.2 Handover Support for Voice and Video in IEEE 802.16

Dong and Dai [6] introduced an improved HO algorithm taking into account the five scheduling services. For each scheduling service, they gave some suitable options of HO based on QoS requirements so as to minimize the control traffic requirement while still the HO delay performance is maintained. For example, for Best Effort (BE) scheduling service, which has no data rate or delay requirement, association level 0 or 1 is encouraged to be used, while association level 2 and Fast BS Switching (FBSS) are suggested to use for Real-Time Polling Service (rtPS) that requires a more stringent delay performance.

Chellappan *et al.* [2] later gave a more careful study and proposed a scheme which supports five QoS scheduling classes in both layer 2 and layer 3 HO. The authors defined five QoS modes corresponding to the five scheduling services in WiMAX. In general, this paper had a wider scope compared to the work described in [6] by considering, in addition, the standby mode, which has no data traffic. Furthermore, it provided several important features such as the support of both micro- and macromobility, both predictive and reactive modes, and mode selection.

To support real-time applications during HO, [13] proposed the Passport Handover scheme which addressed the problem of connection identifier (CID) confliction when the MN migrates to the coverage of a different BS. In the scheme, 3 reserved bits among the 16 CID bits are used to create a pattern at the beginning of a CID. Except the state of "111" that is saved for multicast CID or padding CID, there are 7 different states of 3-bit pattern that can help to reduce the chance of CID conflict. The paper however focused only on link-layer HO, and the scheme supported only a limit number of neighboring BS that made it impractical for real-time services.

To evaluate video communication [7], simulated video traffic, including video conferencing, and evaluated how FMIPv6 supports IEEE 802.11 HO. The simulation is established using a real VLC (VideoLAN Client) video streaming and a real video conference with the help of Gnomemeeting application. Instead of evaluating the benefit of FMIPv6 for layer 3 (network) HO, the paper, however, demonstrated its results on reducing packet loss and HO delay mainly on layer 2 (data link) HO scenarios.

Recently, [16] evaluated FMIPv6 for VoIP support over IEEE 802.16 networks. The authors evaluated both FMIPv6 and MIPv6 protocols using ns-2, and compared their performance in supporting VoIP, where the traffic is generated using constant bit rate (CBR), assuming the G729a codec. This was the first published paper that successfully simulated FMIPv6 over WiMAX networks. In the simulation, the authors considered important metrics of VoIP services such as HO delay, packet loss, and R-factor, and provided a careful evaluation of those metrics under different HO scenarios.

Inspired by the work in [16], the authors of reference [10] experienced the support of FMIPv6 handover for video streaming, another aspect of real-time applications in WiMAX networks. In addition, they proposed some improvements to FMIPv6, and evaluated the improved schemes for supporting video streams and VoIP traffic. The enhanced cross-layer HO to be discussed in Section 6.4 is based on this paper [10], which may be viewed as the preliminary work of this chapter.

6.3.3 Fast MIPv6 (FMIPv6) and Enhancements

This section presents a brief overview of FIMPv6 [12] and two enhancements. Each of them will be described more detailed in Section 6.4. To address the long disruption time during the HO process, many studies on enhancing the existed schemes have been conducted. It has been generally agreed that one major factor of the long delay in HO process is the separation in layer 2 and layer 3 HO executions. When a MN moves from one network to another, the layer 3 HO process is initialized only after the link layer HO has completed. This has introduced an unnecessary gap from the time when layer-2 connection is up until layer 3 HO is initialized. Therefore, the FMIPv6 cross-layer-based HO mechanism for WiMAX has been proposed and evaluated [12]. The main improvement of this scheme is concurrently conducting layer 2 and layer 3 HO steps by using event trigger messages.

To further improve the support of FMIPv6 for real-time applications, Chen and Hsieh proposed the idea of merging layer 2 and layer 3 control messages that have similar functionalities or attributes [4]. Hence, the traffic over the air interface is reduced, and as a result, the scheme can support more mobile nodes with higher mobility speed. On the other hand, [19] proposed an algorithm which guarantees with very high probability that the new CoA generated at the MN will not be duplicated when using in the new network. This algorithm helps to significantly decrease the layer-3 HO delay and lost packets. The above two enhancements have both been adopted in our proposed cross-layer HO mechanism. They will therefore, along with the FMIPv6 scheme, be described in detail in the next section.

6.4 Two Cross-Layer Handoff Mechanisms

6.4.1 Major Features of FMIPv6-Based Cross-Layer Handoff

The FMIPv6 cross-layer HO was proposed to reduce the HO disconnection period, so that the time the receiver waits for a packet after HO is reduced [12]. The main idea is to obtain a new CoA while MN is still in the old network area so that it can use the new CoA as soon as the new MAC layer link is established. To further reduce packet loss, FMIPv6 also uses packet tunneling during channel establishment so that there will be less loss during HO. There are two modes of FMIPv6: predictive and reactive. Most notably, the predictive mode has better performance in term of HO delay because it does not include the time needed for layer-3 HO preparation. Figure 6.5 illustrates the message flow of FMIPv6 scheme in predictive mode.

The predictive mode exploits the advantage of using four new messages: NEW_BS_FOUND (NBSF), LINK_GOING_DOWN (LGD), LINK_SWITCH (LSW), LINK_UP (LUP). These messages are sent internally between layers 2 and 3 of the MN. Particularly, after the MN's link layer receives MOB_SCN-RSP from SBS, it sends NBSF message to the IP layer to trigger layer 3 sending Router Solicitation Proxy (RtSolPr) message to Previous AR (PAR). Concurrently, MN's link layer sends MOB_MSHO-REQ to SBS and waits for MOB_BSHO-RSP to proceed to the next step. The next event LGD is then triggered by layer 2 right after MOB_BSHO-RSP reaches the terminal, which forces MN's layer 3 to send an FBU to PAR indicating the binding update between old and new CoA (nCoA). Upon reception of FBU, PAR communicates with next AR (NAR) about the binding process and also waits for NAR's validation before using nCoA.

After PAR receives the Handover Acknowledgment (HAck) from NAR indicating that the nCoA is valid, PAR sends FBAck to both the terminal and the NAR to start using nCoA in transferring packets. As soon as the MN receives the FBAck from PAR, it will send LSW command from layer 3 to trigger layer 2 to send out MOB_HO-IND message. From this point on, the connection between MN and SBS is corrupted, and no more traffic will be exchanged between them. After that is the period for the terminal to conduct HO to the TBS and perform IEEE 802.16e network reentry procedure to the new network. The last event, LUP, is triggered by layer 2 after the network re-entry and registration is completed. As soon as MN's layer 3 receives LUP command, the Fast Neighbor Advertisement (FNA) message is sent to the NAR, which will then flush all the buffered packets to the terminal.

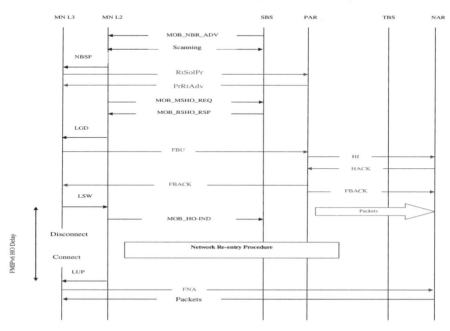

FIGURE 6.5

FMIPv6 handover mechanism in predictive mode.

The reactive mode happens when the FBU cannot reach the PAR before the MN loses its connection with the current BS. When this happens, the MN's layer 2 has to send MOB_HO-IND before nCoA is approved to use. In this case, the layer 2 HO process still takes place. The only difference is that, after finishing the network reentry procedure at the new network, MN sends FNA with an encapsulated FBU message to the TBS. Upon receiving that message, NAR will verify the validity of nCoA and forward the encapsulated FBU to PAR to establish a packet tunnel. PAR then sends FBAck to NAR as a reply of FBU and starts tunneling packets destined for the current CoA to the nCoA, and these packets will, therefore, be sent to the terminal that is currently inside the new network.

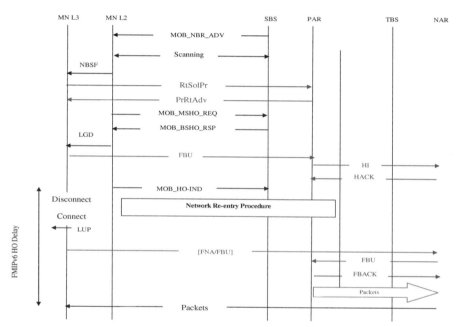

FIGURE 6.6
FMIPv6 handover mechanism in reactive mode.

6.4.2 Proposed Handoff Scheme

In this section, we first explain two drawbacks of the original FMIPv6. Next, we present two major features of the proposed scheme addressing the drawbacks. Finally, a detailed description of the proposed scheme, including message flows, is presented in the last subsection. Although the FMIPv6 HO has been improved in term of HO delay, which is very important for real-time applications, its overall delay is still considered high [4]. Note that in FMIPv6, the number of control messages exchanged during the HO process is still many, especially when there are multiple mobile terminals moving at the same time. That would make the air interface congested, and may cause the loss of control messages, which will lead to wrong behavior of seriously degrade the HO process.

Furthermore, the NAR needs to execute the Duplicate Address Detection (DAD) procedure to verify if the CoA of the incoming MN duplicates with any of the existing CoA being used. The time needed to perform the DAD is relatively long compared to other types of delay occur during the HO process. According to Lee *et al.*, the average time required for binding update is around 140 ms, and the time for layer 2 HO is about 250 ms; the DAD process, on the other hand, takes almost 1 second and therefore becomes the major factor of the overall HO delay [20]. As a result, avoiding the DAD becomes an attractive option.

For the former problem, [4] has proposed to combine correlated layer 2 and layer 3 messages to reduce potential congestion over the air-interface. To address the issue of DAD delay, [20] introduced the idea of using a temporary CoA (tCoA) during the HO process. They also proposed an algorithm to generate nCoA which has a very small possibility to be duplicated when a mobile node migrates a new network area. We propose a new scheme that combines the above two features, as described in detail below.

6.4.2.1 Combination of Control Messages Method

To reduce control message overhead during HO, two pairs of messages are combined, referring to Figure 6.7. The first pair of messages to be combined are MOB_NBR-ADV and Proxy Router Advertisement (PrRtAdv) [4]. Both of them are sent periodically by the BS to the MN to provide necessary information for HO procedure. By combining them, instead of waiting for the MN to send RtSolPr message, the SBS takes the active move to send its router advertisement along with MOB_NBR-ADV of layer 2. Thus, the MN can obtain the needed information to conduct FMIPv6 in advance.

Another pair that can be combined are those sent right before the HO is initiated. They are MOB_HO-IND message of layer 2 and FBU message of layer 3 [4]. In the original FMIPv6, the FBU is sent first to PAR to initialize layer 3 HO. After some time, whether the FBU has reached the BS, the MOB_HO-IND message is sent from the MN to SBS to terminate all packet exchange between them. If the MN moves too fast, or if there is too much traffic over the air interface that prevents the FBU from reaching PAR in time, HO procedure may end up running in the reactive mode, or more seriously, some control messages might be lost and jeopardize the entire HO procedure.

The integration of MOB_HO-IND and FBU is named as FBU_MOB_HO-IND. As suggested, one among six reserved bits in the original layer 2 MOB_HO-IND message is used as a flag to enable the FBU capability in layer 3. By combining these two messages, we could reduce the HO delay. This is because SBS would start tunneling packet as soon as it receives FBU.

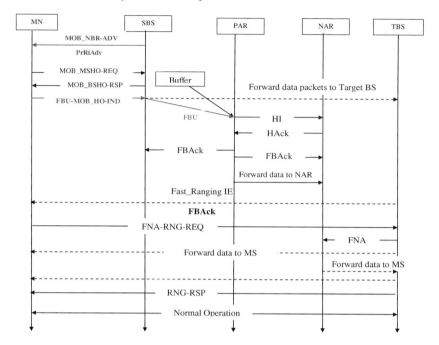

FIGURE 6.7
Message combination scheme.

Therefore, some packets will be held and MN will not receive any packet until it reconnects with TBS. Sending FBU at the last moment before disconnecting with SBS will keep the connection alive with SBS as much as possible. Table 6.1 illustrated the message format of FBU_MOB_HO-IND message [4].

As may be seen in Section 6.5, performance evaluation, combining the two pairs of messages as described above has achieved a significant improvement; some of its major benefits are explained below:

- Reducing the number of control messages over the air interface. It also removes the sending of PrRtAdv during handover period. This is significant if the downlink channel is overloaded by other traffic because PrRtAdv might not be able to reach MN. Therefore, this scheme can concurrently support more mobile nodes performing HO process at the same time.

- With the help of FBU_MOB_HO-IND, we can reduce the packet buffering time. As a result, HO delay is improved.

- As a result of the above two benefits, this scheme can support high mobility speed and overloaded network that makes it more suitable to use in practice.

TABLE 6.1 Format of FBU_MOB_HO-IND message.

Syntax	Size	Notes
Management message type	8 bits	Value = 59
FBU	1 bit	0b00: FBU disable 0b01: FBU enable
Reserved	5 bits	Reserved: will be set to zero
Mode	2 bits	0b00: HO 0b01: MDHO/FBSS: Anchor BS Update 0b10: MDHO/FBSS: Diversity BS Update 0b11: Reserved

6.4.2.2 New CoA Generation Method

To address the time needed to perform DAD algorithm by NAR, an advance method for the MN to generate new CoA has been suggested [20]. The authors also verified that their algorithm would guarantee the new CoA, with very high probability, has almost no duplication with any other existing CoA in a network area. The 16 extra bits inserted into the IEEE 802.16 address of the MN, creates the Interface Identifier (IID) of that MN in the new network. More specifically, the method has the following steps:

- Change the last bit of the first byte in IEEE 802 address to 0, which indicates that the IID is a unicast address.

- Change the second to the last bit of the first byte in IEEE 802 address to 0, which indicates that it is locally controlled within that particular subnet.

- Insert 8 bits of current Cell ID (cID) in the middle of the 48-bit-modified IEEE 802 address.

- Insert 8 more bits of the target cID after the current cID 8 bits.

The result is then used as the IID of the MN, and will be appended to the 64-bits Target subnet prefix to form the nCoA (called tCoA in the simulation section) for the MN in the new network. Figure 6.8 shows the illustration of this method.

The benefit of using this method is that the probability of nCoA to be invalid (due to duplication) is reduced to almost zero. Thus, the chance for HO procedure to run in Predictive Mode is significantly increased since little or no time is spent processing DAD. As a result, this procedure can support a much higher mobility speed. Furthermore, because this method guarantees that the generated nCoA are hardly duplicated, there is no need for the MN to wait for FBAck. Therefore, both layer-3 HO delay and the number of lost packets during handover are significantly decreased due to the fact that it runs in Predictive Mode.

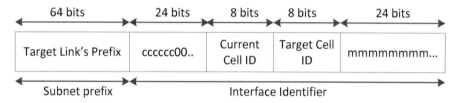

FIGURE 6.8
Temporary CoA creation method.

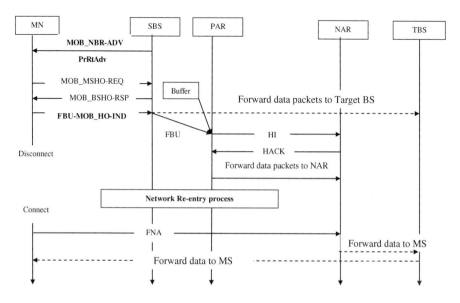

FIGURE 6.9
Message flow of proposed scheme with separated BS and AR.

6.4.2.3 Description of the Proposed Scheme

Stated above are the two key improvements that can be applied to original FMIPv6 to improve its support of real time applications and greater number of concurrent MN. Our proposed scheme exploits the advantages of both. The message flow is given in Figures 6.9 and 6.10. The first one is used when BS and AR are separated while the second one is used when they are co-located with each other.

From the Figure 6.9, the SBS periodically sends the MOB_NBR-ADV along with the PrRtAdv message to the MN. By receiving these messages, the MN updates necessary information of all BSs in range. When the signal strength between MN and SBS is weak, the MN exchanges MOB_MSHO-REQ and MOB_BSHO-RSP with the SBS. After that, it triggers layer 2 HO by sending the combined FBU_MOB_HO-IND message to SBS. Upon receiv-

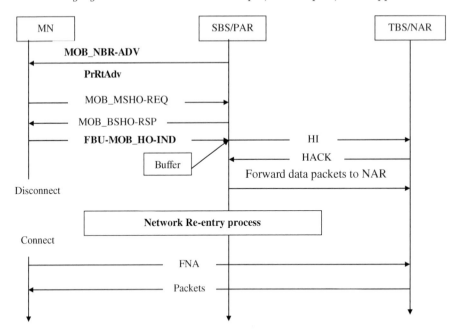

FIGURE 6.10

Message flow of proposed scheme with integrated BS/AR.

ing this combined message, the SBS forwards FBU message to PAR. The PAR starts to buffer packets destined to the current CoA, and sends the Handover Indication (HI) message to NAR indicating the proposed nCoA of the terminal. Note that in this scheme, the nCoA is generated at the MN using the new algorithm described above [20], so it is hardly to be duplicated at the new network. Therefore, after the PAR sends HI to NAR, NAR, upon receiving, will return with HAck as usual, according to FMIPv6 process. However, PAR does not need to send FBAck to MN since the generated address is unique. The process after that is similar to that of the original FMIPv6.

Overall, this scheme omits two messages compared to the original FMIPv6: RtSolPr from the MN to the PAR and the FBAck from the PAR to the MN. Furthermore, the HO delay is significantly decreased because the scheme works mostly in predictive mode, and the DAD process is eliminated.

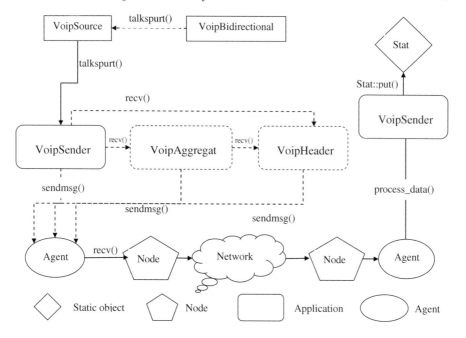

FIGURE 6.11
Modules for VoIP simulation in NS-2.

6.5   Performance Evaluation

6.5.1   VoIP and Video Quality Evaluation Framework

6.5.1.1   VoIP Modules Supported in NS-2

Figure 6.11 shows the VoIP simulation module implemented in an extension of NS-2 [26], which has been described by [1]. On the sender side of the VoIP application, the sender generates speech frames (samples-based or frames-based) either in a periodic interval or, more commonly, the frames being modulated by voice activity detection (VAD), which capitalizes the alternation of talkspurts and silence periods in a single stream of a bi-directional conversation. This module suggests that during the transmission of VoIP packets, a number of speech frames can be multiplexed into the same packet payload to reduce the overhead of transport, network and MAC headers, yet accept the trade-offs of increasing delay. The VoIP payload is then encapsulated into RTP (Real-time Transport Protocol)/UDP (User Datagram Protocol)/IP packets and sent out the Internet. Note that the two objects VoipAggregate and VoipHeader are optional in this module; users are free to choose to enable these two functions in their simulation [1].

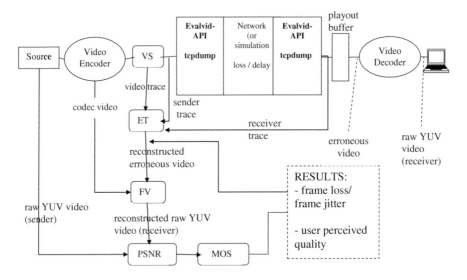

FIGURE 6.12
Scheme of evaluation framework.

6.5.1.2 Video Quality Evaluation Framework

To study video streaming performance of WiMAX network HO process, Klaue *et al.* used a tool-set called EvalVid [17], which has been later extended [15]. Figure 6.12 demonstrates the scheme of this framework [15].

This tool is capable of coding any raw *.yuv* extension file into a codec video file, which becomes the source transmission file. This codec file, however, cannot be sent directly into in NS-2 simulator. We need to transform this file into the trace file containing fragmentation and frame type data, instructing NS-2 while indicating when and how to send packets. This trace file is then used as the source file in the NS-2 simulator.

After running the simulation, we will obtain two more trace files, sender and receiver trace files. The sender trace file contains timestamp, sequence number, and the bytes transmitted at the sender, originally available from EvalVid. We develop a software module so that the receiver would also write the corresponding information into the receiver trace file. Obviously, these two trace files would be different depending on packet loss and network performance. Video may then be reconstructed and be viewed by users using the *etm4* (Evaluate Traces of MP4-file transmission) tool [15].

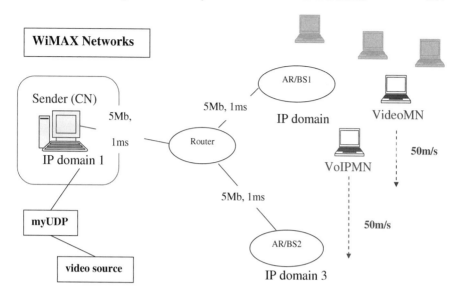

FIGURE 6.13
Simulation network topology.

6.5.2   Simulation Settings

6.5.2.1   Network Topology

As shown in Figure 6.13, the network has three domains. Domain 1 contains CN, which plays as the sender to send video, VoIP packets, and CBR traffic. The receivers, which are located in the other domain (Domain 2), include a mobile node (VideoMN) to receive video traffic, a mobile node (VoIPMN) to receive VoIP traffic, and other fixed nodes for other background CBR traffics. The mobile nodes will move from Domain 2 to Domain 3 with a constant speed throughout a simulation experiment. For the ease of simulation, we assume that each BS has all the functionalities of an AR integrated within.

Referring to Figure 6.13, all the links between wired nodes are configured identically as duplex links with 5 Mbps channel capacity and 1 msec propagation delay. The buffer size of intermediate nodes is assumed to be infinity. Network Simulator version 2.28 (NS-2.28) is used for the simulation. Additionally, the NIST WiMAX patch [23] is exploited to simulate the behavior of PHY and MAC layers in WiMAX networks. Finally, MobiWAN patch [9] has been adopted for MIPv6 functionalities. Based on these patches, we then develop a new module which simulates FMIPv6 proposed in [12], including message sequence, packet tunneling, and cross-layer triggering of messages. We also simulate the two improvements described in Section 6.4. Major simulation parameters are summarized in Table 6.2 and Table 6.3.

TABLE 6.2 Simulation parameters.

Antenna height	1.5 m
Carrier frequency	3.486 GHz
Coverage radius	750 m
Distance between BSs	1 Km
Mobility speed	Varied (30, 40, 45, 50 m/s) 50 m/s (or 180 Km/h) by default
Link and MAC layer queue size	50 packets each
Receiver sensitivity	4×10^{-3}dB
Bandwidth	2 Mbps

TABLE 6.3 Codec parameters and packets size values.

Codec	Video	Voice	Background Traffic (CBR)
	200 Kbps 25 frames/sec	G.711 Sampling frequency: 8 KHz Bit rate: 64 Kbps	Bit rate: 448 Kbps
Packet size	**Video**	**Voice**	**CBR**
	512 bytes	134 bytes	210 bytes

TABLE 6.4 Description of simulated schemes.

Scheme	Description
FMIPv6	The original cross-layer based FMIPv6 (supports both Predictive and Reactive modes), which includes 1 second of DAD
FMIPv6+tCoA	FMIPv6 plus tCoA created by the new CoA Generation Method (without 1 second of DAD process)
FMIPv6+merge	FMIPv6 plus Message Combination Scheme
Proposed scheme	FMIPv6, adopts both CoA and Message Combination Schemes

6.5.2.2 Protocol

There are three types of agents attached to the sender: (1) an UDP agent that has an video application attached and sends out video packets with constant packet size and bit rate, (2) another UDP agent which is attached a VoIP source to generate voice traffic, and (3) one or more UDP agents for generating background CBR traffic. Each MN is equipped with appropriate UDP sink agent to receive incoming packets (video, VoIP). Except for the experiments shown in Section 6.5.3.2, each simulation experiment runs for 20 seconds; the video used is 12 seconds long. A video application starts sending out packets at 0.2 sec while the VoIP source starts to create speech frames at 0 sec. The mobile node starts to move right at the beginning of the simulation. For simplicity, only one MN moves across a domain in every simulation experiment.

Four HO schemes are simulated: the original FMIPv6, FMIPv6+tCoA, which is FMIPv6 that uses the new CoA Generation Method [20], FMIPv6+merge, which is FMIPv6 with the message combination scheme [4], and finally the proposed, which is FMIPv6 with both new CoA Generation Method [20] and message combination [4], as summarized in Table 6.4.

6.5.2.3 Performance Metrics

There are three groups of metrics measured in this study: general network performance, video quality, and VoIP quality.

- Network performance:
  - *Handover Delay*: this is the period between last packets received from PAR and the first packets received from NAR. The delay is measured in the period between LINK_SWITCH and LINK_UP event (refer to Figure 6.5).
  - *Throughput*: The cumulative number of bits received per second during simulation.
  - *Packet Loss*: The number of lost packets during the HO process.

TABLE 6.5

PSNR versus MOS for video quality.

PSNR	MOS
> 37	5 (Excellent)
31-37	4 (Good)
25-31	3 (Fair)
20-25	2 (Poor)
< 20	1 (Bad)

- Video Quality:

  ○ *PSNR (Peak Signal-to-Noise Ratio)*: Evaluated based on the PSNR between the luminance component Y of source image s and destination image d. Theoretically, PSNR can be calculated as below [17]:

  $$PSNR(s, d) = 20 \log \frac{V_{peak}}{MSE(s, d)} dB \qquad (6.1)$$

  where $V_{peak} = 2^k - 1$, k-bits color depth, $MSE(s, d)$ mean square error of s (source image) and d (destination image).

  ○ *MOS (Mean Opinion Score)*: a subjective metric commonly used to judge video quality when displayed at the receiver's application. Usually, MOS measures the human impression of the sound or image at the application layer. MOS can be derived from PSNR, as summarized in Table 6.5 [17].

- Voice Quality:

  ○ *R-Factor*: A value derived from metrics such as latency, jitter, and packet loss that helps to evaluate the quality of VoIP calls on a network. Its value ranges from 50 (bad) to 90 (excellent). The general formula to calculate the R-Factor is given below [25]:

  $$R = 100 - Is - Ie - Id + A \qquad (6.2)$$

  where Is is the signal-to-noise impairment factor, Ie is the equipment impairment factor (loss), Id is the impairment factor (delay), and A is the expectation factor, decided based on communication system condition. Note that only Ie and Id are considered in the typical VoIP simulation. Therefore, the following formula is adopted to obtain the R-Factor [25]:

  $$R = 94.2 - Ie - Id \qquad (6.3)$$

  where Id is the end-to-end packet delay or one-way-mouth-to-ear delay, and Ie is the one-way packet loss.

TABLE 6.6

R-Factor versus MOS for voice quality.

R-Factor	MOS	Quality
90	4.3	Excellent
80	4.0	Good
70	3.6	Fair
60	3.1	Poor

○ *MOS (Mean Opinion Score)*: a subjective metric to judge voice quality when played out at the receiver side. MOS value can be determined by the R-Factor, as shown on Table 6.6 [19].

6.5.3 Simulation Results

6.5.3.1 Network Performance

1. **Handover delay**

   Figures 6.14 and 6.15 show the HO delay of video stream and VoIP while varying background traffic load, keeping default mobility speed of 50m/s. It is clear that the performance of our proposed scheme is superior especially when the background traffic load is heavy. Note that when the network is in ideal status with no background traffic load, the HO delay of FMIPv6+tCoA, FMIPv6+merge, and the proposed scheme are all much smaller than that of FMIPv6. However, when the network becomes busier, only the proposed scheme can sustain and maintain a good delay performance. The reason is that in FMIPv6, as well as in FMIPv6+tCoA, the MN needs to wait for PrRtAdv before it sends the HO request to SBS (Figures 6.5 and 6.6). However, the heavy traffic load in the network may cause that message to be lost during transmission while the MN is waiting, and finally finds out that it has lost the connection with the current BS. This will introduce a long delay in the HO process. Both FMIPv6+merge and the proposed scheme does not have this problem because the PrRtAdv is periodically broadcasted along with the MOB_NBR-ADV, so the MN has many chances to get that message regardless of the traffic load.

   Note that even though FMIPv6+merge does not have to wait for PrRtAdv, it still has a large HO delay. This may be surprising since FMIPv6+merge runs in the predictive mode. The problem, however, is on the predictive mode itself. After sending FNA/FBU message, the TAR must process DAD before it can send back FBACK. As a result, the 1 second delay of DAD adds directly to the HO delay of FMIPv6+merge scheme.

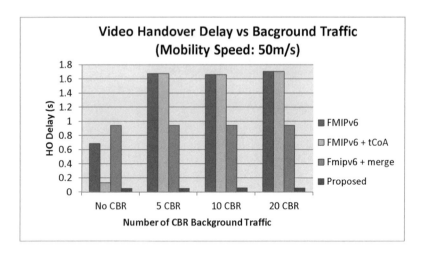

FIGURE 6.14
Video handover delay versus background traffic.

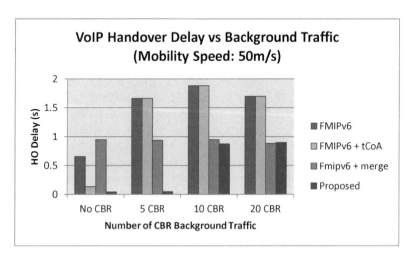

FIGURE 6.15
VoIP handover delay versus background traffic.

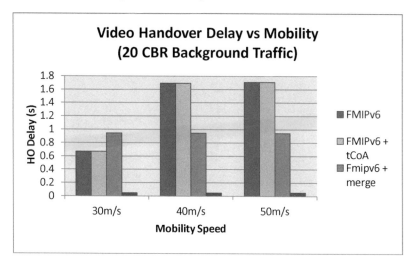

FIGURE 6.16
Video handover delay versus mobility.

Figures 6.16 and 6.17 evaluate HO delay performance while varying the mobility speed, keeping 20 CBR background traffic load by default. As the network traffic is very heavy, both long delay at the BS scheduler and possible buffer overflow (at either BS or AR) may prevent the FBAck message from reaching the MN, both the FMIPv6 and FMIPv6+tCoA scheme will end up in reactive mode, and thus have the same HO delay, while the proposed scheme yields relatively small value of HO delay regardless of mobility speed. This happens because it does not require the reception of FBAck. The same applies to FMIPv6+merge, yet, as discussed, the HO delay of FMIPv6+merge in predictive mode still suffers from DAD delay.

2. **Handover Packet Loss**

Figures 6.18 to 6.21 are the results of HO Loss of video stream and VoIP while varying background traffic load and mobility speed. Again the difference between FMIPv6 and FMIPv6+tCoA can only be seen on a network with a few CBR background traffic load. With 5 or more CBR background traffic load, these two schemes tend to have similar poor loss performance (up to 145 packets for video and 186 packets for voice). The reason is that the MN has not received any PrRtAdv message because the downlink is extremely busy, thus, they end up in the reactive mode. Both FMIPv6+merge and the proposed scheme give a better result. They do not suffered from the above problem since PrRtAdv is integrated with MOB_HO_ADV. This message will be sent

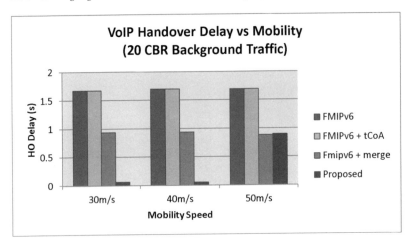

FIGURE 6.17
VoIP handover delay versus mobility.

periodically. However, as the background traffic is increased, the loss in FMIPv6+merge also increases due to buffer overflow because it takes longer to finish HO. The proposed scheme has the smallest number of lost packets during HO (at most 6 packets for video and 55 packets for voice). This is because it has eliminated both problems by integrating the two schemes together.

3. **Network Throughput**

Because of the significant improvements in HO delay and packet loss, the proposed scheme has achieved much better performance in term of network throughput compared to the FMIPv6 and FMIPv6+tCoA as shown in Figures 6.22 to 6.25. Note that its performance is close to FMIPv6+merge. This is because both protocols have successfully avoided large packet loss using tunneling. When the MN moves fast and the network is getting heavier, the proposed scheme has the best throughput since it has the top delay performance and therefore best avoids packet loss.

6.5.3.2   Video Quality

In this section we evaluate video quality in terms of PSNR and MOS. To clearly demonstrate the differences in MOS, this set of experiments is each run for 10 (instead of 20) seconds. In the first part, we vary the mobility speed while the background traffic load is set at 10 CBR; Figures 6.26 and 6.27 show the PSNR and the corresponding MOS, respectively. It is clear that as MN moving speed increases, both the original FMIPv6 and FMIPv6+tCoA are

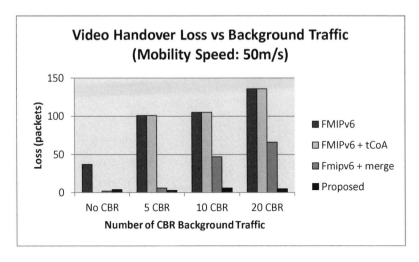

FIGURE 6.18
Video handover loss versus background traffic.

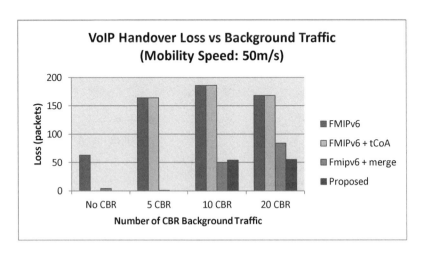

FIGURE 6.19
VoIP handover loss versus background traffic.

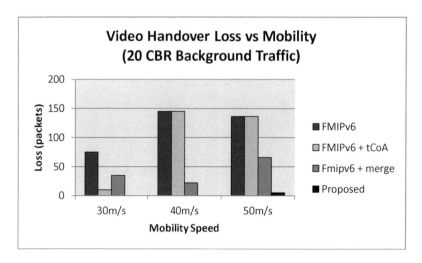

FIGURE 6.20
Video handover loss versus mobility.

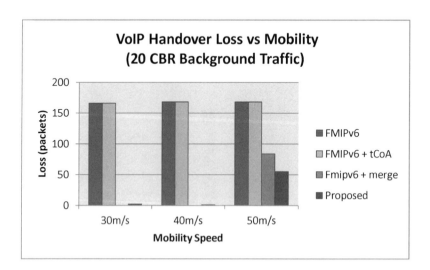

FIGURE 6.21
VoIP handover loss versus mobility.

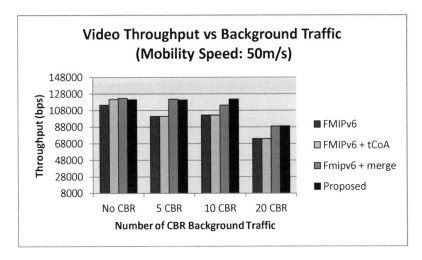

FIGURE 6.22
Video throughput versus background traffic.

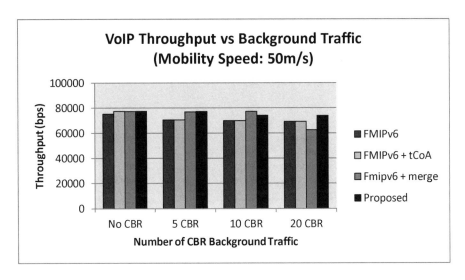

FIGURE 6.23
VoIP throughput versus background traffic.

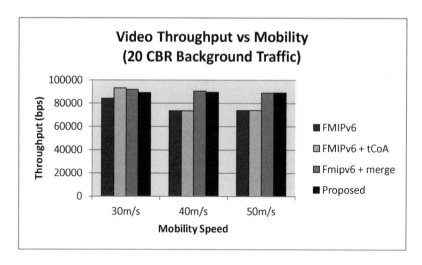

FIGURE 6.24
Video throughput versus mobility.

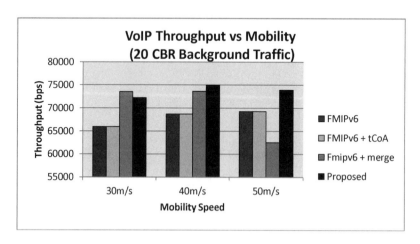

FIGURE 6.25
VoIP throughput versus mobility.

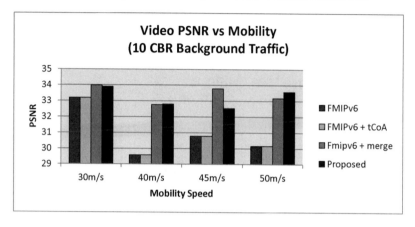

FIGURE 6.26
Video PSNR versus mobility.

not able to maintain PSNR (Figure 6.26), causing a lower MOS (Figure 6.27). PSNR depends solely on image quality differences between the source and the destination. The two improved scheme, FMIPv6+merge and the proposed, are able to maintain small HO delay and low packet loss during HO, and therefore achieve higher PSNR and better MOS. Among the four, the proposed scheme clearly gives the highest PSNR at higher speed due to its small HO delay and small packet loss.

In the second half of this set of experiments, the background traffic load is varied while the mobility speed is kept at 40 m/s. Figures 6.28 and 6.29 show the PSNR and the corresponding MOS, respectively. We see that in the first two schemes, FMIPv6 and FMIPv6+tCoA, both have began to suffer lower PSNR and lower MOS when background traffic load increased to 5 CBR, whereas the other two improved schemes, FMIPv6+merge and the proposed, achieve high PSNR and MOS until 20 CBR background traffic load. Again this is due to their ability to maintain smaller HO delay and low packet loss even during heavy network condition.

6.5.3.3 Voice over IP (VoIP) Quality

In this section we show the voice quality in the experiments reported in Section 6.5.3.1, including R-Factor and MOS. In the first set of experiments, the background traffic load varies while the mobility speed is kept at 50 m/s; results are shown on Figures 6.30 and 6.31, respectively. As described in Section 6.5.2.3, R-Factor is derived from latency and loss of voice data. From the simulation, the proposed scheme has a smaller number of one-way packet loss and a lower end-to-end packet delay than the other three, its R-Factor, as calculated using Equation (6.2), is therefore the highest (see Figure 6.30), especially during busy network condition. The corresponding MOS values,

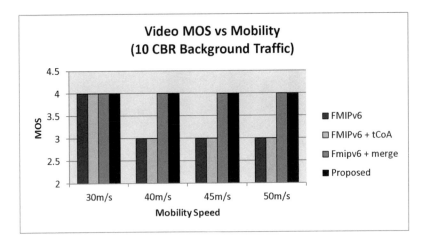

FIGURE 6.27
Video MOS versus mobility.

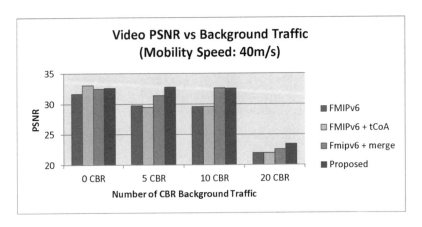

FIGURE 6.28
Video PSNR versus background traffic.

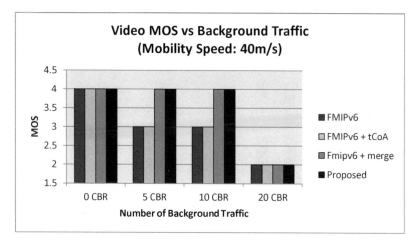

FIGURE 6.29
Video MOS versus background traffic.

however, do not show as much different, since the MOS value (representing user perspective) is defined according to a range of values of the R-Factor (see Table 6.6).

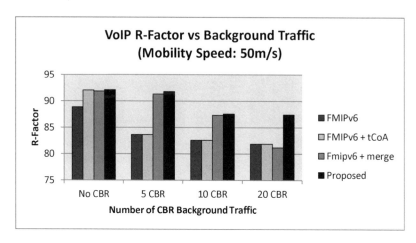

FIGURE 6.30
VoIP R-Factor versus background traffic.

In the second half of the set of experiments, the background traffic load is kept the highest (20 CBR) while varying mobility speed. The R-Factor results are shown in Figure 6.32. Again, it is clear that the proposed scheme has maintained the highest R-factor due to its ability to keep both HO delay and packet loss low. The differences, however, are not significant. Notice that

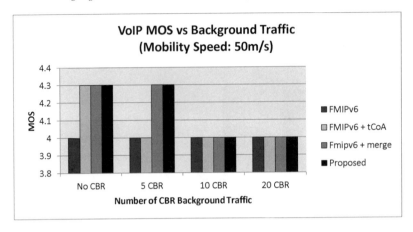

FIGURE 6.31
VoIP R-Factor versus mobility.

all the results of R-Factor are in the range of 80 and 90, corresponding to MOS value of 4 (Table 6.5). Therefore, all four cross-layer schemes are able to maintain a similarly good value of MOS for VoIP under heavy traffic load regardless of the mobility speed.

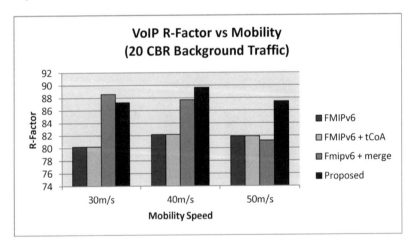

FIGURE 6.32
VoIP MOS versus background traffic.

6.6   Conclusion

As a promising emerging wireless metropolitan-area network technology, it is crucial for WiMAX to support seamless mobility for real-time services. The handover of WiMAX suffers long delay due to both layer-2 and layer-3 HO processes. Cross-layer HO combines and integrates HO in the two layers and improves HO performance. This book chapter describes the FMIPv6 cross-layer HO [12], and a newly proposed enhancement that combines two features that have been previously proposed [4, 20]. Performance evaluation has been conducted using computer simulation, comparing four schemes: FMIPv6, FMIPv6+tCoA [20], FMIPv6+merge [4], and the proposed scheme. We found that the proposed scheme has successfully eliminate several control messages and the DAD delay, therefore, performs the best in terms of HO delay, loss, and user perspectives for video and voice. Future works may include extending the proposed scheme for supporting all the five scheduling QoS classes [2], combining the proposed scheme with link-layer retransmission schemes [22], and applying the scheme for high-speed vehicular networks [21].

Bibliography

[1] A. Bacioccola, C. Cicconetti, and G. Stea. User-level performance evaluation of VoIP using ns-2. In *ACM International Conference Proceeding Series*, page 321, 2007.

[2] B. Chellappan, T. Moh, and M. Moh. Supporting Multiple Quality-of-Service Classes in IEEE 802.16e Handoff. In *Proc. of the International Conference of Computing, in Engineering, Science and Information*, Fullerton, CA, April 2009.

[3] J. Chen, C. C. Wang, and J. D. Lee. Pre-coordination mechanism of fast handover in WiMAX networks. In *Wireless Broadband and Ultra Wideband Communications (AusWireless)*, page 15, Aug. 2007.

[4] W. Chen and F. Hsieh. A cross layer design for handover in 802.16e network with IPv6 mobility. In *Proc. of IEEE Wireless Communications and Networking Conference*, pages 3844–3849, March 2007.

[5] S. Cho, J. Kwun, and C. Park. Hard handoff scheme exploiting uplink and downlink signals in IEEE 802.16e systems. In *Proc. of IEEE Vehicular Technology Conference (VTC-Spring)*, volume 3, pages 1236–1240, May 2006.

[6] G. Dong and J. Dai. An improved handover algorithm for scheduling services in IEEE802.16e. In *Mobile WiMAX Symposium*, volume 3, pages 38–42, March 2007.

[7] Ivov E., J. Montavont, and T. Noel. Thorough empirical analysis of the IETF FMIPV6 protocol over IEEE 802.11 networks. In *Proc. of IEEE Wireless Communications and Networking Conference (WCNC)*, pages 65–72, April 2008.

[8] C. Eklund, R.B. Marks, and K.L. Standwood. IEEE Standard 802.16: A technical overview of the WirelessMAN air interface for broadband wireless access. *IEEE Communication Magazine*, pages 98–107, June 2002.

[9] Ernst, T. MobiWAN: NS-2 extensions to study mobility in Wide-Area IPv6 Networks, `http://www.inrialpes.fr/planete/mobiwan/`, May 2002.

[10] P.Q. Huynh, P. Jangyodsuk, and M. Moh. Supporting video streaming over WiMAX networks by enhanced FMIPv6-based handover. In *Proc. of the Fourth International Conference on Information Systems, Technology and Management (ICISTM'10)*, Bangkok, Thailand, March 2010.

[11] IEEE. IEEE 802.16e Standard for Local and metropolitan area networks. Part 16: Air interface for fixed broadband wireless access systems - Amendment for physical and medium access control layers for combined fixed and mobile operation in licensed bands. Technical report, IEEE, Dec. 2005.

[12] H. Jang, J. Jee, and Y. Han. Mobile IPv6 fast handovers over IEEE 802.16e network. IETF RFC 5270, Nov. 2007.

[13] W. Jiao, P. Jiang, and Y. Ma. Fast handover scheme for real-time applications in mobile WiMAX. In *Proc. of IEEE International Conference on Communications(ICC)*, pages 6038–6042, June 2007.

[14] D. Johnson, C. Perkins, and J. Arkko. Mobility Support in IPv6. IETF RFC 3775, June 2004.

[15] C. Ke, C. Shieh, W. Hwang, and *et al.* An evaluation framework for more realistic simulations of MPEG video transmission. *Journal of Information Science and Engineering*, pages 98–107, Aug. 2006.

[16] H.J. Kim and M. Moh. Performance study of FMIPv6-based cross-layer WiMAX handover scheme for supporting VoIP service. In *Proc. of IEEE Pacific RIM Conference on Communications, Computers and Signal (PACRIM)*, Victoria, B.C., Canada, Aug. 2009.

[17] J. Klaue, B. Rathke, and A. Wolisz. EvalVid – A framework for video transmission and quality evaluation. In *Proc. of the 13th International Conference on Modelling Techniques and Tools for Computer Performance Evaluation*, pages 255–272, Sept. 2003.

[18] R. Koodli. Fast Handovers for Mobile IPv6. IETF RFC 4068, July 2005.

[19] D.H. Lee, K. Kyamakya, and J.P. Umondi. Fast handover algorithm for IEEE 802.16e broadband wireless access system. In *Proc. of the 1st International Symposium on Wireless and Pervasive Computing*, page 6, July 2006.

[20] J.S. Lee, S.Y. Choi, and Y.I. Eom. Fast handover scheme using temporary CoA in mobile WiMAX systems. In *Proc. of 11th International Conference Advanced Communication Technology (ICACT)*, pages 1772–1776, Feb. 2009.

[21] M. Moh, B. Chellappan, T.-S. Moh, and S. Venugopal. Handoff mechanisms for IEEE 802.16 networks supporting intelligent transportation systems. In *Proc. of 5th IEEE Consumer Communications and Networking Conference (CCNC)*, Las Vegas, NV, Jan. 2009.

[22] M. Moh, T.-S. Moh, and Y. Shih. On Enhancing WiMAX HARQ: A multiple-copy approach. In *Proc. of 5th IEEE Consumer Communications and Networking Conference (CCNC)*, Las Vegas, NV, Jan. 2008.

[23] NIST. Seamless and Secure Mobility Project. An IEEE 802.16 Model for NS-2.

[24] C. Perkins. Mobility Support in IPv4. IETF RFC 3320, Jan. 2002.

[25] S. Sengupta, M. Chatterjee, and S. Ganguly. Improving quality of VoIP streams over WiMAX. *IEEE Transaction on Computers*, 57(2):45–156, Feb. 2008.

[26] The Network Simulator (NS-2). http://www.isi.edu/nsnam/ns/.

7

Service Continuity Support in Self-Organizing IMS Networks

Christian Makaya

Telcordia Technologies, Piscataway, NJ
Email: chrismak@ieee.org

Satoshi Komorita

KDDI R&D Labs, Saitama, Japan

Ashutosh Dutta

NIKSUN Inc., Princeton, NJ

Hidetoshi Yokota

KDDI R&D Labs, Saitama, Japan

F. Joe Lin

Telcordia Technologies, Piscataway, NJ

CONTENTS

The Next-Generation Networks (NGN), which can provide advanced multimedia services over all-IP based networks, was the subject of a lot of attention many years ago. While there have been tremendous efforts to develop its architecture and protocols, especially for IMS (IP Multimedia Subsystem), which is a key technology of the NGN, it is far from being widely deployed. However, efforts to create an advanced signaling infrastructure realizing many requirements have resulted in a large number of functional components and interactions between those components. Thus, carriers are trying to explore effective ways to deploy IMS while offering the value-added and rich communication services. As one of such approaches, we have proposed self-organizing IMS (SOIMS). The self-organizing IMS enables IMS functional components and corresponding physical nodes to adapt dynamically and automatically, based on situations like network load and available system resources while continuing the IMS operation. To realize this, service continuity provisioning to end-users is an important requirement when a reconfiguration occurs during the operation. This chapter proposes a novel mechanism that provides service continuity to end-users. The proposed mechanism has been implemented in a real system (prototype), and its performance evaluation analyzed in terms of number of control signaling and processing time during reconfiguration.

7.1   Introduction

In recent years, the NGN, which is a standard platform for providing highly advanced network services by integrating fixed and mobile communication networks, has been attracting attention. The IMS [3] is its key technology, and has been standardized by 3GPP [6]. The IMS has standard specifications for a multimedia services platform operating over an all-IP network, which provides quality of service (QoS) control, accounting, and various services by third-party service providers. The IMS is currently deployed and operated on a large scale by telecommunication network operators. Their core network is equipped with IMS components such as several call control servers, subscriber database servers, and various application servers. A user terminal or UE (User Equipment), connects to the IMS core network through access networks such as wireless and wired networks. However, the IMS is a complex system with many components. In addition, the operator needs to take care of the reliability and redundancy of the IMS. Thus, cost effective implementations and deployment of the IMS are desired.

Self-organization offers an attractive way to reduce the deployment and operational costs of such a complex system. A self-organizing approach has been studied for network configuration, such as mobile ad-hoc networks (MANET) and zero-configuration networks. Furthermore, in IMS, components are defined as functional components and can be considered separately from physical nodes. Thus, dynamic adaptation with the nodes merging and splitting IMS functional components can realize effective IMS operation in terms of resource usage and availability. In its current form, IMS architecture and protocols do not have the mechanisms that can easily help IMS components to self-organize. Thus, we have proposed additional features and mechanisms to support the self-organizing capability for IMS [9].

In addition to configuration before bootstrapping IMS functionality, the efficiency of the IMS reconfiguration while already in operation is also possible, depending on the network and resource usage situation. However, this has a huge impact on registered UEs with IMS components that have established sessions. Thus, this self-organizing mechanism also needs to support service continuity to allow the UEs to continue their services seamlessly after the reconfiguration. In this chapter, we propose a method to guarantee service continue by effective notification of the affected UEs. This method allows the UEs to reregister seamlessly with IMS and maintain the established session. Furthermore, we implement our proposed method on real systems and assess its efficiency and performance.

The rest of this chapter is organized as follows: Section 7.2 provides an overview of the IMS and the SOIMS. Section 7.3 describes related work dealing with the self-organization in IMS environment. Section 7.4 presents the proposed method for service continuity provisioning in the SOIMS environment. Section 7.5 describes the prototype implementation and performance results. Finally, Section 7.6 concludes the chapter.

7.2 Overview of IMS and Self-Organizing IMS

7.2.1 IP Multimedia Subsystem Concepts

Figure 7.1 shows the basic IMS core network configuration. The following IMS components are located in the IMS core network that an operator manages: the HSS (Home Subscriber Server) is a database server for managing subscribers, the S-CSCF (Serving CSCF) is the main session initiation protocol (SIP) server for call control, the P-CSCF (Proxy CSCF) communicates with the UE directly and establishes a secure connection to it, the I-CSCF (Interrogating CSCF) is a kind of resolver of SIP message routing and a gateway to other IMS domains, the PCRF (Policy and Charging Rules Function) conducts QoS control, the MGCF (Media Gateway Control Function), and the BGCF (Breakout Gateway Control Function) allows interconnection to the existing circuit-switched networks. Each operator has these IMS facilities and interconnects their own IMS to another operator's one by a secure and quality guaranteed route (e.g., via a dedicated line). A UE connects with the IMS through the IP core network and wireless access networks such as EV-DO

[5] and LTE (Long Term Evolution) [2], and wired access network such as packet cable networks [8]. All IMS components are functional entities, thus, they can run on any physical nodes except specific components, which require the dedicated lines.

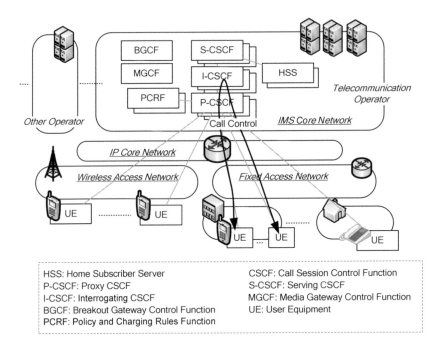

FIGURE 7.1
IP Multimedia Subsystem (IMS) overview.

In IMS, the SIP [18] is used for call control between UEs and IMS components. First, a UE registers with IMS before availing IMS services. The UE sends a registration request message to the S-CSCF via the P-CSCF of the IMS to which the UE belongs. The S-CSCF verifies the UE based on the UE information stored in the HSS, and if successful, the S-CSCF registers the UE. After that, the UE establishes a secure IPsec (IP Security) connection to the P-CSCF and uses the IMS services by sending and receiving SIP messages to and from a correspondent UE and Application Servers (ASs) via the P-CSCF and the S-CSCF. For example, when a UE makes a call, the UE sends an INVITE message to a correspondent UE and establishes active session to communicate. CSCFs handling the active session are decided when the session is established and cannot be changed during runtime.

7.2.2 Self-Organizing IMS (SOIMS)

The basic concept of the SOIMS is shown in Figure 7.2. In the resource pool, there are physical nodes that are capable of running several IMS components. According to the load and network conditions, the nodes can adapt the components and run one or multiple instances of them. They negotiate with each other and take over missing components and allows IMS to remain available in case of failure or overload. When the physical resource is not needed any more, the processing is moved to other nodes and the unnecessary node is removed. Thus, the system realizes efficient redundancy and effective resource usage.

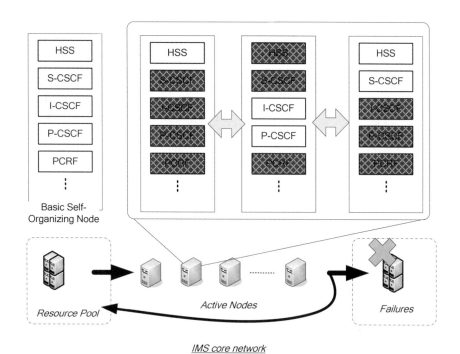

FIGURE 7.2
Concept of the self-organizing IMS (SOIMS).

The SOIMS can be structured in one of the two modes: centralized or distributed. For sake of simplicity and easy optimization of the entire system, we consider a centralized approach that has a central control node called a *master node* that maintains operator policy and state information for all nodes under its control. On the basis of the information, the master node assigns IMS functionalities to different nodes. The SOIMS require additional features such as node discovery, role request, and assignment, and interaction protocols between the nodes and the master node have been proposed in [9]. In addition, after the IMS configuration is changed, those IMS components assigned to each node need to take care of active sessions that were established before the reconfiguration. Otherwise, services are interrupted and it degrades the user's experience.

7.3 Related Work

With the emerging 3GPP LTE (Long Term Evolution) technology, self-organizing networks are envisioned as the new model for the next-generation OSS/BSS (Operations and Business Support System). The 3GPP [3] and Next Generation Mobile Networks Alliance [16] have standardized a set of capabilities known as self-organizing network to improve O&M (Operation and Maintenance). Since IMS is envisioned as a main component for the deployment of LTE or 4G wireless networks, for example, in order to provide real-time applications and other value-added and rich communication services, it is important to extend the self-organizing network concepts to the IMS core networks rather than only to the LTE RAN (Radio Access Network) and EPC (Evolved Packet Core).

The concept of SOIMS networks has not been widely studied to the best knowledge of authors. Here, it is important to understand that the self-organizing IMS is different from the P2P-SIP concept [10] on which significant research have been done. Bessis [7] describes performance analysis and benefits of running multiple SIP servers on the same host. That paper shows how to design the IMS networks in order to maximize IMS server colocation and explains which types of SIP calls can benefit from the colocation of IMS servers. Fabini *et al.* [11] describe an optimal IMS configuration with respect to architecture and QoS aspects. It demonstrates the feasibility of an IMS system implementation within a single virtual device (all-in-one). In [15], a distributed IMS architecture has been proposed by representing network functional elements in DHT (Distributed Hash Tables) overlay networks. The main focus was to distribute S-CSCF functionalities by using an overlay network where these functionalities are merged in one node, which is called IMS DHT. A common issue with DHT overlay networks is related to the number of operations or message in DHT to retrieve information or data.

In a SOIMS, efficiency requires the reconfiguration of the IMS to be executed while the IMS is in operation. From the perspective of users, it is important to continue their services without interruption. The SCC (Service Centralization and Continuity) [4] is defined as continuing an IMS session when an UE executes a handoff by establishing new session related to the old one. In [19], a method was proposed to use the same session by updating SIP servers responsible for active sessions. This method sends a specific message to SIP clients, then updates session information to continue to use the old ones. In the self-organizing IMS, it is important to take care of the impact on UEs and how to manage those affected. This chapter deals with service continuity from the perspective of a self-organizing IMS networks.

7.4   Service Continuity in SOIMS

7.4.1   Requirements for Service Continuity

In the SOIMS, the IMS components are dynamically assigned and transferred to appropriate physical nodes when IMS reconfiguration happens. However, in the IMS, each component has its own state and information for call and session control. A UE also keeps its connection state to the IMS network and the state of the call. In particular, if an inconsistency occurs between the UE and the P-CSCF that communicates with the UE directly, and the S-CSCF that manages registration information of the UE, the service will be interrupted and terminated.

To solve the inconsistency between those IMS components and the UE, we have proposed two types of approaches: UE non involved and UE involved. The former tries to hide the reconfiguration from the UE. However, that will require a hiding mechanism at IP layer, which requires new equipment such as SIP proxy or SBC (Session Border Controller) in front of the P-CSCF. In addition, a handoff procedure for the state registered in the IMS is needed. This approach requires a significant modification of the existing IMS, including operational modes. The latter tries to inform the UE of the reconfiguration and make the UE updates information affecting service continuity. In this method, the UE that is informed about the reconfiguration registers with the IMS again, and continues the processing of the call. The drawbacks of this method is the signaling overhead due to UEs notification when IMS network has been reconfigured.

This chapter proposes a method based on the latter approach in order to guarantee service continuity and reduce signaling overhead. The proposed approach has the following requirements. First, the method figures out the UEs affected by the reconfiguration effectively because a lot of UEs connect to the IMS and relationships between CSCFs and UEs are dynamically assigned when UEs register. The method needs to determines which UE connects to which CSCFs and notifies the UE if the CSCFs are changed. Second, a notification mechanism to the UE is needed. In current IMS, there is no way to notify the reconfiguration information to UEs. Third, one is to minimize the service interruption time from the reconfiguration to the end of appropriate procedure at UEs. During the interruption time, control messages for the call cannot be processed. If a correspondent UE sends a message to the UE, the message is dropped. This failure affects service continuity. Finally, impact on the existing IMS should be minimized because IMS has been already deployed in some operators' networks. Furthermore, IMS is a well defined technology, then major change must be avoided.

7.4.2   Proposed SOIMS Method

In this chapter, we focus on the SUBSCRIBE session that is established between a UE and S-CSCF through P-CSCF in order to notify the UE's registration state to the UE. We propose a method to notify the reconfiguration to the UE by an extension of this session. In this method, the P-CSCF and S-CSCF, which are responsible for the UE, learn and store information about each other in the establishment of this session. Then, in case of any change in one of them, the other can recognize it and notify the change to the UE. This allows the IMS to inform only the affected UEs. The UEs execute registration and session handoff procedure [4] through the P-CSCF designated by the notification message. The P-CSCF asks the HSS for the S-CSCF of the UE in the same way as basic IMS and processes the UE registration. The assignments of P-CSCF and S-CSCF to UEs are done and provided to the appropriate CSCFs and HSS by the master node.

Figure 7.3 shows the overview of the proposed notification mechanism. In this figure, UE#1 and UE#2 register and establish a SUBSCRIBE session with S-CSCF#2 through P-CSCF#1 and P-CSCF#2. UE#K also registers with S-CSCF#M through P-CSCF#2. In the case (A) where the master node tries to move all UEs using S-CSCF#2 to S-CSCF#1, the master node sends commands to all P-CSCFs and HSS. Thus, P-CSCF#1 and P-CSCF#2 send a SIP NOTIFY message to UE#1 and UE#2. The HSS also updates the assignment of S-CSCF to UEs based on the master node information. In the case (B) where the master node tries to move all UEs using P-CSCF#2 to P-CSCF#1 and P-CSCF#N, the master node sends commands to all S-CSCFs, and then S-CSCF#2 and S-CSCF#M send a SIP NOTIFY message to UE#2 and UE#K. In the both cases, the UEs register with the indicated CSCFs and continue their ongoing services.

This method is also applicable by sequential execution when both S-CSCF and P-CSCF are changed. In addition, this is valid when some failure takes place at P-CSCF and S-CSCF because the failed CSCF is not involved with this notification sequence. This method can send notifications effectively and quickly to the affected UEs, and can be realized by an extension of subscribe session and an additional master node compared with the basic IMS. Thus, it satisfies the requirements described in previous section.

7.4.3   Message Exchange with SOIMS

In this section, the proposed call flows for notification and session continuity are described. Figure 7.4 shows the first registration flow when using an AKA

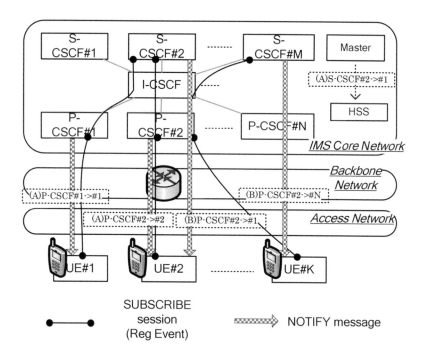

FIGURE 7.3
Overview of notification method.

(Authentication and Key Agreement) [1] as the authentication procedure. This
call flow is almost the same as a standard IMS registration flow. However, with
the standard IMS flow, the P-CSCF and S-CSCF obtain and store the information required for notification procedure in (17), (19) and (20). First, a UE
sends a REGISTER to P-CSCF (1) and P-CSCF forwards it to I-CSCF (2).
The I-CSCF asks HSS which S-CSCF is assigned to the UE (3) and forwards
REGISTER to that S-CSCF (4). The S-CSCF asks HSS for authentication
information (5) and sends back a 401 Unauthorized message to the UE via
the I-CSCF and the P-CSCF (6, 7, 8). The UE sends REGISTER again with
authentication information to the S-CSCF (9, 10, 11). The S-CSCF verifies
the authentication credentials, and it gets and updates the UE's information
on the HSS (12), then sends back a 200 OK message to the UE (12, 13, 14).
After that, the UE sends a SIP SUBSCRIBE message to the S-CSCF through
the P-CSCF (15, 16) in order to subscribe to IMS network reconfiguration
event. Here, the S-CSCF stores the P-CSCF of the UE (17), and sends back a
200 OK to the UE (18). The P-CSCF also stores the S-CSCF of the UE (19)
and session information to ensure the P-CSCF sends a NOTIFY (20), then
forwards the 200 OK to the UE.

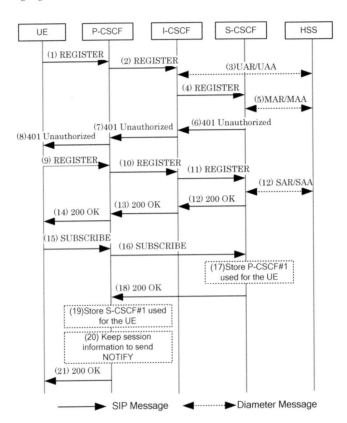

FIGURE 7.4
Signaling message during registration.

Figure 7.5 shows the P-CSCF change flow from P-CSCF#1 to P-CSCF#2. First of all, the UE is already registered to the S-CSCF through P-CSCF#1 and establishes an active call session with an INVITE through them (1). Let us assume for a given reason the master node decides a reconfiguration of the IMS network and informs all S-CSCFs about the change. If P-CSCF#2 is not running yet, the master node also sends commands to launch it at the same time (2). The S-CSCF finds out which registered UEs use P-CSCF#1 based on the stored information, and sends a NOTIFY to the UE including the new P-CSCF information via the new P-CSCF, that is, P-CSCF#2 in Figure 7.5 (4). The P-CSCF#2 forwards the NOTIFY to the UE (5). Upon receipt of the notification from the S-CSCF, the UE sends back a 200 OK (6, 7), registers and subscribes through the P-CSCF#2 (8). The UE also takes care of the active session. The UE (i.e., the caller) sends an INVITE with a session transfer identifier that associates the previous active session with the new session generated by this INVITE (9, 10, 11). The correspondent UE (i.e., the callee) sends back a 200 OK (12, 13, 14) and the caller UE sends an ACK (15, 16, 17). Finally, the UE sends a BYE to disconnect the previous active session.

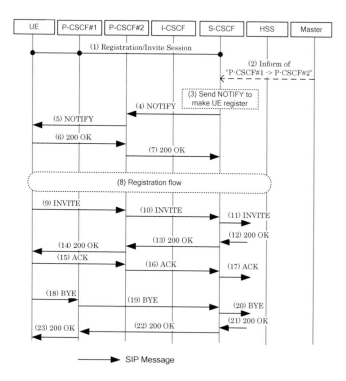

FIGURE 7.5
Call flow when P-CSCF changes.

Figure 7.6 shows the S-CSCF change flow from S-CSCF#1 to S-CSCF#2. This flow is similar to the P-CSCF change case. However, in addition to the notification to the UE, the master node needs to inform the HSS of the change because the HSS needs to assign another S-CSCF when the UE registers. The UE is already registered to S-CSCF#1 through P-CSCF and establishes an active call session by sending an INVITE through them (1). Let us assume for a given reason the master node decides a reconfiguration of the IMS network and informs all P-CSCFs about the change. If S-CSCF#2 is not running yet, the master node also sends commands to launch it at the same time (2). The master node also informs the HSS of the change to update assignments of S-CSCF (3, 4). The P-CSCF finds out which UEs were associated with S-CSCF#1 based on the stored information, and sends the NOTIFY message (6). Here, the NOTIFY indicates the same P-CSCF because P-CSCF does not change. The UE sends back a 200 OK (7), registered and subscribed through the P-CSCF. Then, the HSS informs the I-CSCF of the new S-CSCF in the registration sequence, thus the UE can register with the new S-CSCF (i.e., S-CSCF#2). This way, the session continuity can be guaranteed similarly to the P-CSCF change scenario.

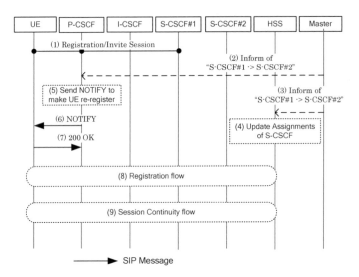

FIGURE 7.6
Call flow when S-CSCF changes.

7.5   Prototype and Results

7.5.1   Experimental Configuration

Figure 7.7 shows our experimental network configuration for verifying the proposed SOIMS method's behavior and performance. The core network is composed of four IMS nodes (node#i, $i = 1, ..., 4$) and the master node. The HSS and I-CSCF are running on node#1, and the S-CSCF function is running on node#2 and node#3, while the P-CSCF function is running on node#3 and node#4, respectively. The UEs and UE Emulator connect to the IMS through hubs and a router by using the Ethernet. A monitor machine captures packets passing through the hubs for statistics.

The proposed method has been implemented in the master node and the different IMS entities, namely, CSCFs, HSS, and UE. The CSCFs and HSS are built based on NIST SIP [17] implementation, which is an open-source SIP stack. The master node and slave middleware, which are our original

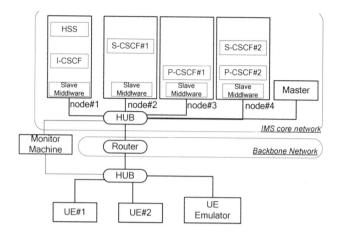

FIGURE 7.7
Experimental network configuration.

software, communicate to get information of each node and dictates which IMS functions are running on the node. An extended SIP Communicator [20] is used as IMS client (i.e., UE) or user agent. The UE Emulator, which can handle a thousand of sessions but only control signaling, is built based on IMS Bench SIPp software [13]. Regular PCs (CPU: Intel Atom 230 1.6 GHz, Memory: 2 GB, HDD: 80 GB, NIC Intel Pro/100) with Fedora 10 as operating system (OS) are used for the testbed.

7.5.2 Measurements and Evaluations

In the experiment, first UE#1 and UE#2 register with the IMS through P-CSCF#1 and S-CSCF#1, and then establish a VoIP session. For this scenario, P-CSCF#2 and S-CSCF#2 on the node#4 are not running. During the call, the master node decides to add a new P-CSCF function to increase the processing capacity of P-CSCF in the IMS core network, and then sends out a command to node#4 to launch P-CSCF#2 and also send out commands to S-CSCF#1 to change P-CSCF of UE#1 from P-CSCF#1 to P-CSCF#2. With our proposal, the call session is not interrupted due to IMS network reconfiguration. In a similar way, the master node decides to add a new S-CSCF on the node#4 to increase processing capacity of S-CSCF in the IMS core network and changes the S-CSCF of UE#1 from S-CSCF#1 to S-CSCF#2. The performance of the proposed SOIMS is evaluated for this two scenarios.

Moreover, by using the UE Emulator, several instances of UEs ranging from 10 to 1000 registers with the IMS to S-CSCF#1 through P-CSCF#1. For a given reason or network policy, the master node decides to change their P-CSCF to P-CSCF#2 and their S-CSCF to S-CSCF#2, respectively. As measurement metrics, we use the processing time for service continuity which is the time from when the master node sends out commands to each IMS node until all UEs complete the registration and session handoff of active sessions. For the use case where the UE Emulator is used, we measure the total time from when the master node sends out commands until all instance of UEs complete their registration. These performance metrics are computed based on the captured packets at the monitor machine during the experiments.

7.5.3 Performance Results

The desired behavior of our proposal was verified in this prototype. Figure 7.8 and Figure 7.9 show the signaling and VoIP traffic for an ongoing session when a reconfiguration happens in the IMS core network. In both cases, the UEs communicate with each other after call establishment, and then the P-CSCF and S-CSCF of UE#1 are changed, respectively. Although the CSCF is changed, the VoIP session continues without disruption and terminates correctly. In the IMS, VoIP traffic is not transferred through CSCFs. Thus, VoIP can continue seamlessly unless there is a failure of control signaling due to the change of CSCFs. However, if a new UE tries to establish a session with previously registered UEs, this attempt will fail, since the CSCF (either P-CSCF or S-CSCF) has changed. With our proposal, previously registered UEs are still reachable since the state of their registration has been transferred to the new CSCF. The new session establishment will be supported seamlessly.

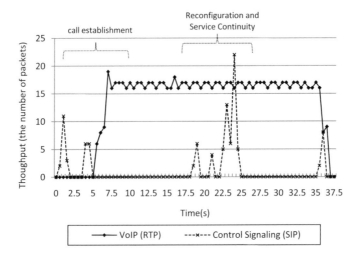

FIGURE 7.8
Traffic volume when P-CSCF has been changed.

Figure 7.10 shows the processing time during IMS network reconfiguration

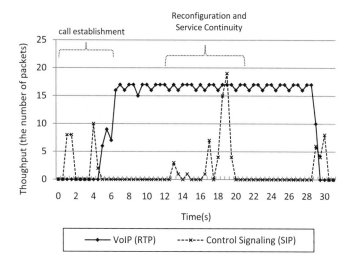

FIGURE 7.9
Traffic volume when S-CSCF has been changed.

when the S-CSCF and P-CSCF of UE#1 were changed. It takes about 6 seconds to restore session state for the S-CSCF change case and about 7 seconds for the P-CSCF change case. In this experiment, the processing times are due to the processing and waiting at each IMS node since there is no transmission delay and congestion in the testbed network. In the former case, the dominant factor that affects the processing is the registration, because it takes several seconds to launch the new S-CSCF on node#4 and then complete the registration. In contrast, the dominant factor due to processing is the notification from the IMS core network to UEs in the latter case. However, this is also attributed to the start-up time of a new P-CSCF on the node#4 because the notification is sent through the new P-CSCF.

Figure 7.11 shows the total processing time when the P-CSCF and S-CSCF are changed by using the UE Emulator as the IMS client. In both cases, the processing time increases linearly, but processing time when the S-CSCF changed takes six times longer than when the P-CSCF is changed. Figure 7.12 shows the success rate for service continuity. The rate when the S-CSCF is changed was always 100%. However, the success rate when the P-CSCF (Figure 7.13) is changed decreases as the number of UEs increases. Here, the failure of service continuity means that the UE cannot complete the processing of the service continuity and its service is terminated.

The difference between both cases seems to be due to the processing of notification and registration. When the P-CSCF of UEs was changed from P-CSCF#1 on node#3 to P-CSCF#2 on node#4, S-CSCF#1 on the node#2 sends notifications to all affected UEs through P-CSCF#2 regardless of pro-

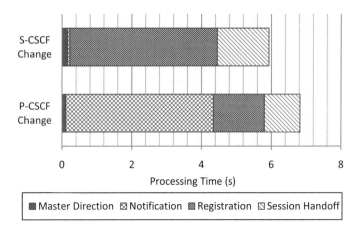

FIGURE 7.10
Processing time for session state restoration.

cessing capability of other nodes, such as node#4. This behavior results in the retransmission of notifications and exceeds the processing speed of registration at P-CSCF#2 on node#4. Thus, the success rate decreases. On the other hand, when the S-CSCF of UEs was changed from S-CSCF#1 on node#2 to S-CSCF#2 on node#4, the P-CSCF#2 on node#3 sends notifications to all affected UEs directly. However, all registration comes to the P-CSCF#2 and exceeds its processing power so that it cannot send the needed notifications.

7.5.4   Consideration

In this prototype, we were able to verify the service continuity after the IMS core network reconfiguration by using our proposed SOIMS method. The overall time required for IMS reconfiguration is about 7 seconds. We think this time is short enough to continue the service because a UE can try to continue its service if the UE recognizes the reconfiguration before retransmission of other signaling messages expires. When there are many affected UEs, quick notifications within the limited time become more important. The master node needs to take care of the capability and the processing power of CSCFs.

However, some open issues remain. In our SOIMS method, the master node does not directly deal with the affected UEs and each CSCF sends notifications based on its stored information. This can improve the performance for sending notifications quickly without the dedicated management nodes. On the other hand, each CSCF does not help with or take into account the processing capability and bandwidth of other CSCFs. This behavior results in congestion of control signaling and failure of service continuity. Thus, in future, some mechanisms are required to control the rate of sending the noti-

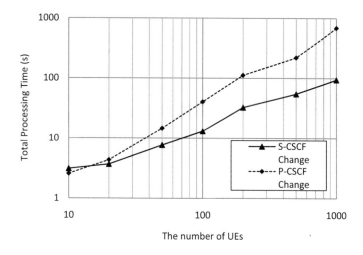

FIGURE 7.11
Total processing time.

fications. For example, the master node can inform CSCFs of their capability. It is also possible that related CSCFs communicate with each other about their capability.

Security is another issue. A UE uses IPsec to connect to P-CSCF in order to secure communication. If the P-CSCF of the UE changes, the notification is sent out over the secure connection. In particular, in case that a previous P-CSCF crashes, the UE does not have any choice to receive the unsafe notification from another P-CSCF. Thus, it is possible that a malicious attacker could send bogus notification to the UE and lead the UE to a fake P-CSCF. In SOIMS, it is difficult to send a bogus notification to the UE and redirect the UE to a fake P-CSCF. In our proposal, it is difficult to send a bogus notification because the notification includes Call-ID and tags that are unique keys generated in the previous secure registration procedure. However, there is still the probability for some attacks, such as TCP session hijacks [12]. Thus, additional security mechanisms for notifications might be needed.

7.5.5 Self-Organizing IMS and Load Balancing

In the above proposed SOIMS approach, the UE is aware of IMS node changes in the core IMS network, and can be refereed as the *UE involved method*. With the UE involved method, when any change happens in IMS networks, the UEs are notified. This will induce a higher signaling traffic and scalability issues when the number of UEs increases. In the wireless environment where radio resources are limited, these issues are critical. The CSCF must notify all UEs affected by the change. On the other hand, all of these UEs must refresh their

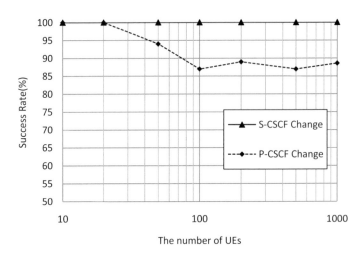

FIGURE 7.12

Success rate when S-CSCF and P-CSCF change.

ongoing IMS session. The CSCF will then receive simultaneously a higher number of request. This might be interpreted as a denial of service (DoS) attack by a firewall deployed in the network. To avoid these issues, change in the IMS core network should be transparent to UEs. In other words, the UEs should not be notified about the IMS node failure or reconfiguration. This approach is called *UE non involved*. In case of the UE non involved method the UE is completely unaware of the IMS node changes in the core network. With this method, no additional message will be sent to UEs over the wireless link, and the UEs do not take any action after reconfiguration of the IMS core network.

To allow support of the UE non involved method, we proposed a solution based on load balancing (LB) concept. Load balancing is a technique used to distribute network load across different network components in order to get optimal resource utilization, scalability, high availability and flexibility. The proposed solution satisfies these LB features while handling support of session continuity IMS networks, topology hiding, and distributed IMS functions in an efficient manner. In the proposed SOIMS with LB approach, the LB appears as a virtual P-CSCF to UEs. In other words, from the UE's perspective, the LB is the P-CSCF and the UE sends all request or SIP messages to the LB's virtual IP address (VIP).

Moreover, rather than establishing an IPsec security association (SA) between UEs and P-CSCF as specified by IMS standard, the IPsec SA is established between UEs and LB. Figure 7.14 illustrates our proposed SIP-based load balancing procedure for IMS networks, called SOIMS-LB. When the LB receives SIP messages from the UE, it forwards these messages to the selected

P-CSCF. The selection of P-CSCF is done by a scheduling algorithm defined in LB and by using information provided by the Master Node. There is no direct communication between UEs and the real P-CSCF since all communications are handled by the LB. In order to correctly route the SIP packets to P-CSCFs and maintain the session persistence, the LB needs to intercept the SIP packets and modify the headers accordingly. In fact, since the LB is acting as an outbound SIP proxy, it processes this message and adds itself in (the topmost) Via and Record-Route headers of SIP message before forwarding the message to the selected P-CSCF. By adding its information to the Via header, the LB will receive the response to the request. In the opposite direction, the LB removes the Via header, which contains its information, before sending the message to the UE.

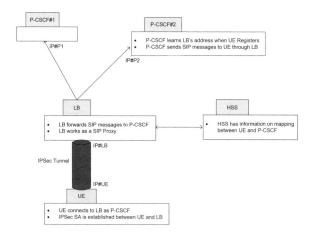

FIGURE 7.13
Load balancing with SOIMS.

In this section, we will describe session continuity when IMS role change event (e.g., IMS node failure, role assignment) occurs during ongoing session. Let us assume that the P-CSCF's (e.g., P-CSCF#1 in Figure 7.14) role changes due to failure, the LB will be notified by the Master Node about network configuration change or since LB gets list of P-CSCFs from the Master Node, it will discover role assignment change or failure. When the Master Node detects failure of P-CSCF#1, it notifies the S-CSCF about this event. Upon this notification, the S-CSCF can retrieve information of the new P-CSCF (e.g., P-CSCF#2) from the HSS. At the same time, the S-CSCF updates registration status (e.g., association and mapping) of LB and UE through P-CSCF#2. Then P-CSCF#2 can restore registration information and update mapping between LB and UE for subsequent SIP messages.

After restoration of registration information, the S-CSCF sends a message to P-CSCF#2 with information about media negotiation (i.e., SDP), new and old SIP Route. This information exchange allows restoration of the ongoing SIP session state in the S-CSCF and P-CSCF#2, and reconfiguration of IMS core network. When P-CSCF#2 completes the update of session information, it informs the LB and HSS about the new SIP Route. The old and new SIP Routes are stored in the LB and HSS to allow mapping of previously established SIP dialog with P-CSCF#2.

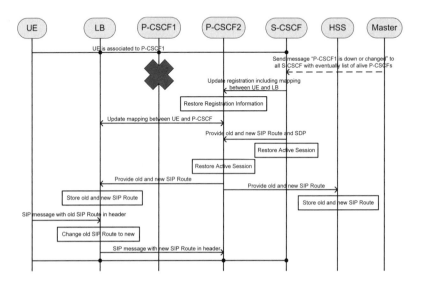

FIGURE 7.14
Signaling when P-CSCF changes.

All of these changes are transparent to the UEs. In fact, the LB hides the change and reconfiguration of IMS core network and UEs have no direct communication with the IMS components. Any subsequent SIP message sent by the UE will be set with the old SIP Route. It is the LB's responsibility to change the old SIP Route (e.g., Service-Route) in SIP message to the new SIP Route before to forward this message to the first IMS entity (i.e., P-CSCF). Since any change in the core IMS network is transparent to UEs, with the proposed solution, there is no need for UEs to subscribe for events notification. In other words, there is no need to exchange SIP SUBSCRIBE/NOTIFY messages, leading to minimal signaling overhead and network resources usage. Similar behavior can be observed when S-CSCF change, hence we don't show all signaling message for this scenario. More details about SOIMS and LB integration can be find in [14].

7.6 Conclusion

In this chapter, we introduced the self-organizing capability for IMS, in short SOIMS, that provides cost effective operation and makes efficient usage of resources. To realize this, service continuity for UEs affected by the IMS reconfiguration is an important aspect. We proposed an approach to realize it without a large impact on specification and implementation of the IMS. The proposed SOIMS method sends notifications to the UEs effectively by using the subscribe session and make the users to reregister or update their registration information and help the session handoff procedures. Further, we implemented the method and demonstrated its behavior and performance. The performance results show the realization of IMS reconfiguration completes in less than 10 seconds without any media disruption. This time seems to be short enough considering retransmission time of control signaling. Although the required time increases with the number of the affected UEs and capability of CSCFs, the performance could be improved by appropriate control and distribution of the notifications according to the capability. On the other hand, the proposed SOIMS-LB approach allows to hide any change in the core IMS networks to end-users (i.e., user equipment doesn't participate in the IMS network reconfiguration), minimize IMS networks reconfiguration delay, and reduces signaling overhead. As a result, the IMS networks efficiency and scalability has been improved.

Acknowledgment

The authors would like to thank the other project members: Dana Chee, Subir Das, Suren Alathurai, Manabu Ito, and Tsunehiko Chiba for their contributions to this work.

Bibliography

[1] 3GPP. 3G Security; Security Architecture. 3GPP TS 33.102 v9.0.0, Sept. 2009.

[2] 3GPP. Evolved Universal Terrestrial Radio Access (E-UTRA) and Evolved Universal Terrestrial Radio Access Network (E-UTRAN); Overall Description; Stage 2. 3GPP TS36.300, 2009.

[3] 3GPP. IP Multimedia Subsystem (IMS); Stage 2. 3GPP TS 23.228 v9.0.0 23.228 v9.0.0, Sept. 2009.

[4] 3GPP. IP Multimedia Subsystem (IMS); Stage 2. 3GPP TS 23.237 v9.0.0, Sept. 2009.

[5] 3GPP2. cdma2000 Wireless IP Network Standard: Introduction. 3GPP2 TS X.S0011-001-C v3.0, 2006.

[6] 3rd Generation Partnership Project (3GPP). http://www.3gpp.org.

[7] T. Bessis. Improving Performance and Reliability of an IMS Network by Co-locating IMS Servers. *Bell Labs Technical Journal*, 10(4):167–178, 2006.

[8] CableLabs. http://www.cablelabs.com/.

[9] A. Dutta, C. Makaya, S. Das, D. Chee, F. J. Lin, S. Komorita, T. Chiba, H. Yokota, and H. Schulzrinne. Self-organizing IP multimedia subsystem. In *Proc. of IEEE Int'l Conf. on Internet Multimedia Systems Architecture and Application (IMSAA'09)*, Bangalore, India, Dec. 2009.

[10] E. Marocco, A. Manzalini, M. Sampo, and G. Canal. Interworking between P2PSIP Overlays and IMS Networks – Scenarios and Technical Solutions, http://www.p2psip.org.

[11] J. Fabini, P. Reichl, A. Poropatich, R. Huber, and N. Jordan. IMS in a Bottle, Initial Experiences from an OpenSER-based Prototype Implementation of the 3GPP IP Multimedia Subsystem. In *Proc. of Int'l Conf. on Mobile Business (ICMB'06)*, June 2006.

[12] B. Harris and R. Hunt. TCP/IP security threats and attack methods. *Computer Communications*, 22(10):885–897, 1999.

[13] IMS Bench SIPp. http://sipp.sourceforge.net/ims_bench/.

[14] C. Makaya, A. Dutta, S. Das, D. Chee, S. Komorita F. J. Lin, H. Yokota, and H. Schulzrinne. Service continuity support in self-organizing IMS networks. In *Proc. of 2nd Wireless VITAE (WVITAE'11)*, Chennai, India, Feb. - Mar. 2011.

[15] M. Matuszewski and M. Garcia-Martin. A distributed IP multimedia subsystem (IMS). In *Proc. of IEEE Int'l Symposium on a World of Wireless, Mobile and Multimedia Networks (WoWMoM'07)*, pages 1–8, 2007.

[16] NGMN Alliance. http://www.ngmn.org.

[17] NIST. http://www-x.antd.nist.gov/proj/iptel/.

[18] J. Rosenberg, H. Schulzrinne, G. Camarillo, A.R. Johnston, R. Sparks J. Peterson, M. Handley, and E. Schooler. SIP: Session Initiation Protocol. IETF RFC 3261, June 2002.

[19] T. Hasegawa S. Komorita, T. Kubo and H. Yokota. Network-controlled SIP server switching methods for active SIP sessions. In *Proc. of IASTED Parallel and Distributed Computing and Networks (PDCN'09)*, Feb. 2009.

[20] SIP Communicator. `http://sip-communicator.org`.

8

Vehicular Communication Networks : Challenges, Solutions, and Services

Robil Daher

University of Rostock, Germany
Email: robil.daher@uni-rostock.de

Djamshid Tavangarian

University of Rostock, Germany

CONTENTS

Vehicular Communication Networks (VCNs) provide a promising technology for improving traffic safety and traffic efficiency in different road systems. Based on DSRC/WAVE Technology VCNs offer an efficient communication platform for Intelligent Transportation Systems (ITS) and related ITS services as well as value-added services such as multimedia and data services. However, different types of services have different requirements of network performance. The rapidly varying network topology in access network, where vehicles may travel with high speeds of more than 200 Km/h, creates several challenges for QoS provision and efficient routing integration. To investigate possible solutions for those challenges, related issues have to be addressed in conjunction with VANETs and roadside networks. Consequently, relevant solutions are developed for QoS mechanisms and routing protocols for both VANETs and roadside networks. This chapter introduces some of those most important solutions and proposes an overview of future ITS service, called "Start-to-Destination Driving Route Reservation (S2D-DRR)." Finally, some open research issues will briefly be introduced with an objective to spark new research interests in the field of VCNs.

"The more your car knows, the safer you - and everyone around you - will be." [3]

8.1   Introduction

Vehicular Communication Networks (VCNs) and related Intelligent Transportation Systems (ITS) provide a promising technology for improving traffic safety and traffic efficiency in different road systems. A large increase in R&D activities in area of VCNs was achieved in the last few years [15]. Several factors have led to this rapid development of VCNs, among others:

1. The increase deployment of wireless networks, which leads to increasing use of wireless Internet services;

2. The awareness of national and regional governments of the importance of integrating ITS into their transport infrastructure;

3. The embrace of vehicle manufacturers (VW, Daimler, GM, BMW, Ford, Toyota, Nissan, etc.) of information technology to address safety, environmental, and comfort issues of their vehicles;

4. The constructive cooperation between vehicle manufacturers, national and international organizations (IEEE, ASTM, ISO, etc.), and public transport authorities (US Department of Transportation and equivalent transport authorities in Europe and Japan, etc.) for developing standards for VCNs.

Consequently, various projects have been initiated: AKTIV, COOPERS, etc., are underway; FleetNet, ASV 4, VSC, etc., were completed just recently. In addition, several consortia (C2C-CC, VSCC, etc.) were established to explore the potential of VCNs and related ITS [15]. Those projects are funded substantially by national governments and involve several constituencies, including the vehicle manufacturers, the road administrators and operators, telecom and other service providers. Figure 8.1 presents a scheme that summarizes the main players in the field of VCNs. Furthermore, several national and regional governments have allocated licensed spectrum for vehicular wireless communication, generally in the 5.8/5.9 GHz band [33]. In that respect, the U.S. Federal Communications Commission (FCC) has assigned 75 MHz of radio spectrum at 5.9 GHz band for Dedicated Short-Range Communications (DSRC) [12], while the European Electronic Communications Committee (ECC) has allocated 50 MHz (5875-5925 MHz) to DSRC in Europe [5].

Because of their advantages not only for ITS, but also for a wide range of future multimedia applications, e.g., video/audio, and data applications, e-maps and road/vehicle as well as traffic/weather information [34], VCNs are expected to provide an efficient wireless platform for on-the-way Internet, as well as for future pervasive roads. Therefore, VCNs should support several applications and services, particularly real-time applications such as that for active safety and VoIP (voice-over IP) services [4]. Thus, to investigate challenges and solutions related to development and employment of VCNs, this chapter deals with mechanisms, protocols, and models in the field of routing and quality of service (QoS) for VCNs and concentrates on DSRC as the leading technology for vehicular communications [3], especially that based on IEEE 802.11(p) and 1609 standards. Moreover, some open research issues are also discussed with an objective to spark new research interests in this field.

The rest of this chapter is organized as follows. First, we present an overview of the VCNs, ITS, and related network technologies, as well as projects initiated worldwide for inspecting VCNs. After that, we briefly describe the specifications of emerging DSRC/WAVE and related network architectures. In addition, we also present the most important emerging vehicular applications in relation to ITS and value-added services. We also discuss the main challenges for routing, and QoS provisioning in conjunction with specifications of vehicular environments. Accordingly, we propose several solutions

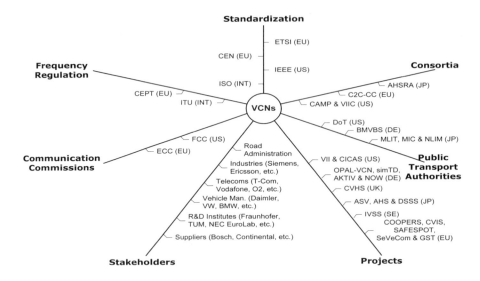

FIGURE 8.1
Main players participating in the development of VCNs.

for each of those challenges in order to reflect the state of the art in this field. Then, we introduce a future ITS service called "Start-to-Destination Driving Route Reservation (S2D-DRR)". After that we present some related open research issues. Finally, we conclude with a summary in the last section.

8.1.1 Intelligent Transportation Systems (ITS)

The IEEE Intelligent Transportation Systems Society (ITSS) defines the ITS as "those utilizing synergistic technologies and systems engineering concepts to develop and improve transportation systems of all kinds," including road, rail, water, and air transport, as well as navigation systems. However, although the ITS are not restricted to road transport, we concentrate in this chapter on ITS for only road transport systems. In that respect, the main focus of ITS from wireless communication point of view lies on Telematics and all types of communications in vehicles, between vehicles (vehicle-to-vehicle – V2V), and between vehicles and infrastructure (vehicle-to-roadside – V2R) [11].

8.1.2 Communication Technologies Used in VCNs

Several wireless technologies such as WLAN, WiMAX, and UMTS could be used as a basic technology for VCNs. However, the nature of VCNs and related ITS applications and services have certain requirements of specifications in media access control (MAC) and physical layers of those technologies. The dynamic topology of VCNs at high speeds and the nature of active safety applications, especially concerning the message dissemination that requires very low latency of connection initiation and link operation, make the adoption of those wireless technologies, especially WiMAX and UMTS standard families, for ITS very difficult. Therefore, several organizations such as ASTM and ETSI, as well as several projects such as FleetNet and VCS, have investigated the available wireless technologies to find out whether one or more of these technologies could satisfy the requirements of VCNs, or whether some kind of enhancements to those technologies would be sufficient to enable them meeting the expectations.

The WMAN and WWAN technologies such as WiMAX and UMTS provide QoS provision within a relatively long range and are already deployed in several regions, especially the UMTS technology. Nevertheless, due to the nature of active safety applications, especially because of their message dissemination nature (instead of data flow in usual real-time applications such as VoIP) that requires very low latency of connection initiation, both technologies WiMAX and UMTS, without additional enhancements, cannot satisfy the requirements of ITS services, especially that of active safety. Similarly, since the physical layer properties in VCNs are rapidly changing, when vehicles are traveling at high speeds of up to 200 Km/h, and where very short-duration communications exchanges are required [26], the current IEEE 802.11 WLAN cannot satisfy the requirements of ITS services, especially that of active safety.

In that respect, the American Society for Testing and Materials (ASTM) has concentrated on the DSRC-based solutions and developed the standard ASTM E2213-03 that describes the MAC and physical (PHY) layer specifications for wireless connectivity using DSRC services in 5 GHz band. This ASTM standard is based on and refers to the IEEE 802.11 and 802.11a. Accordingly, the IEEE 802.11 working group has been developing the standard IEEE P802.11p as amendment to integrate the ASTM E2213-3 specifications into the current MAC and PHY layers of IEEE 802.11 and 802.11a. In addition, the IEEE 1609 working group for Wireless Access in Vehicular Environments (WAVE) has already published four trial-use standards for the higher layers (above the MAC layer) of DSRC. The DSRC/WAVE currently represents the most completed standard suite specified for VCNs and related ITS services and is considered as the leading technology for vehicular communications [3].

In contrast, the ISO TC204 WG16 (International Organization for Standardization Technical Committee 204 Working Group 16) has standardized the ITS communication architecture, which is termed as CALM (Continuous Air Interface for Long and Medium range). CALM is intended to be platform-independent, and therefore to avoid the battles over regional standards that have dogged existing ITS standards like DSRC (ITU 07). For this purpose, the CALM architecture supports several wireless network technologies: 2G (ISO 21212) and 3G (ISO 21213) cellular systems for long-range, CALM Infrared (ISO 21214) and CALM 60 GHz (ISO 21216) for directional communications in short and medium rang, CALM M5 (ISO 21215), compatible to IEEE 802.11p, for omnidirectional in short and medium ranges.

However, the use of DSRC/WAVE for ITS-services, especially that of active safety, is recommended and supported by all participants in the R&D activities in the field of VCNs. Other wireless technologies, especially 2G and 3G due to their wide deployment, are foreseen as complementary infrastructure for DSRC-based VCNs.

8.1.3 VCN-Related Projects in Europe, US, and Japan

Several projects in the field of VCNs were initiated by several national and international organizations (IEEE, ASTM, ISO, etc.), public transport authorities (US Department of Transportation and equivalent transport authorities in Europe and Japan, etc.), and vehicle manufacturers (i.e., Daimler, BMW, General Motors, Renault, Toyota, etc.) or a combination thereof since the 1980s [22]. The advanced development and the wide deployment of wireless networks since the 1990s have led to more investment in the field of VCNs and ITS, especially since 2000, when many projects have been initiated in several countries worldwide, especially in the USA, Europe and Japan. However, while the first generation of VCN-related projects have dealt with the basic communication platform and service models and systems, the second generation projects focus mostly on the enhancement and optimization of concepts and solutions developed in the first-generation projects.

The first generation of projects and R&D activities in the field of VCNs have concentrated on the communication platform to be used for ITS, where investigations of challenges and solutions for adopting known communication technologies such as 2G and WLAN or for developing new communication technologies represent the main focus of those projects such as FleetNet (2000-2003), NoW – Network-on-Wheels (2004-2008) in Germany and more recently COOPERS (2006-2010) in Europe and VSC (2002-2004) and VSC-2 (2005-2009) in the United States. The results achieved in those projects form a basis for developing ITS-related communication standards such as IEEE 802.11p and 1609. Besides the communication platform, the development of ITS-related service models and systems are also important topics in the first generation of VCN-related projects such as WILLWARN (2004-2007), PRe-VENT (2004-2008) and more recently INTERSAFE-2 in Europe. Table 8.1 presents a brief list of the most known projects worldwide.

The second generation of projects and R&D-activities in the field of VCNs concentrates on the field trial of standards and services, developed within the first generation projects, in real-world vehicular environments. The main goal of the field trial is to investigate, among others: (1) the interoperability between already developed standards and related software and hardware components; (2) the performance of communication platform; and (3) the functionality of ITS services. Those projects serve finding the gabs and problems in the available standards and technical solutions, and thus addressing the

TABLE 8.1

List of the most important projects in the US, Europe and Japan.

	Project Name	Duration	Abbreviation/Website
EU	PReVENT	2004-2008	Preventive and Active Safety www.prevent-ip.org
	NoW	2004-2008	Network-on-Wheels www.network-on-wheels.de
	AKTIV	2006-2010	Adaptive und Kooperative Technologien für den Intelligenten Verkehr www.aktiv-online.org
	CVIS	2006-2010	Cooperative Vehicle Infrastructure Systems www.cvisproject.org
	COOPERS	2006-2010	Vehicle to Infrastructure, Traffic Management www.coopers-ip.eu
	INTERSAFE 2	2007-2013	INTERSAFE 2 www.interafe-2.eu
	simTD	2008-2012	Safe and Intelligent Moility - Test Field Germany www.simtd.de
	OPAL-VCN	2010-2014	Open Air lab for Vehicular Communication Networks www.opal-vcn.org
US	IVI	1998-2004	Intelligent Vehicle Initiative
	VII	2004-2009	Vehicle Infrastructure Integration www.vehicle-infrastructure.org
	VSC I & II	2002-2009	Vehicle Safety Communication
JP	ASV I to IV	1991-2007	Advanced Safety Vehicle Program
	AHS	1998-2008	Advanced Cruise-Assist Highway Systems (AHS) www.ahsra.or.jp

necessities and possibilities of enhancement and optimization tasks required for achieving an appropriate ITS functionality and sufficient performance. In that respect, those projects have been dealing with the last step of checking the developed VCN technology before launching the markets. OPAL-VCN (2010-2014) and simTD (2008-2012) in Germany and VII (2004-2009) in the United States represent the most important second generation VCN-related projects. Moreover, simTD is considered as a predecessor of the European Field Operational Tests (FOTs) in Europe [22].

8.2 Emerging DSRC/WAVE-Technology for VCN

Dedicated Short Range Communications (DSRC) describes a general purpose wireless link between the vehicle and the roadside, or between vehicles. The standards developed to support DSRC provide a short- to medium-range communications service for a variety of ITS applications. However, there are two sets of standards supporting DSRC and operating in different frequency bands: 915 MHz and 5.9 GHz. The standards developed for 5.9 GHz DSRC are designed to support a larger variety of ITS services, in the area of active safety and traffic efficiency, than that of 915 MHz DSRC. Moreover, the 5.9 GHz DSRC standards provide higher information capacity than DSRC in 915 MHz band, and have also a longer range. Therefore, the 5.9 GHz DSRC represents the leading technology for adopting ITS. This chapter focuses mainly on the 5.9 GHz DSRC and related technologies. In the rest of this chapter the term DSRC refers to 5.9 GHz DSRC, unless mention to the contrary.

The increasing demand on broadband wireless services for ITS, besides the wide employment of IEEE 802.11, has led to the adoption of the DSRC/WAVE DSRC in accordance with IEEE 802.11p and IEEE 1609 Wireless Access in Vehicular Environments (WAVE) standards [12]. For purpose of message exchange in ITS applications, the Society of Automotive Engineers (SAE J2735 – DSRC Message Set Dictionary) provides the ITS-related data dictionary and message sets to be used in DSRC. In general, several ITS standards for DSRC are already published or well under development [10], among others:

- ASTM E2213-03, which describes the specifications of MAC and PHY layer for wireless connectivity in vehicular environments using DSRC services.

- IEEE 1609 standards family for WAVE, which defines the network architecture, communications model, management structure, security mechanisms, and physical access in vehicular environments.

- IEEE P802.11p, which represents an extension of IEEE 802.11 technology specified for vehicular environments. P902.11p describes the required modifications of 802.11 and 802.11a for DSRC implementations.

8.2.1 DSRC/WAVE Background

The DSRC/WAVE is designed specifically for VCNs and related ITS. DSRC/WAVE specifies using IEEE 802.11p for physical and MAC layers, while using IEEE 1609 for the upper layers. Accordingly, IEEE 1609 family of WAVE comprises specifications for the network architecture, communications model, management and security structure and mechanisms, and network access for wireless communications in vehicular environments.

The IEEE P1609.0 draft standard describes the WAVE architecture and related services necessary for DSRC/WAVE devices to communicate in a high-dynamical vehicular environment. The IEEE 1609 family of standards consists of four trial-use standards which, have full-use drafts under development [32]:

1. IEEE 1609.1 (2006) for specifying the services and interfaces of resource manager applications;

2. IEEE 1609.2 (2006) for defining security services for applications and management messages, including message format and processing;

3. IEEE 1609.3 (2007) for defining networking services, including network and transport layer services;

4. IEEE 1609.4 (2006) for providing multichannel operations as enhancements for 802.11 for supporting WAVE.

Moreover, new IEEE 1609 members namely P1609.5, 1609.6, and P1609.11 are still in process; P1609.5 (Communication Manager) enhances the communication management services and is expected to be published in 2012 [19]; P1609.6 (Facilities) provides a combination of the layers nominally referred to as Session, Presentation, and Application [30]; P1609.11 (Over-the-Air Data Exchange Protocol for ITS) defines the services and secure message formats necessary to support ITS interoperability with secure electronic payments and is expected to be published in 2013 [19]. The ITS Standards Fact Sheets [32] reported that IEEE P1609.0 is still under development. IEEE 1609.1, 1609.2, 1609.3, and 1609.4 standards are trial-use published, while draft standards are still under development.

IEEE 1609 standards are based on IEEE P802.11p which represents an extension to IEEE 802.11 and IEEE 802.11a for vehicles traveling at high speeds. 802.11p introduces modifications to parameters such as adjacent/non adjacent channel rejection and receiver minimum input sensitivity. Furthermore, IEEE 802.11p supports two different network stacks: IPv6 for service channels and WAVE Short Message Protocol (WSMP) on any channel. While WSMP is mostly used for safety applications and services, IPv6 stack is planned for data services. Figure 8.2 shows the protocol suite DSRC/WAVE. The ITS Standards Fact Sheets [32] reported that IEEE P802.11p is an active unapproved draft.

Different frequency bands are used for DSRC in different countries. While the FCC has allocated 75 MHz of radio spectrum at 5.9 GHz (5.850–5.925 GHz) band in the US [12], the ECC has assigned 50 MHz (5875–5925 MHz) to DSRC in Europe [5]. Besides that, the ECC has recommended additional 20 MHz (5855–5875 MHz) for DSRC in Europe [6]. In Japan 80 MHz (5770–5850 MHz) were allocated to DSRC [5], whereas other countries worldwide have been considering the 5.9 GHz band for DSRC [5]. Table 8.2 summarizes the properties of channel assignment in regard with ECC and US FCC 90.377.

DSRC/WAVE defines two types of channels: Service Channel (SCH) and

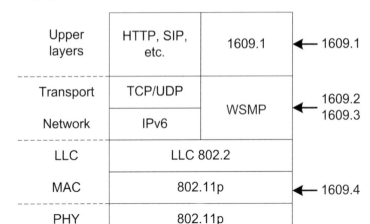

WSMP: WAVE Short Message Protocol
LLC: Logical Link Control
MAC: Medium Access Control
PHY: Physical

FIGURE 8.2
DSRC/WAVE protocol suite [8].

Control Channel (CCH). Although DSRC mainly uses 10 MHz channels (7 channels: even channel numbers between 172–184), it also specifies using 20 MHz channel (2 channels: 175 and 181). In that respect, in relation to 802.11a and due to the channel bandwidth, DSRC/WAVE provides transmission bitrates of 3, 4.5, 6, 9, 12, 18, 24, and 27 Mbps for 10 MHz-channels and up to 54 MHz for 20 MHz channels. Similar to the conventional 802.11 technology, the bit rate selected for data transmission depends on several parameters, especially the link quality for instance between OBUs (Onboard Units) in V2V communications and between OBU and RSU in V2R communication. However, since the traffic safety applications require a high reliability, a low rate (e.g., 6 Mbps) will most likely be chosen for such cases [4].

In general, DSRC enables communications over line-of-sight (LOS) with distances of up to 1000 meters between RSUs (RSU class D and max. output power of 28.8 dBm) and mostly high speed (up to 200 Km/h), occasionally also for stopped and slow moving OBUs [12].

8.2.2 VCN Architecture

The DSRC defines two main network components in VCN: On-board Unit (OBU) and Roadside Unit (RSU). The OBU is installed in the vehicle and connected to the vehicle's board computer and sensors; it acts as a networking

TABLE 8.2 DSRC channel assignment in US and EU [5].

	Channel No.	Frequency range (MHz)	Channel use	Bandwidth (MHz)	Bitrate (Mbps)
US	170	5850–5855	Reserved	5	
	172	5855–5865	SCH	10	3–27
US and EU (recommended)	174	5865–5875	SCH	10	3–27
	175	5865–5885	SCH	20	6–54
	176	5875–5885	SCH	10	3–27
	178	5885–5895	CCH	10	3–27
US and EU (assigned)	180	5895–5905	SCH	10	3–27
	181	5895–5915	SCH	20	6–54
	182	5905–5915	SCH	10	3–27
	184	5915–5925	SCH	10	3–27

platform for the vehicle. The RSU is established on the roadside as part of the access network and acts as relay, on the one hand, and as a gateway to the roadside backbone (i.e., to the Internet), on the other hand. Accordingly, two main communication forms are distinguished in a vehicular communications system [15]: Vehicle-to-Vehicle (V2V) communication for direct communication between OBUs, and Vehicle-to-Roadside (V2R) communication for communication between OBUs and RSUs. However, a vehicular communications system specifies two levels of communications network in its infrastructure [39, 15, 33], as revealed in Figure 8.3:

1. Vehicular Ad-hoc Network (VANET), which provides multihop networking, based on V2V and V2R communications in access network [15].

2. Roadside network, which consists of twofold: (a) Roadside Access Network (RAN), which comprises the RSUs and offer interfaces to enable V2R-communications to the backbone; (b) Roadside Backbone Network (RBN), which represents the backbone network of RSUs, and in which RSUs can communicate with each other and with Internet [20].

Since the functionality of RAN forms a part of VANET, only RBN is considered in this chapter under the roadside network. While a VANET has a high dynamic nature, the roadside network has a static nature. However, providing a stable performance of roadside network for rapidly moving vehicles in VANETs represents an essential challenge for the development of VCNs.

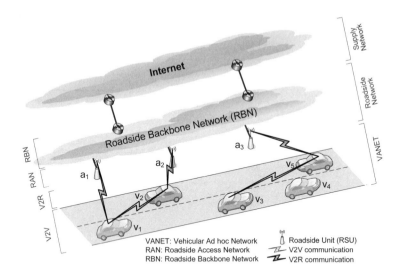

FIGURE 8.3
Vehicular Communication Networks (VCN) architecture.

8.3   Emerging Applications for VCNs

The DSRC/WAVE-based VCNs provide a basic communication platform for a wide range of applications and services in the field of ITS and Internet-related services. Although VCNs are developed as a platform for operating ITS services mainly, the large scale of VCNs' deployment as well as the efficient employment of VCNs' resources are the main factors that lead to operate conventional services such as entertainment applications over VCNs. In that respect, Schoch *et al.* [31] has classified applications and services, which could be operated over VCNs, in four groups: active safety, public service, improved driving, and business and entertainment. However, two main categories of services can mainly be addressed for VCNs as a communication platform:

1. ITS services, which include safety and traffic efficiency related applications

2. Value-added services (VASs), which include all other types of applications, especially Internet-related applications, such as business and entertainment

Most of the ITS services, especially that of active safety, is based on the communication concept signaler (warner)/responder, for example, in case of crash warning. In contrast, value-added services follow different communication concepts, which are mostly based on client/server principle, for example, in case of web-based services.

8.3.1   Intelligent Transportation Systems Services

There are different classification schemes of ITS-services, for example, the VSC project (US) addressed the intelligent safety applications enabled by DSRC, where five categories are classified: intersection collision avoidance, public safety, sign extension, vehicle diagnostics, and maintenance, and information from other vehicles [7]. In general, the ITS services can be classified into three main categories of applications:

1. Active safety for critical situations, such as dangerous road features, abnormal traffic and road conditions, danger of collision (crash), imminent crashing, and occurred incident (e.g., low bridge and curve speed warning, road condition warning, lane change warning, pre-crash sensing, and breakdown warning, respectively [31]).

2. Public services for the purpose of emergency response and support for authorities (e.g., approaching emergency vehicle warning and electronic drivers license, respectively). Furthermore, applications for improving driving on the road are also included for the purpose of enhanced driving such as left turn assistant and in-vehicle signage [31].

3. Traffic efficiency for increasing the performance of road systems. Applications such as enhanced route guidance and navigation, intelligent traffic flow control, and parking spot locater service help reducing the traffic load in cities and on highways.

Most of the ITS services, particularly active safety and public services, rely on the multihop and message-oriented communication concept "signaler/responder." In this concept, the sender (signaler) and the receiver (responder) exchange information between each other over messages, whose content, transmission period and life time and distance is dependent on the warning type – Table 8.3 provides a list of some applications and related message-properties. The signaler represents the vehicle who signals the neighbors (vehicles or roadside units) his critical situation, such as in case of danger of collision, by sending a suitable message through V2V or V2R communication or a combination thereof. The responder is a neighboring vehicle – or roadside unit – that has to react appropriately on the signal received. The reaction of the responder can be, among others: (1) forwarding the received message to other neighbors, to a certain service-related server, or to both; (2) informing/warning the driver; (3) triggering a physical reaction of responder-related vehicle such as braking or changing the lane. For instance, in case of "danger of collision" when a vehicle changes its lane on the highway, it signals this action

TABLE 8.3 Example of messages for some ITS-related applications [1, 7].

Application	Comm. type	Required data into message	Mode	Frequency (Hz)	Life time (sec)	Life distance (m)
Curve speed warning	R2V	Curve location, curve speed limits, curvature, bank and road surface condition	Periodic	1	1	100
Lane change warning	V2V	Position, velocity, heading, acceleration and turn signal status	Periodic	10	100	150
Approaching emergency vehicle warning	V2V	Emergency vehicle position, speed, lane information, and intended path/route	Event-driven	1	1	1000
Left turn assistant	V2V and R2V	Traffic signal status, timing, and direction; road shape and intersection information; etc.	Periodic	10	100	300

to other driving vehicles with broadcasting a "lane change message", which has a determined life-time and life-distance, and include information such as position, speed, and acceleration. Other vehicles (as responders) in around receive this message and each one reacts appropriately such as warning the driver to start braking or to keep eye on adjacent vehicles.

The message sets used for ITS applications are provided by the Society of Automotive Engineers (SAE J2735 – DSRC Message Set Dictionary), where several data elements were uniquely defined for V2V and/or V2R communication. In general, the SAE J2735 standard defines message sets, data frames, and data elements used by ITS applications to exchange data over DSRC/WAVE, as well as other communication protocol [16]. J2735 contains several message categories: general, safety, geolocation, traveler information, and electronic payment, where all messages, data frames, and data elements are presented in ASN.1 and XML encoding format [16].

8.3.2   Value-Added Services

Value-added services include all other wireless services such as business and entertainment services for the purpose of vehicle maintenance, mobile services, enterprise solutions, and e-payment (e.g., just-in-time repair notification, VoIP telephony, fleet management, and gas payment, respectively [31]). In that respect, two main categories of value-added applications can be addressed:

1. Real-time applications such as voice- and video-related services, classified as soft real-time applications, can tolerate an extent degradation of the promised QoS [23]. For instance, the VoIP-applications can tolerate an E2E delay of up to 150 ms in case of using G.711 codec [14]

2. Non-real-time applications including other Internet and business-related applications such as Internet service provisioning and wireless diagnostics

8.4   QoS Models and Solutions in VCNs

The high demand on real-time capability of applications used in VCNs reflects the requirements on QoS provision in such networks. In that respect, the active safety applications represent the most time-critical applications under ITS services, where certain extents must not be exceeded such as the E2E-delay that must not exceed 50 ms [36]. This is very important for avoiding critical situations such as crash. For this reason, the active safe applications are considered to have hard QoS requirements. In contrast, the VoIP and multimedia applications, considered as the main real-time applications of value-added services, require certain level of QoS provisioning. For instance, VoIP applications require an E2E-delay of up to 150 ms in case of using G.711 codec [14] in order to guarantee an acceptable speech communication quality.

However, other QoS parameters such as bandwidth, jitter, and packet loss are also very important for the determination of QoS provision. While the QoS requirements for the most real-time applications in value-added services have been standardized such as in case of VoIP applications, the QoS requirements for the most real-time applications under ITS services still have to be standardized. Table 8.3 provides an attempt to investigate some boundaries of ITS services. Moreover, we addressed in a previous study [9] the QoS parameters for both hard and soft real-time applications in VCNs in relation with specific configurations. Accordingly we found that while active safety applications require 16 Kbps bandwidth, lower than 25 ms jitter and much less than 1% packet loss, VoIP applications require 64 Kbps bandwidth, lower than 50 ms jitter and less than 1% packet loss concerning codec G.711. This section addresses the issues and challenges for providing QoS in VCNs, and proposes a classification of most known solutions, where some of those solutions will be briefly discussed.

8.4.1 Challenges for QoS Support in VCNs

Beside the characteristics of road systems and related roadside networks, the characteristics of VANETs affect the QoS provision in VCNs. While the vehicles' velocity, movement patterns, and density determine the main characteristics of mobility in VANETs [31] and thus have a direct influence on QoS provisioning in VCNs, the different types of road systems and related topography, as well as the roadside networking infrastructure, represent the key basis for providing the required QoS levels for several applications, among others, in VANETs.

8.4.1.1 Challenges for QoS Support in VANETs

Although VANETs have similar issues to those of Mobile Ad-Hoc Networks (MANETs) such as error-prone-shared-radio channel, dynamically changing network topology, and imprecise state information [23], the high speed network nodes and the related rapidly varying network topology in VANETs cause new challenges for QoS provisioning in comparison to that of MANETs, especially in relation with channel resource utilization and QoS-oriented routing. The main issue for supporting QoS-based channel resource utilization in DSRC/WAVE is to control the medium access especially for SCHs, which are based on resource sharing of wireless medium [12]. Since safety applications are prioritized over non-safety applications like value-added services (e.g., Internet), the QoS-provision in such environments could be a real challenge for value-added services. On the other hand, the rapidly varying network topology represents the main challenge for adopting QoS-oriented routing methods. In general, the higher the speed differences among vehicles building a VANET are, the more difficult it is to provide a stable routing inside this VANET. Several challenges can be addressed in this context such as topology discovery and updates, route reestablishment/maintenance, and mobility-aware routing.

8.4.1.2   Challenges for QoS Support in Roadside Networks

The real challenge for designing a roadside network in relation with QoS requirements is to adapt the RBN performance to the rapidly varying network topology of related VANET, especially in respect to V2R-communications over RBN and to/from Internet. In accordance with value-added services, in which V2R communications to/from Internet are required, two main issues can be addressed: seamless RSU-change (handoff) and mobility-aware routing. The RBN must provide seamless RSU-change through low-latency data stream switching from previous RSU to the posterior one along the driving way of the vehicle, where several layer-2 and layer-3 mechanisms could be required in RAN.

The RBN must provide a mobility-aware routing capability through (1) adapting the routing inside RBN to follow the vehicles' mobility, and (2) predicting the future path of vehicles and thus forwarding data packets to reach related RSUs before those target vehicles [17]. Finally, beside the network architecture and related networking mechanisms and protocols there are two main types of RBNs to be considered in this field: wireless and wired networks, or a combination thereof. The decision about which type should be used for RBN determines the type of issues to be solved for QoS provision. For example, while the wireless RBN offers more flexibility and scalability for network design, the wired RBN offers higher bandwidth and more reliability [9].

8.4.2   Classifications of QoS Solutions in VCNs

There are several criteria to be considered in the classification of QoS-solutions in VCNs. However, for purpose of simplification, we deal in this chapter only with the ISO OSI layer-wise classification scheme. The layer-wise classification scheme is essential for understanding which characteristics of those QoS solutions are related to which layers of the network protocol. In this chapter we consider only the QoS solutions specified for VCNs. That is, we do not consider solutions developed for other networks and used for VCNs, such as using QoS solutions of MANETs for VANETs. Figure 8.4 shows layer-wise classification of considered solutions [9].

8.4.3   QoS Models and Solutions for VANETs

The unique characteristics of active safety applications determine the data traffic pattern required for providing the related functionality. In that re-

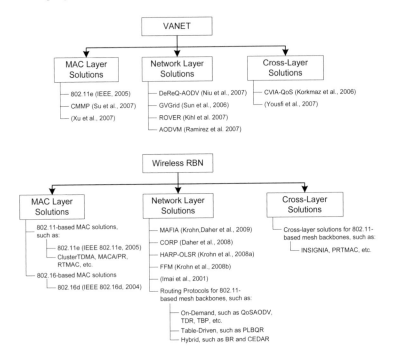

FIGURE 8.4
Layer wise classification of QoS solutions in VCNs [9].

spect, while usual value added services such as Internet employ the packet flow communication, active safety applications employ the message dissemination to transfer safety-related notification among OBUs and RSUs, where very low latencies are required in this case [36]. In general, the most important QoS solutions specified for VANETs are based on MAC layer and/or IP layer. While MAC-layer-based QoS solutions such as by Su and Zhang [34], and IEEE 802.11e focus mostly on active safety applications, the IP-layer-based solutions concentrates on value-added services due to their packet flow nature.

8.4.3.1   MAC Layer Solutions: 802.11e and 802.11p

IEEE 802.11e extends the functionality of 802.11 MAC-layer with QoS capability. IEEE 802.11e defines a new medium access method (Hybrid Coordination Function, HCF), in order to provide a prioritization mechanism within the 802.11 MAC layer as a basis for supporting differentiated QoS. While HCF defines two access methods: Enhanced Distributed Channel Access (EDCA) for contention based access and HCF Controlled Channel Access (HCCA) for contention free access, WAVE/802.11p uses only the EDCA access method. EDCA provides a priority-based, differentiated and distributed access to the wireless medium. For this reason, 802.11e defines eight different User Priorities (UPs) and four Access Categories (ACs). While UPs can be obtained from IEEE 802.1D capable devices in conjunction with the transferred frames, ACs must be mapped to UPs levels for each frame, as explained in details in IEEE 802.11-2007 documents. Accordingly, ACs determine the access priority to the wireless medium.

WAVE/802.11p enhances the EDCA through the integration of AC-queues on a per-channel basis in relation with channel coordination [38]. In that respect, CCH and each SCH channel has four priority queues that reflect the required ACs. While WSMP datagrams can only be served by CCH, SCHs serve both WSMP and IPv6 datagrams. To coordinate the functionality of those channels WAVE/802.11p defines a channel router that distribute ITS-related WSMP datagrams coming from the higher layer on suitable channels depending on the related priority and channel number specified in WSMP header. The channel router distribute the IPv6 datagrams on appropriate SCHs depending on the channel number specified in the sender profile registered by related IP Application, while the related priority is selected in accordance with the transferred UP. Consequently, beside the internal contention within each channel among frames of different ACs, an external contention between outputs of devices working on same channels is controlled in accordance with related 802.11 access method. Based on that, WAVE/802.11p guarantee that ITS-related frames can be sent before IPv6 frames over SCHs, while active safety-related WSMP frames can be sent before, for instance, traffic-efficiency related WSMP.

8.4.3.2 IP Layer Solutions

The addressed VANET-specific QoS solutions on IP layer proposed several routing concepts and protocols. While Ramirez and Fernandez [27] presented multipath routing functionality to provide QoS for VANETs, other studies enhance the routing concepts and protocol in VANETs with a position-based routing functionality, such as [24, 35]. Since the addressed QoS solutions for layer-3 describe only routing protocols, we discuss those solutions in the section "Routing Models and Solutions in VCNs."

8.4.4 QoS Models and Solutions for Roadside Networks

The used network architecture and related communication technologies in a roadside network determine the kind of required QoS solutions for this network. Therefore, we concentrate in this section only on QoS solutions developed specifically for RBNs.

8.4.4.1 QoS in Wired RBNs

The wired RBNs usually provide higher bandwidth and lower latency than that of wireless RBNs, and thus a higher QoS satisfaction. The most known

VCNs-related projects and studies provide marginal details about their used RBNs [9], such as in simTD. However, we suppose that most of those projects mainly integrate conventional wired communication technologies in their RBNs. That means that "conventional QoS solutions, such as those based on a combination of IEEE 802.1D, DiffServ, IntServ, MPLS, etc. are expected to be in use" [9].

The study of Okanishi *et al.* represents one of the few studies on using wired RBN for highways [25]. Okanishi *et al.* [25] proposed an IP-based infrastructure as an alternative to the currently used dedicated communication network system of highway administrators. However, although the authors mentioned that the QoS provision must be provided, they did not give any details about possible QoS solutions.

8.4.4.2 QoS in Wireless RBNs

The wireless RBNs, in spite of their lower QoS satisfactions and reliability in comparison with wired RBNs, provide higher architecture flexibility and scalability as well as lower decision-to-installation time and more reduced costs [9]. Although the advantages of wireless RBNs are addressed by early studies such as by Lamm *et al.* [21] and Bengsch *et al.* [2]; nevertheless, the QoS provision was not seriously considered.

The addressed QoS solutions developed specifically for wireless RBNs focus on layer-3, where QoS-oriented routing protocols and pre-fetching/preloading mechanisms represent the main topics. Moreover, due to the use of additional small range technology such as 802.11 WLAN as RAN for value-added services, additional topic could also be addressed, namely, seamless handoff mechanisms. While the addressed layer-3 QoS solutions concerning routing protocols are covered in the section "Routing Models and Solutions in VCNs"; only the prefetching/preloading solutions will briefly be discussed in this section. The main idea behind the Packets Pre-Fetching Mechanisms (PPFMs) is to minimize the effects of frequent handoffs on the QoS provision through dynamically adaption of routing mechanism to forward data packets to RSUs, in which destination vehicles are expected to be at certain time points. For this purpose, the cooperation between a location service and a routing/forwarding mechanism is very necessary in order to enable a kind of mobility-aware routing, as discussed in [17]. Krohn *et al.* [17] proposed a novel mechanism, called Mobility Aware Forwarding in Advance (MAFIA), for improving the packet delivery at the RSU over a wireless RBN through tracking and localizing the moving vehicles at the level of access networks [17].

The MAFIA is proposed in conjunction with a mesh wireless RBN, but MAFIA as a concept is principally independent of the used RBN technology. In addition, two types of MAFIA are defined: best efforts MAFIA (BE-MAFIA) and QoS-oriented MAFIA (QoS-MAFIA). In general, MAFIA proposes an essential mechanism for increasing the RBN performance, as well as the QoS provision experienced by clients. The main difference of MAFIA to other preloading mechanisms is concentrated on the packet-level-dependent forwarding mechanism, where the preloading server does only observe the traffic and accordingly decide when which packet must be forwarded to which network access point. In that respect, location service architecture is specified for MAFIA mechanism; however, the MAFIA concept is independent of the location service and can also be used in other network environments. Although the MAFIA mechanism was proposed in a special VCN-based scenario, a first implementation of BE-MAFIA was expected in 2010.

8.5 Routing Models and Solutions in VCNs

Routing is a rapidly developing field of research in VCNs. Due to the differences in data traffic characteristics between ITS and value-added services, two main research fields can be distinguished in relation with routing principles: message dissemination and routing. The former considers message forwarding (broadcasting) such as in case of warning message (active safety applications in ITS). The latter considers routing between certain nodes such as in case of flow-based data traffic (infotainment and some kinds of ITS services) and represents the main focus in this section.

Various techniques were developed or are still in development for routing in VCNs. However, due to the rapidly changing network topology and related channel characteristics in VANETs, establishing and maintaining routes over V2V and/or V2R communications forms a huge challenge for routing protocols, where the higher is the variety of network topology in a vehicular environment, the higher it is the cost of establishing and maintaining of a stable route. Accordingly, several routing protocols and concepts have already been developed specifically for VANETs' environments. Also, several already existing MANET routing protocols have been additionally enhanced with new routing concepts or algorithms in order to suite the requirements of VANETs. On the contrary, only very few routing solutions for RBNs could be found in the literature, actually; only two routing protocols specifically developed for RBNs could be addressed. This lies on the currently existing gap in the research and development of RBNs, since known VCNs' solutions suppose that a backbone infrastructure of the roadside network is already well established and provide the required performance for vehicular communications [9].

The Internet community "Routing-Lexicon.org"[1] we established at the beginning of 2009 focuses on the collection, classification, and evaluation of routing solutions of different technologies and in different fields. Accordingly, we have developed a novel scheme of Routing Protocols Classification Criteria (RPCC) as a basis for understanding, analyzing and evaluating of routing

---

[1]www.routing-lexicon.org

solutions. Thus, the results presented in this section represent a part of the working results of the Routing-Lexicon. To simplify working with routing protocols in the rest of this section, we list all used routing protocols and related references in Figure 8.5 which also shows all found routing solutions classified in accordance with VANET and RBN. In the rest of this section we mean by the terms "routing protocol and routing concept/algorithm" different things. The former indicates a routing protocol that is completely specified and developed for VCNs (e.g., CAR [37] and VADD [40] protocol). The latter refers to a routing algorithm or mechanism that is specified for VCNs but implemented into an already existed routing protocol (e.g. integrating DeReQ [24] algorithm into AODV protocol).

8.5.1   Challenges for Providing Routing in VCNs

Similar to issues and challenges proposed in Section 8.4, the characteristics of VCNs as well of vehicle's mobility form a real challenge for design and development of efficient routing protocols for both VANETs and RBNs. Moreover, different types of road systems (city and rural roads and highways) have different vehicles movement's patterns and thus different requirements on routing performance and reliability. In that respect, while the roads inside cities have straight streets and a relatively high density of traffic, the rural roads have a lower density of traffic and more curves. However, the traffic in both cities and rural roads is mostly unordered compared to highways, where vehicle movement is approximately one-dimensional. These different patterns pose special challenges, especially for routing [31]. Finally, the variation of vehicle density along the roads inside cities, on rural roads, or highways form a key challenge for resource management of VANET, as well as roadside networking. For instance, the required bandwidth in some segments of the highways (e.g., in case of a traffic congestion), could be very high in comparison to the rest of the highway. Thus, network overloading in such cases may not be avoidable, which degrades the provided QoS drastically for all participating RSUs and OBUs. On the contrary, with low traffic density and absence of roadside network such as in rural areas, the interruption of communication among vehicles is unavoidable, which represents a real challenge not only for routing, but also for several other networking mechanisms.

To benefit from experience in similar technologies, several studies such as Ferreiro-Lage *et al.* [13] and Ros *et al.* [29] have attempted to apply some routing schemes of MANETs, especially that of AODV and OLSR, on VANETs due to the similarity of both topologies and mobility environments. However, there are essential differences between both technologies, especially in accordance with the rapidly varying topology of VANETs, as well as the high demand on reliability and real-time capability of the message-based communications of safety applications. These differences make the use of most MANETs routing protocols into VANET's environments relatively inefficient for the most applications, particularly if certain modifications concerning VANETs' requirements are not carried out.

8.5.2 Classifications of Routing Solutions in VCNs

Due to the different network and operation characteristics of VANETs and roadside networks, different routing concepts and solutions have been developed for each of these networks. Therefore, as a first classification level of routing solutions, we consider the VCN-level-wise classification, as revealed in Figure 8.5(a), which simplifies the classification process of routing solutions. However, many other classification schemes may also be applied in relation to the several VCN-characteristics and related routing properties such as proactive/reactive, unicast/broadcast/geocast, city/rural area/highways, vehicle position, QoS, security, etc.

The most found routing solutions of VANET are routing protocols, whereas only a few routing algorithms/concepts could be addressed. Figure 8.5(b) shows the number of emerging routing protocols and concepts per year since 2000; we could not find in the literature any routing solution specified for VANETs at the time before 2002. However, as revealed in Figure 8.5(b), the most routing solutions for VANETs are published after 2004, especially in 2006, where VANETs began to gain a lot of attention at universities and in research centers due to the increasing support from industry and politic.

8.5.3 Routing Concepts and Protocols for VANETs

Due to the diversity of VANET routing protocols and related properties we select only two routing solutions to be proposed for VANETs. Both routing solutions are developed for supporting QoS in VANETs. We select those routing solutions as complementary for the section "IP Layer Solutions."

8.5.3.1 GVGrid: QoS Routing Protocol for VANET

Sun *et al.* [35] introduced a QoS routing protocol, called GVGrid, specified for VANETs. GVGrid represents a reactive and position-based routing protocol that initiates and manages routes between a source as fixed node and vehicles driving or stopping in a destination region. Additionally, the geographical area in GVGrid is partitioned into uniform-size squares called grid. The main goal of GVGrid is to employ the vehicles' mobility characteristics as a basis for analyzing their effects on the stability of network routes. In that respect, the stability of network routes initiated over V2V communications will be higher if related vehicles are driving in the same direction, and even much higher if those vehicles are driving at the same speed. Accordingly, the stable network routes provide higher reliability and thus can be employed for providing QoS-oriented routing.

For achieving position-based routing in GVGrid, participating vehicles

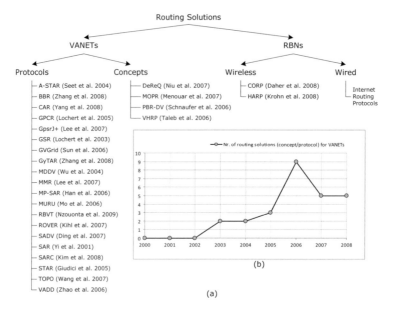

FIGURE 8.5
(a) VCN-level wise classification of routing solutions; (b) emerging routing solutions for VANETs (routing solutions of 2009/2010 are not included).

must be equipped with same ranged wireless devices, as well as a car navigator (GPS and digital map). The network nodes can communicate with nodes of neighboring grids, where information concerning, among others, vehicle's position and driving direction will be exchanged periodically. Simulation results showed that GVGrid achieves a better routing performance, concerning lifetime, and packet arrival ratio, in comparison with GPCR. Consequently, although GVGrid brings several advantages for the high quality communication data transfer in VANETs compared to existing methods, some critical issues still have to be dealt with such as the combination of geographic- and link-state based routing. Also, a performance comparison between GVGrid and another QoS-based routing protocol specified for VANETs would be necessary for better estimation of the GVGrid capability.

8.5.3.2 Delay and Reliability Constrained QoS Routing Algorithm

Niu *et al.* [24] proposed a routing algorithm, called link Delay and Reliability-constrained QoS (DeReQ), for multimedia communications in VANETs. The goal of developing DeReQ algorithm was to find a best route in relation with reliability and QoS requirements with focus on delays. The essence of this solution can be summarized in the following points: (1) considering the influence of link duration, traffic density, average speed, and traffic flow for obtaining a macroscopic knowledge scale about node mobility pattern; (2) considering two key QoS metrics satisfactions – link reliability and link delay; (3) reducing space complexity through setting up time-to-live factor in terms of hops number for each routing message. Niu *et al.* [24] did not develop any routing protocol. Instead, they integrated their DeReQ algorithm as a module into the routing discovery process of AODV (Ad-hoc On Demand Distance Vector) routing protocol. Although, Niu et al. presented DeReQ in combination with AODV, they assert that DeReQ could be integrated into other known MANET routing protocols. The simulation results showed a clear improvement of routing performance of DeReQ-AODV in comparison to the original AODV and its QoS extensions. However, AODV is a routing protocol developed for MANETs and not for VANETs; therefore, a results comparison with a VANET-specific routing protocol would be valuable. Consequently, DeReQ provides an important step towards adopting QoS requirements in routing protocols to be specified for VANETs. Nevertheless, the proposed study concentrated only on link delay; other QoS parameters such as packet loss, jitter, and bandwidth were not considered. Moreover, the integration of DeReQ in a VANET-specific routing protocol such as CAR for instance should be investigated.

8.5.4 Routing Concepts and Solutions for Roadside Networks

The routing in roadside networks is essential for services that require certain connection characteristics, especially between VANETs and the Internet, such as in case of value-added services. The routing solutions for roadside networks usually depend on the network design and use communication technologies. In other words, the network architecture and network technology used for roadside networks are very important factors in deciding which routing concepts and protocols should be adopted for which RBN. That means, routing solutions can mostly be considered only in accordance with the used RBN infrastructure. That is, the routing solution in a RBN can be efficiently analyzed if, and only if, a sufficient knowledge is available about the related RBN infrastructure. Therefore, we concentrate in this section only on routing solutions for RBNs that were specifically developed for VCNs. In the rest of this section, only RBNs will be considered, since the functionality of RAN forms a part of VANET.

8.5.4.1 Routing Solutions for Wired RBNs

The well-designed wired RBNs usually provide a reliable platform for routing protocols in comparison to wireless RBNs. Since it is supposed that the already existing VANETs use conventional wired RBN solutions, mostly known routing solutions for wired networks such as OSPF, RIP, etc., are envisaged to be used in such RBNs.

8.5.4.2 Routing Solutions for Wireless RBNs

Though some limitations exist with wireless RBNs, such as low reliability and bandwidth restrictions, the wireless infrastructure of such RBNs has essential features, such as high flexibility, low decision-to-installation time, and reduced costs, in comparison to wired RBNs [9, 38]. The attempts to develop routing solutions for wireless RBNs are concentrated on (1) compensating the influence of bandwidth differences between RAN and Internet, on the one side, and wireless RBN, on the other side, on the network performance in V2R-VANETs; (2) compensating the influence of the rapidly varying topology in VANETs on the routing in RBNs through providing an intermediate layer over RBN, which enables addressing clients (vehicles) in relation to their associated RSUs. Only two routing solutions could be found in the literature: Cluster-Oriented Routing Protocol (CORP) and Host Abstraction Routing Platform (HARP). Both solutions are developed for the same wireless network architecture that is specifically developed as RBN for VCNs [9]. In the following we introduce only the CORP protocol.

CORP provides a QoS-oriented routing protocol and is basically designed for hierarchical backbone infrastructures of roadside networks, which are developed within the project Wi-Roads [9] and used as a basis for the RBN of the following project of Wi-Roads, called "Open Air Lab for Vehicular Communication Networks (OPAL-VCN)." The project OPAL-VCN deals with the R&A activities of design, installation and operation of VCN infrastructure for a test field of at least 30 Km length on the highways A19/A20 around Rostock city in northeastern Germany; more information could be found on the project website (www.opal-vcn.org).

However, CORP is specifically developed for the hierarchical multilayer backbone infrastructure proposed by Krohn *et al.* [18], as shown in Figure 8.6. This backbone infrastructure represents a three-layer physical hierarchy, in which two main components are defined: cluster and domains, where the roadside network is constructed of basis of multiple domains. The related network topology defines three types of nodes: Mesh Point (MP), Cluster Head (CH), and Domain Head (DH). Accordingly, CORP distinguishes between two main types of routing: vertical routing and horizontal routing. The vertical routing consists of downwards routing, from DH to MP in case of downloads, and upwards routing from MP to DH in case of uploads. The horizontal routing is between MPs in case of inter-vehicle communications over roadside network. CORP specifies two types of routing mechanisms: relayed routing and QoS-oriented routing. The QoS-oriented CORP provides (1) an end-to-end resource reservation and admission control, and (2) a fairness-oriented balanced routing mechanism, which improves the resource utilization within RBN. Those properties of CORP help to reduce the effect of the bandwidth problem (bandwidth difference between RAN and Internet) on the whole performance of RBN.

Moreover, CORP employs a Vehicle-to-MP resolution mechanism, which

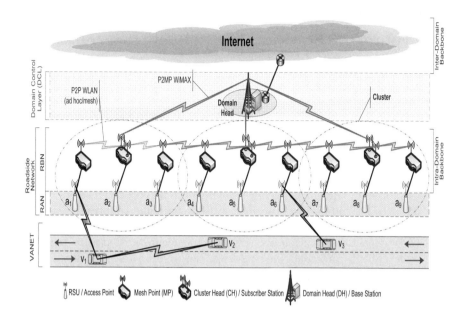

FIGURE 8.6

Multilayer backbone infrastructure used for CORP (Wi-Roads/OPAL-VCN project).

enables addressing vehicles through the MPs connected to their access portal (RSU). That is, each MP acts in this case as a gateway between CORP and VANET's routing protocol. In that respect, routing occurs between DH and MPs, in case of vertical routing, or among MPs in case of horizontal routing. Thus, converting the routing problem of rapidly varying network topology into a routing in a quasi stationary network environment, that is, the negative influence of rapidly varying network topology of VANET on routing in RBNs, should be reduced drastically. The routing concept of CORP and HARP is similar; however, CORP provides a complete routing protocol. The concept and development of CORP and HARP are still a work in progress, and we expect a first complete specification and implementation of both protocols in the second half of 2010 [9].

8.6 Future ITS Services: Driving Route Reservation Service

Modern road networks such as urban roads and highways provide an efficient transport platform within and between cities. However, other than transportation methods based on rails and airlines, the road networks offer more flexible mobility for people, not only through the deep integration of road networks into the people's living space, but also through the dynamic provided for driving (almost) anytime to anywhere. In contrast to railways and airlines, the road networks in their current status are not capable of providing a guarantee, in terms of traveling (driving) time, to the drivers or travelers [28]. In that respect, the public bus transport systems in some countries such as in Germany represent an exception in this case, where the departure and arrival time at each bus station are usually very accurate, that is, a relatively high punctuality is provided. Nevertheless, the timely varying traffic and the flexibility provided by driving anytime to anywhere make the applying of mechanisms, which are successfully experienced by bus transport systems, on traveling with other vehicular travel and transport systems such as private cars and trucks unfeasible. Ravil *et al.* provides one of the first studies about providing driving time reservation through a lane reservation for highways [28]; however, they have concentrated only on differentiated services and only for highways; other road systems were beyond the scope of [28] study.

However, inspired by QoS techniques of integrated services (IntServ) a similar concept to that of end-to-end resource reservation can be adopted in order to support a Start-to-Destination Driving Route Reservation (S2D-DRR) for all types of vehicles in different types of road networks. The main goal of such a mechanism is to provide a driving time reservation along the route from start to destination, where a time slot can be assigned to the related vehicle at each road segment and each junction, independent of the type of roads and junctions, at certain time points or within a certain time range along the path from start to destination. To achieve this vision, the road network' resources and the vehicles' behavior on the roads must be controlled interactively by an intelligent traffic management system (ITMC) that manages the utilization of roads' resources in accordance with the required driving route reservations and, vice versa, the ITMC can be considered as a centralized or distributed ITS service running somewhere into the related RBN. For this purpose, the road network resources (including, among others, driving-in/out times to/from certain roads, opening/closing times at junctions, speed limits for road segments, opening/closing driving lanes, and controlling driving times on lanes and road segments) must be controllable through that ITMC. Also the vehicles' behavior including driving time, driving route, and driving behavior (driving speed, stopping times, overtaking other vehicles, etc.) must be controllable either directly through the related ITMC that trigger the onboard driving control system, or indirectly through the driver himself who will necessarily be notified by the related ITMC in order to adapt his driving behavior accordingly. Consequently, efficient S2D-DRR concepts have to deal with issues and challenges of controlling (1) the resource utilization of road networks, and (2) vehicles' behavior.

The control of the resource utilization of a road system in accordance with the requirements of driving route reservations requires new methods for managing the utilization of, among others, driving lanes, speed limits of road segments, and junctions' switching times. Therefore, new approaches are necessary for adapting the functionality and performance of current road systems to requirements of S2D-DRR concepts. It is worth noting, that we do not intend to present new concepts or models for road systems, but to enhance the current road systems with new services, which should increase the provided service quality and service experience in available as well as future road systems.

The control of vehicles' behavior in relation with the resource utilization of a road system requires interactions between vehicles/drivers as well as related driving behavior and the ITMS used to control the resource utilization of the road system. For this purpose, new dimension of ITS services is required for operating S2D-DRR mechanisms. In other words, ITS services used for traffic efficiency and active safety must be coordinated cooperatively in relation with S2D-DRR related applications in order to enable an effective S2D-DRR service. Therefore, VCNs provides an essential platform not only for enabling the communication between vehicles and related ITMC, but also for operating others ITS services required for supporting and operating S2D-DRR mechanisms.

We have already dealt with those challenges and accordingly developed a novel concept for developing and operating S2D-DRR service in accordance with VCNs and for several types of road systems. However, this concept is beyond the scope of this chapter and will be introduced in another scientific publication.

8.7 Open Research Issues in VCNs

The most studies about VCNs focus on VANETs functionality and related ITS services. However, there is an obvious lack of studies about roadside backbone infrastructures in spite of their importance for providing high reliability in VCNs. Similarly, though the importance of QoS for safety applications, only few studies could be found about VCN-specific QoS solutions [9]. Thus, several topics such as message prioritization and forwarding, end-to-end QoS and QoS-oriented routing still provide an important field of research. Furthermore, we believe that the constructive interactions between VANETs and related RBNs are very necessary for improving the whole network performance from user/vehicle point of view. Therefore, issues such as interactions between routing protocols of VANETs and RBNs as well as inter-RSU communication over RBN present very hot topics for research. Also challenges and issues for designing and installing reliable, modular, and real-time capable roadside backbone infrastructures represent important research topics in the future [9].

On the other side, the technology of VCNs has been developing rapidly, where the first commercial products are expected to be on the market in a few years. However, it cannot be expected that all vehicles will be equipped with this technology a short time. The reasonable scenario for integrating VCN technology into the current road and traffic systems is to start with certain road types and segments as well as with new manufactured vehicles, and thus gradually developing a solid basis for future successful operation of ITS services. However, this creates new challenges for the reliability and functionality of the whole system of VCNs and related ITS. Therefore, issues such as distributed service architectures, effects' compensation of coverage gaps, and routing and handoff between VCN's islands represent important topics for future research. Also, the integration of future network technologies such as that of Delay Tolerant Networks (DTNs) forms a promising step toward pervasive vehicular environments.

8.8 Conclusion

This chapter introduced vehicular communication networks (VCNs) as well as Intelligent Transportation Systems (ITS) and related ITS services, including value-added services such as multimedia and data services. Accordingly, the emerging DSRC/WAVE as a leading technology for VCNs was presented and related vehicular applications including ITS and value-added applications have been introduced. After that, issues and challenges for provisioning QoS for real-time applications of ITS and multimedia services were discussed, where only few VCN-specific QoS solutions could be found in the literature. The VCN-specific QoS solutions are classified in relation to VCN level, and some of those solutions are discussed briefly. Similarly, issues and challenges for providing routing protocols in VCNs are discussed and the addressed routing solutions are classified according to VCN level. Afterwards, several routing solutions for VANETs and roadside network were proposed, and the future ITS service "Start-to-Destination Driving Route Reservation (S2D-DRR)" was introduced. Finally, we proposed a brief description about some open research issues in VCNs.

Bibliography

[1] M. S. Almalag. Safety-related vehicular applications. In S. Olariu and M. C. Weigle, editors, *Vehicular Networks: From Theory to Practice*. CRC Press, Boca Raton, FL, 2009.

[2] A. Bengsch, H. Kopp, A. Petry, Daher R., and D. Tavangarian. Efficiency of wireless local area networks in high speed moving clients. In *Innovative Internet Community Systems (I2CS)*, Guadalajara, Mexico, 2004.

[3] I. Berger. Standgerards for Car Talk. *IEEE – The Institute*, 31(1), 2007.

[4] R. Bossom, R. Brignolo, T. Ernst, K. Evensen, A. Fritscher, and W. Hifs. Communication for eSafety (COMeSafety). Overall Framework Proof of Concept Implementation, Version 2.0. European ITS Communication Architecture, 2008.

[5] Electronic Communications Committee. ECC Decision of 14 March 2008 on the harmonised use of the 5875-5925 MHz frequency band for Intelligent Transport Systems (ITS). ECC/DEC (08)01, 2008.

[6] Electronic Communications Committee. ECC Recommendation (08)01 Use of the Band 5855-5875 MHz for Intelligent Transport Systems (ITS). ECC/REC (08)01, 2008.

[7] Vehicle Safety Communications. Vehicle Safety Communications Project: Final Report Submitted to NHTSA and FHWA in response to Cooperative Agreement. Dtfh61-01-x-001, 2005.

[8] IEEE P1609.0 D0.2. IEEE Trial Use Standard for Wireless Access in Vehicular Environments (WAVE) - Architecture, 2007.

[9] R. Daher and D. Tavangarian. Qos in vehicular communication networks. In S. Adibi and R. Jain, editors, *Quality of Service Architectures for Wireless Networks: Performance Metrics and Management*. IGI Global, 2010.

[10] DSRC 5GHz: Dedicated Short Range Communication at 5.9 GHz Standards Group. http://www.iteris.com/itsarch/html/standard/dsrc5ghz.htm. Accessed on Dec. 1, 2009.

[11] ETSI TC Intelligent Transport Systems (ITS). http://www.etsi.org/WebSite/technologies/IntelligentTransportSystems.aspx. Accessed on Dec. 1, 2009.

[12] Federal Communications Commission (FCC). Amend Rules Regarding Dedicated Short Range Communications Services and rules for Mobile Service for Dedicated Short Range Communications of Intelligent Transportation Services. FCC 03-324 A1, 2004.

[13] J. A. Ferreiro-Lage, C. P. Gestoso, O. Rubios, and F. A. Agelet. Analysis of unicast routing protocols for VANETs. In *International Conference on Networking and Services*, pages 518–521, 2009.

[14] ITU-T Recommendation G.114. One-Way Transmission Time, 2003.

[15] H. Hartenstein and K. P. Laberteaux. A Tutorial Survey on Vehicular Ad Hoc Networks. *IEEE Communications Magazine*, 46(6):164–171, 2008.

[16] C. Hedges and F. Perry. Overview and Use of SAE J2735 Message Sets for Commercial Vehicles. In *SAE Technical Paper Series*. SAE International, 2008.

[17] M. Krohn, R. Daher, M. Arndt, A. Gladisch, and D. Tavangarian. Mobility-aware preloading mechanism for enhancing QoS in mesh wireless backbones. In *Proceedings of International Congress on Ultra Modern Telecommunications (ICUMT)*, St. Petersburg, Russia, 2009.

[18] M. Krohn, R. Daher, M. Arndt, and D. Tavangarian. Aspects of roadside backbone networks. In *Wireless VITAE 2009, Vehicular Technology*, Aalborg, Denmark, 2009.

[19] T. M. Kurihara. IEEE 1609 Working Group – Project Status Report. IEEE 802.11-09-0093-02-000p, 2009.

[20] K. Kutzner, J.-J. Tchouto, M. Bechler, L. Wolf, B. Bochow, and T. Luckenbach. Connecting vehicle scatternets by Internet-connected gateways. In *Proceedings of Workshop on Multiradio Multimedia Communications (MMC)*, Dortmund, Germany, Feb. 2003.

[21] R. Lamm and S. J. Schneider. Investigation into the Development of a Wireless Ethernet Backbone. Technical Report 10-9233, Southwest Research Institute, http://www.swri.org/3pubs/IRD2001/10-9233.htm, 2001.

[22] L. Le, A. Festag, R. Baldessari, and W. Zhang. Car-2-X Communication in Europe. In S. Olariu and M. C. Weigle, editors, *Vehicular Networks: From Theory to Practice*. CRC Press, Boca Raton, FL, 2009.

[23] S. R. C. Murthy and B. S. Manoj. *Ad Hoc Wireless Networks*. Prentice Hall, Upper Saddle River, NJ, 2004.

[24] Z. Niu, W. Yao, Q. Ni, and Y. Song. DeReQ: A QoS Routing Algorithm for Multimedia Communications in Vehicular Ad Hoc Networks. In *International Conference on Wireless Communications and Mobile Computing*, pages 393–398, New York, NY, USA, 2007.

[25] S. Okanishi, M. Kon, S. Chiku, M. Sugiyama, and H. Sakurai. Traffic network system. *NEC Technical Journal, Special Issue on ITS*, 3(1), 2008.

[26] IEEE P802.11p. Wireless Access in Vehicular Environments (WAVE), 2007.

[27] C. L. Ramirez and V. M. Fernandez. QoS in vehicular and intelligent transport networks using multipath routing. In *IEEE International Symposium on Industrial Electronics (ISIE)*, pages 2556–2561, 2007.

[28] N. Ravil, S. Smaldone, L. Iftode, and Gerla M. Lane Reservation for Highways (Position Paper). In *Proceedings of the 10th International IEEE Conference on Intelligent Transportation Systems (ITSC)*, 2007.

[29] F. J. Ros, P. M. Ruiz, J. A. Sanchez, and I. Stojmenovic. Mobile Ad hoc routing in the context of vehicular networks. In S. Olariu and M. C. Weigle, editors, *Vehicular Networks: From Theory to Practice*. CRC Press, Boca Raton, FL, 2009.

[30] R. Roy. IEEE 1609 Working Group, Revised Architecture Diagram. 2008.

[31] E. Schoch, F. Kargl, M. Weber, and T. Leinmller. Communication patterns in VANETs. *IEEE Communications Magazine*, 46(11):119–125, 2008.

[32] ITS Standards Fact Sheets. `http://www.standards.its.dot.gov/fact_sheet.asp?f=80`. Accessed in Dec. 2009.

[33] L. Stibor, Y. Zang, and H.-J. Reumerman. Neighborhood evaluation of vehicular Ad-hoc network using IEEE 802.11p. In *13th European Wireless Conference*, Paris, 2007.

[34] Z. Su and X. Zhang. Clustering-based multichannel MAC protocols for QoS provisionings over vehicular ad hoc networks. *IEEE Transactions on Vehicular Technology*, 6(56):3309–3323, Nov. 2007.

[35] W. Sun, H. Yamaguchi, K. Yukimasa, and S. Kusumoto. GVGrid: A QoS routing protocol for vehicular ad hoc networks. In *14th IEEE International Workshop on Quality of Service*, pages 130–139, New Haven, CT, 2006.

[36] Q. Xu, T. Mak, J. Ko, and R. Sengupta. Medium access control protocol design for vehicle-vehicle safety messages. *IEEE Transactions on Vehicular Technology*, 56(2):499–518, 2007.

[37] Q. Yang, A. Lim, and P. Agrawal. Connectivity aware routing in vehicular networks. In *IEEE Wireless Communications and Networking Conference*, pages 2218–2223, 2008.

[38] M. Zhang and R. S. Wolff. Routing protocols for vehicular ad hoc networks in rural areas. *IEE Communications Magazine*, 46(11):126–131, 2008.

[39] X. Zhang, H. Su, and H. Chen. Cluster-based multi-channel communications protocols in vehicle ad hoc networks. *IEEE Wireless Communications*, 13(5):44–51, 2006.

[40] J. Zhao and G. Cao. VADD: Vehicle-assisted data delivery in vehicular ad hoc networks. In *Proceedings of 25th IEEE INFOCOM*, pages 1–12, Barcelona, Spain, April 2006.

Part III

Resource Management and Cognitive Networks

9

Network Coding Approach to Improving TCP Throughput in Wireless Networks

Tebatso Nage

Department of Systems and Computer Engineering, Carleton University, Ottawa, Canada

Marc St-Hilaire

Department of Systems and Computer Engineering, Carleton University, Ottawa, Canada
Email: marc_st_hilaire@carleton.ca

F. Richard Yu

Department of Systems and Computer Engineering, Carleton University, Ottawa, Canada

CONTENTS

Compared to traditional routing protocols, network coding is bandwidth efficient and can achieve high throughput gains. By intelligently mixing (coding) packets together, fewer transmissions are required and bandwidth becomes available for new data. The goal of this chapter is to show that when XOR network coding is used in conjunction with an opportunistic scheduling and the Transmission Control Protocol (TCP) window state, higher throughput can be achieved. The necessary motivation and background are provided to enable the reader to acquire the essence of the problem. Finally, a cross-layer approach is proposed followed by simulation results.

9.1 Introduction

Network coding is a new transmission paradigm pioneered by Ahlswede *et al.* [1]. In recent years, it has generated huge research interest especially in wireless communications. The main attraction of this novel concept is that it is bandwidth efficient and achieves high throughput gains [6, 10, 12, 13, 18, 20]. On top of that, network coding can also allow peer-to-peer live multimedia streaming with finer granularity [17]. Network coding can be regarded as an extension of the traditional routing protocols in which relay nodes (R) simply receive, store and forward packets as shown in Figure 9.1. In this case, four transmissions are required in order to transmit two packets between A and B. The basic idea behind network coding, as exposed in Figure 9.2, is that packets are intelligently mixed (or coded) together at an intermediate node into one coded packet which is then broadcast. This process generates fewer transmissions in the network due to information rich transmissions. As we can see from Figure 9.2, only three transmissions are required to exchange the same two packets. As a result, more bandwidth becomes available for new data to be transmitted resulting in high network throughput [3].

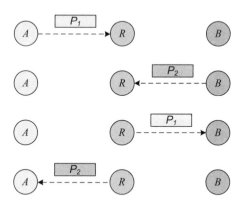

FIGURE 9.1
Traditional routing using store and forward.

It is also important to note that network coding is heavily dependent on network traffic pattern, medium access link scheduling and topology. For example, Figure 9.2 shows a scenario in which nodes A and B first send their messages. This link scheduling facilitates network coding. However, this link scheduling scenario is not always the case in real networks. Consider a link

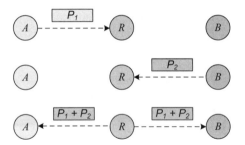

FIGURE 9.2
Information exchange using network coding.

scheduling scenario in which node A first sends its message followed by relay node and then node B. In this scenario, network coding is not possible because the relay node only has one message to forward. To facilitate network coding, some studies have considered what is referred to as coordinated network coding in which traffic flows are made to overlap so as to generate coding opportunities. However, this suggestion has been criticized for being unrealistic [14].

Although network coding was initially designed for wired networks which employ multicast transmission mode, it is currently one of the hottest research topics in wireless communications because of its capability to exploit the broadcast nature of wireless networks. To that end, this chapter presents a cross-layer design approach in order to improve TCP throughput in wireless ad-hoc networks. The rest of this chapter is organized as follows. The remainder of this section presents background information about different methods used to perform network coding and the benefits of cross-layer design. Section 9.2 introduces relevant parameters that are used in the proposed cross-layer design approach. Section 9.3 presents the framework of the proposed approach and the interactions between the different layers of the protocol stack. Finally, Section 9.4 presents simulation results followed by the conclusions in Section 9.5.

9.1.1 Different Flavors of Network Coding

Network coding comes in different forms. Some studies consider what is referred to as physical layer network coding [16, 29] while some others deal with random network coding [3, 6, 10, 11, 17, 27].

9.1.1.1 Physical Layer Network Coding

The broadcast nature of wireless networks is treated as an interference-inducing nuisance in most of today's wireless networks such as IEEE 802.11 [28]. As a result, Zhang *et al.* [28] proposed a Physical Layer Network

Coding (PNC) scheme that embraces interference to improve performance in a multihop network. Instead of performing coding arithmetic on digital bit streams (which is a requirement in straight forward network coding), it makes use of the additive nature of the simultaneously arriving electromagnetic waves by directly mapping the combined signals received to a signal to be relayed. In order to achieve this, two conditions must be met:

1. A relay node must be able to convert simultaneously received signals into sensible output signals to be relayed.

2. A destination node must be able to extract its intended signal from the relayed signals.

Xue and Sandhu [26] proposed a simple adaptive coding scheme that combines channel coding with network coding, which greatly simplifies the complexity in code design and decoding as compared to pure random network coding. Specifically, the scheme first designs codes independently for each channel, then combines them with simple network coding to achieve performance comparable to ideal cases. Part of the motivation for Reference [26] is that in XOR network coding, coded data packet first has to be decoded by intended receivers before being XOR-ed with old known packets. According to information theory [26], this limits information rates in the broadcast stage to the weakest link through which one of the intended receivers receives its version of the coded packet.

9.1.1.2 Random Network Coding

Sundararajan *et al.* [22] came up with a new network coding scheme which is inserted between the transport and the network layers. This scheme allows random linear combination of packets currently in TCP sender's congestion window. They claim that this scheme reacts to packet drops in a smooth manner, resulting in a novel and effective approach for congestion control over networks involving lossy links such as wireless links. Since the receiver receives a linear combination of the packets (and not the original packets of the message), the notion of ordered sequence of packets used by TCP is missing. Therefore, to make this scheme work, there is a need to change the TCP mechanism that acknowledges packets at the receiver.

Due to the resilience and stable transmission offered by random network coding, Jin and Li [10] have proposed adaptive random network coding in Worldwide Interoperability for Microwave Access (WiMAX). The objective is to optimize WiMAX network performance in which it has been established that hybrid automatic repeat request (HARQ) is underutilizing wireless bandwidth because it was designed for point-to-point channel. As an attempt to minimize delay created by waiting for full rank transfer matrix to be formed, Reference [17] has implemented a decoding process using the Gauss–Jordan elimination technique which allows decoding to be carried out while coded packets are progressively received. Although, Gauss–Jordan elimination usually leads to numerical instability [17], it does not affect network coding since the operation is done in the Galois field. It has been found that when the Galois field size is increased, the success probability of decoding is increased [3]. However, this introduces additional system design and computational complexity. In order to address this problem, a special case of random network coding using a Galois field of size 2, called XOR network coding, has been proposed. The latter has been identified as a promising technique for coding packets in future networks [12, 27]. In XOR network coding, packets are coded using bitwise XOR operation into one coded packet. At the receiver, decoding is then carried out by applying bitwise XOR operation again to the received coded packet using packets which were previously overheard from the network or those that were previously sent. In the rest of this chapter, XOR network coding is used because it is the simplest and cheapest method of network coding.

XOR Network Coding Example
As mentioned previously, the secret of network coding is rooted in the fact that the relay node does not only store and forward messages but also codes packets together and then broadcasts a coded version of messages. In the scenario illustrated in Figure 9.2, suppose packet P_1 has bit stream 10101110011010... and packet P_2 has bit stream 010001101010100... in their data portions. As shown in Figure 9.3, when the relay node receives packets P_1 and P_2, it combines them by performing the bitwise XOR operation to produce a new packet $P_3 = P_1 \oplus P_2$, where $\oplus$ stands for the bitwise XOR operation. Note that before the bitwise XOR operation is carried out at the data link layer, XOR headers and MAC headers for packets P_1 and P_2 are unwrapped, and the remainder is only the data portion which includes the Internet Protocol (IP) header. After the operation, P_3 has bit stream 111010011001110... in its data portion as shown in Figure 9.3.

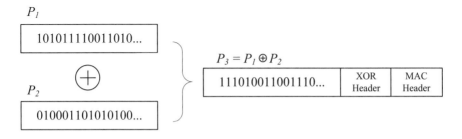

FIGURE 9.3
Coding packets P_1 and P_2 at the relay node.

Upon receiving the broadcast coded message P_3, node A uses a copy of packet P_1 it sent earlier to extract P_2 from P_3 by performing bitwise XOR operation as illustrated in Figure 9.4. Again, XOR header and MAC header are removed from P_3 before the operation. Node B, on the other hand, uses a copy of packet P_2 to extract packet P_1 from P_3 as depicted in Figure 9.5. Since only three transmissions are required instead of four as in traditional routing protocols, the remaining transmission could be used to send new messages, hence more efficient bandwidth utilization and high throughput by network coding.

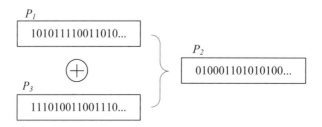

FIGURE 9.4
Extracting P_2 from coded packet P_3 at node A.

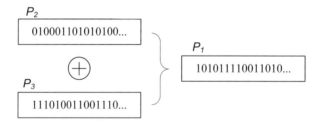

FIGURE 9.5
Extracting P_1 from coded packet P_3 at node B.

9.1.2 Cross-Layer Design: Definition and Benefits

The layered structure in the Open System Interface (OSI) model has been praised for its simplicity in maintaining, managing and optimizing protocols. However, it has been found that this approach introduces inefficiency, and redundancy, and that it can lead to poor performance, especially in applications with hard quality of service constraints. Therefore, a promising cross-layer design approach has been introduced. This approach allows information exchange across layers, which offers substantial gains in efficiency, throughput, and quality of service. It comes handy in wireless networks in which there is mobility, frequent handovers, and fluctuation of channel conditions. For example, suppose a particular node in a wireless mobile ad-hoc network has a TCP sending rate that is increasing gradually, and a low data rate is used yet the current wireless channel conditions are favorable so that the highest data rate could be used. If layered structure is employed, such a node may fail to take advantage of an opportunity to transmit at a high data rate due to lack of efficient coordination among protocol layers. However, when a cross-layer design approach is used, the data link layer could use the TCP and channel information received from the transport and physical layers, respectively, to determine the transmission data rate that can help speed up the TCP sending rate. Furthermore, if wireless channel conditions are so poor that even the lowest modulation scheme cannot be used, by employing channel information from physical layer, opportunistic scheduling scheme in the data link layer could temporarily halt transmission to allow conditions to improve.

9.2 Parameters Considered in the Cross-Layer Approach

When using a cross-layer design approach, several parameters from different layers of the stack can be considered simultaneously in order to improve the decisions made by the algorithm. In this chapter, we consider the channel information in order to perform opportunistic scheduling and to determine what data rate to use depending on the current TCP window state.

9.2.1 Channel Information for Opportunistic Scheduling

In reality, wireless channel conditions fluctuate. At times they are favorable but at other times they are unfavorable, a situation referred to as fading. When a node has channel information, it can take advantage of the stochastic fluctuation of channel conditions by transmitting at high data rates, where applicable, to maximize throughput. Also, a node could halt transmission when wireless links are poor to avoid creating interference to other nodes and to save power. In order to have access to channel information, nodes periodically send each other information of outgoing channel conditions using feedback channels. After receiving this information, a node then makes some statistical analysis from which it makes decisions about its outgoing channel conditions. Although this scheme is promising, the downside of it is that it introduces a stochastic halt of data transfer that can degrade network performance as some packets may be kept at a node for too long or they may be received out of order.

Lin and Vucetic [15] analyzed average capacity for a wireless network with joint opportunistic scheduling and wireless network coding. From their analysis, they found out that when more packets are coded together into a single packet, there is high probability of some intended nodes experiencing deep fading. Therefore, they proposed a scheme which uses power control and rate adaptation in network coding to optimize packet delivery of coded packets.

Sagduyu and Ephremides [19] considered the joint design of network coding and medium access control (MAC) in wireless ad-hoc networks. They also outlined an extension of network coding to operate with arbitrary MAC protocols.

Yomo and Popovski [27] addressed a scheduling problem in which they did not only consider information from neighboring nodes for network coding, but also considered instantaneous wireless link conditions. The idea was to find a set of packets whose destinations have good quality links and optimize data rate for the coded packet. Authors of [4, 21] realized that opportunistic coding scheme (COPE) misses several coding opportunities due to random link access scheduling at the data link layer. They proposed a coordinated network coding scheme to fully utilize coding opportunities missed by COPE by making sure that neighboring nodes transmissions are scheduled such that network coding gain is maximized [14]. Regrettably, according to Koutsonikolas *et al.* [14], these works either provide only theoretical results, making unrealistic assumptions of a slotted MAC layer or only consider a super special traffic pattern, consisting of pairs of perfectly overlapping flows going towards opposite directions.

9.2.2 TCP Window State

Transmission Control Protocol (TCP) is a connection-oriented, reliable transport layer protocol. It is the dominant transport layer protocol in today's Internet. Many Internet applications such as email, file transfer and web browsers use TCP as their transport layer protocol [2, 8, 24]. TCP was originally designed for wired networks in which bit error rate (BER) is lower than 10^{-8} [2, 24]. This protocol was designed based on the assumption that packet losses due to damage in transit are rare, hence most of packets get lost due to network congestion [7, 24]. Even though this assumption is valid for wired networks, it may not be valid for wireless networks in which it is difficult to provide channel quality of bit error rate of 10^{-6} [24]. The reason being that in wireless networks, most packet losses are due to high bit error rate as opposed to network congestion. Therefore, TCP tends to misinterpret packet losses due to bit errors in wireless networks as congestion thereby slowing down its transmission unnecessarily resulting in performance deterioration.

9.3 TCP-Aware Network Coding with Opportunistic Scheduling

In this section, we introduce a framework for TCP-aware network coding with opportunistic scheduling to increase TCP throughput in wireless mobile ad-hoc networks. The proposed scheme is implemented at the data link layer of

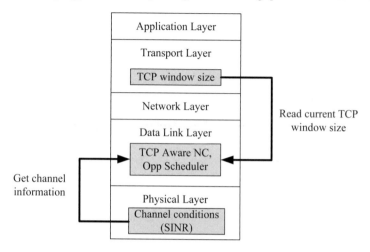

FIGURE 9.6
Interactions between the layers of the proposed scheme.

the protocol stack as shown in Figure 9.6. For full functionality of the system, information about the current TCP window size and channel state has to be known.

9.3.1   Channel Model

The propagation model used is the Two Ray Ground model in a 1Km×1Km area. Packet error rate is set to 0.01 and the number of MAC retransmissions is set to 6. For simplicity and simulation purposes, TCP packet size is set to 100 bytes. The transmit power is 1 watt. Perfect channel estimation is assumed. It is also assumed that the sender has already received channel information from the receivers. Using Equations (9.1), (9.2), and (9.3), we determine the required Signal to Interference and Noise Ratio (SINR) for the different modulation schemes shown in Table 9.1.

$$P_e = 1 - (1 - F_e^{N_{retrans}+1})^{N_{fr}} \tag{9.1}$$

$$F_e = 1 - (1 - BER)^{L_{fr}} \tag{9.2}$$

$$BER = K_1 \exp\left(\frac{-K_2 \gamma P_{tr}}{2^\rho - 1}\right) \tag{9.3}$$

where K_1 and K_2 are constellation and code-specific constants, respectively, $N_{retrans}$ is the number of retransmissions, N_{fr} is the number of frames, L_{fr} is the length of the frame, and BER is bit error rate.

TABLE 9.1

Modulation schemes and required SINR

Modulation scheme	Required SINR (dB)
BPSK	6.549
QPSK	8.471
QAM16	10.989
QAM64	16.130

9.3.2 TCP Model

TCP Reno, which is the most widely deployed version of TCP protocol [5], is used in all simulations. TCP Reno has three phases: slow-start, congestion-avoidance, and fast recovery. When a TCP session is established, TCP is said to be in slow-start phase. In this state, the congestion window size is increased by doubling its size for every acknowledgement (ACK) received. When the slow start threshold is reached, which indicates transition from slow-start to congestion avoidance state, the window size starts increasing linearly. If three duplicate packets[1] are received by the sender due to corrupted, lost, or out-of-order packet delivery at the receiver, fast retransmission of the lost packet is performed without waiting for retransmission timeout (RTO) [2]. This moves TCP Reno into fast recovery state in which the new RTO equals double the previous RTO, and the transmission rate is reduced by half. Reno remains in this state until the entire transmit window has been acknowledged, after which it moves into congestion-avoidance state. In case of timeout, Reno will move into slow-start mode. Even though this strategy is better than the one used in earlier TCP version, it is inefficient in that it only recovers at most 1 lost packet [25]. Therefore, its performance deteriorates a lot in wireless networks where channel errors are frequent [2, 7, 24].

9.3.3 Proposed Scheme

In this scheme, TCP dynamics and link conditions are both considered to determine which packets to code and which data rate to use. This scheme facilitates quick and efficient response of the TCP sender upon realizing any changes in unpredictable link conditions. Specifically, at the sender, let N be the largest TCP window size such that $1 \leq \eta \leq \alpha \leq \beta \leq N$, where η, α, and β are some parameters indicating the state of the TCP congestion window as shown in Figure 9.7.

If $n \in N$ is the current state of the congestion window such that $\beta < n \leq N$, it means that TCP sending rate is very high. Therefore, the scheduler will allocate the highest channel rate (QAM64) irrespective of the link quality [7]. This is done because if the channel quality is low, and low channel rate is

---

[1]It is important to mention that these duplicate packets are actually ACK packets from the receiver.

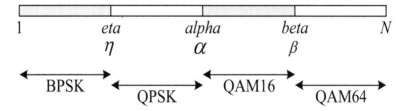

FIGURE 9.7
TCP congestion window size states and associated modulation scheme.

allocated when the TCP sending rate is high, this will inevitably results in packet loss due to congestion. The advantage of transmitting at high data rate in this situation is that the medium can become available quickly to other nodes with good links. This facilitates efficient resource sharing. However, transmitting at high modulation scheme in a channel of low quality can lead to high bit error rate. This is accounted for by employing adaptive coding to ensure that a particular bit error rate is obtained.

Furthermore, in case packets get lost due to low quality channel to the intended receiver, other nodes with good quality links can overhear such packets and could forward them by network coding to the intended receiver. When $\alpha < n \le \beta$, the least channel rate that can be allocated is QAM16. In case the link quality is good such that it allows QAM64 channel rate, the scheduler allocates QAM64 instead of QAM16. This is done to ensure that TCP reacts quickly enough to take advantage of the available bandwidth [9]. For $\eta < n \le \alpha$, QPSK is the least channel rate that can be allocated. Again, if the link quality allows transmission at a higher modulation scheme, the scheduler will allocate the corresponding modulation scheme. The same is also the case for $1 \le n \le \eta$. Figure 9.8 gives a summary of how a self-generated packet is processed in TCP-aware network coding with opportunistic scheduling. Note that a delay of 1 ms is induced in case the wireless link is poor to allow channel conditions to improve, after which transmission resumes.

At the intermediate nodes, information on TCP sending rate annotated on XOR header of a packet is used to determine data rate to be used for either coded packet or non coded packet. In case there is coding opportunity, the intermediate node uses the strategy discussed previously to determine a set of packets to code. Note that the final set of packets to be coded may have packets that require different data rates for transmission. This scheme chooses the lowest data rate required by one of the native packets in the set. Even though this decision does not favor packets that require high data rates, it ensures that both TCP dynamics and link conditions requirements are met to enhance performance.

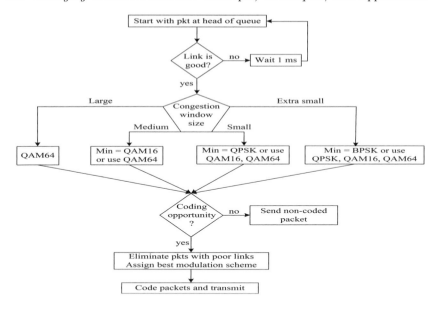

FIGURE 9.8

TCP-aware network coding with opportunistic scheduling.

9.4   Simulation Results

Network Simulator 2.33 was used to perform all simulations. The length of each simulation was set to 30 seconds since there was no significant change in the simulation results after 30 seconds. Note that the transmission range was determined by NS-2.33 and that the network was stable. We implemented our scheme in IEEE 802.11 and WirelessPhy extended versions developed by Mercedes-Benz Research and Development North America and University of Karlsruhe for NS-2.33 [23]. For simulation purposes, η, α and β were set to 20, 80 and 140. A total of 300 simulations were carried out for each case and an average was computed. Two simulation scenarios were defined: network traffic increase and increasing network nodes. For network traffic increase scenario, there were 19 nodes in the network, were placed at fixed predefined locations. In the second simulation scenario, nodes were added in the network by being placed at fixed predefined locations. Five TCP sessions were generated.

Figure 9.9 illustrates that network traffic increase is directly proportional to network throughput. However, it is expected that when the network traffic reaches saturation point, TCP throughput will stabilize or may even diminish as intermediate nodes become overwhelmed by high network traffic to be routed through the network. It can be seen from Figure 9.9, when there is low network traffic, there is insignificant performance improvement by any scheme. This is attributed to unavailability of coding opportunities. Also, with low network traffic, the wireless medium is readily available thereby presenting opportunities to all schemes to perform MAC retransmissions without affecting TCP performance. However, if network traffic is high, performing retransmission introduces delay because of time spent waiting to access the medium. Despite this, TCP-aware network coding with opportunistic scheduling achieves the highest performance as compared to other schemes. This indicates that when the traditional network coding is upgraded to TCP-aware network coding and combined with opportunistic scheduling in wireless mobile ad-hoc networks, performance can significantly improve.

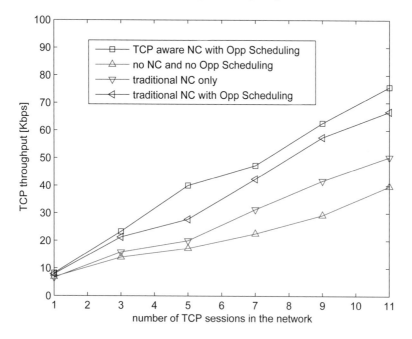

FIGURE 9.9
Impact of traffic increase on TCP throughput.

Figure 9.10 exhibits no performance improvement when nodes are added in the network. Although additional nodes in the network create options for routing packets, there are more routing messages that are being generated, resulting in few chances of a node being granted channel access per unit time.

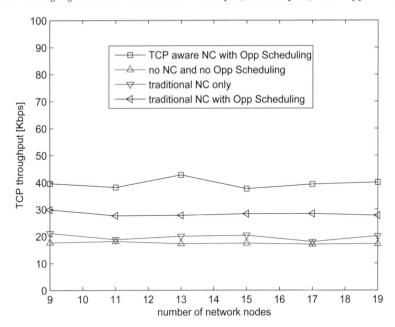

FIGURE 9.10
Impact of node increase on TCP throughput.

9.5 Conclusion

In this chapter, an efficient TCP-aware network coding with opportunistic scheduling, which is capable of enhancing TCP performance in wireless mobile ad-hoc networks, has been presented. Simulation results show that when traditional network coding is upgraded to TCP-aware network coding and combined with opportunistic scheduling in wireless mobile ad-hoc network, there is significant TCP performance improvement as compared to other schemes when network traffic increases and when the number of network nodes increases.

Bibliography

[1] R. Ahlswede, N. Cai, S.-Y.R. Li, and R.W. Yeung. Network information flow. *IEEE Trans. Inform. Theory*, 46(4):1204–1216, Jul. 2000.

[2] H. Balakrishnan, V.N. Padmanabhan, S. Seshan, and R.H. Katz. A

comparison of mechanisms for improving TCP performance over wireless links. *IEEE/ACM Trans. Netw.*, 5(6):756–769, Dec 1997.

[3] A.T. Campo and A. Grant. Robustness of random network coding to interfering sources. In *Proc. 7th Australian Communications Theory Workshop*, pages 120–124, Feb. 2006.

[4] P. Chaporkar and A. Proutiere. Adaptive network coding and scheduling for maximizing throughput in wireless networks. In *Proc. of ACM MOBICOM'07*, 2007.

[5] K. Daoud and B. Sayadi. HAD: A novel function for TCP seamless mobility in heterogeneous access networks. In *Proc. IEEE VTC'07 (Fall)*, pages 1451–1455, Sept. 30–Oct. 3 2007.

[6] E. Fasolo, M. Rossi, J. Widmer, and M. Zorzi. On MAC scheduling and packet combination strategies for practical random network coding. In *Proc. IEEE ICC'07*, pages 3582–3589, June 2007.

[7] M. Ghaderi, A. Sridharan, H. Zang, D. Towsley, and R. Cruz. TCP-aware channel allocation in CDMA networks. *IEEE/ACM Trans. Netw.*, 8(1):14–28, Jan. 2009.

[8] T. Hasegawa and M. Lagreze. A mechanism for TCP performance enhancement over asymmetrical environment. In *Proc. IEEE ISCC03*, 2003.

[9] K. Igarashi and K. Yamazaki. Flight size auto tuning for broadband wireless networks. *IWCMC 2009 Next Generation Mobile Networks Symposium*, 2009.

[10] J. Jin and B. Li. Adaptive random network coding in WiMAX. In *Proc. IEEE ICC'08*, pages 2576–2580, May 2008.

[11] L. Kai and W. Xiaodong. Cross-layer design of wireless mesh networks with network coding. *IEEE Trans. Mobile Comput.*, 7(11):1363–1373, Nov. 2008.

[12] D. Katabi, S. Katti, H. Wenjun, H. Rahul, and M. Medard. On practical network coding for wireless environments. In *Proc. Int'l Zurich Seminar on Communications*, pages 84–85, 2006.

[13] S. Katti, H. Rahul, H. Wenjun, D. Katabi, M. Medard, and J. Crowcroft. XORs in the air: Practical wireless network coding. *IEEE/ACM Trans. Netw.*, 16(3):497–510, June 2008.

[14] D. Koutsonikolas, Y.C. Hu, and C.-C. Wang. An Empirical study of performance benefits of network coding in multihop wireless networks. In *Proc. IEEE INFOCOM'09.*, 2009.

[15] Z. Lin and B. Vucetic. Power and rate adaptation for wireless network coding with opportunistic scheduling . *IEEE ISIT'08*, pages 21–25, July 2008.

[16] K. Lu, S. Fu, and Y. Qian. Capacity of random wireless networks: Impact of physical-layer network coding. In *Proc. IEEE ICC'08*, pages 3903–3907, May 2008.

[17] W. Mea and L. Baochun. Lava: A reality check of network coding in peer-to-peer live streaming. In *Proc. IEEE INFOCOM'07*, pages 1082–1090, May 2007.

[18] D. Nguyen, T. Tran, T. Nguyen, and B. Bose. Hybrid ARQ-random network coding for wireless media streaming. In *Proc. ICCE'08*, pages 115–120, June 2008.

[19] Y.E. Sagduyu and A. Ephremides. Cross-layer design for distributed MAC and network coding in wireless ad hoc networks. In *in Proc. IEEE ISIT'05*, pages 1863–1867, Sept. 2005.

[20] L. Scalia, F. Soldo, and M. Gerla. PiggyCode: A MAC layer network coding scheme to improve TCP performance over wireless networks. In *Proc. IEEE GLOBECOM'07*, pages 3672–3677, Nov. 2007.

[21] B. Scheuermann, W. Hu, and J. Crowcroft. Near-optimal coordinated coding in wireless multihop networks. In *Proc. of ACM CoNEXT'07*, 2007.

[22] J.K. Sundararajan, D. Shah, M. Medard, M. Mitzenmacher, and J. Barros. Network coding meets TCP. In *Proc. IEEE INFOCOM'09*, April 2009.

[23] The Network Simulator (NS-2). `http://www.isi.edu/nsnam/ns/`.

[24] J. Wang, X. Fang, N.Q. Cheng, M. Jia, and P. Zhang. A cross-layer Wireless TCP Enhancement Scheme in OFDM network. In *Proc. IEEE Region 10 Conference*, pages 1–4, Nov. 2006.

[25] Y. Wu, Z. Niu, and J. Zheng. A network-based solution for TCP in wireless systems with opportunistic scheduling. In *Proc. IEEE PIMRC'04*, volume 2, pages 1241–1245, 2004.

[26] F. Xue and S. Sandhu. PHY-layer network coding for broadcast channel with side information. In *Proc. IEEE ITW '07*, pages 108–113, Sept. 2007.

[27] H. Yomo and P. Popovski. Opportunistic scheduling for wireless network coding. In *Proc. IEEE ICC'07*, pages 5610–5615, June 2007.

[28] S. Zhang, S. Chang Liew, and L. Lu. Physical layer network coding schemes over finite and infinite fields. In *in Proc. IEEE GLOBECOM'08*, pages 1–6, 30 2008-Dec. 4 2008.

[29] H.-M. Zimmermann and Y.-C. Liang. Physical layer network coding for unicast applications. In *Proc. IEEE VTC'08 (Spring)*, pages 2291–2295, May 2008.

10

Network Selection and Spectrum Sharing for Heterogeneous Cognitive Radio Systems

Pengbo Si

College of Electronics Information and Control Engineering, Beijing University of Technology, Beijing, P.R. China
Email: sipengbo@bjut.edu.cn

F. Richard Yu

Department of Systems and Computer Engineering, Carleton University, Ottawa, ON, Canada

Hong Ji

Key Laboratory of Universal Wireless Communication, Ministry of Education, Beijing University of Posts and Telecommunications, Beijing, P.R. China

Yanhua Zhang

College of Electronics Information and Control Engineering, Beijing University of Technology, Beijing, P.R. China

CONTENTS

The integration of different radio access technologies becomes attractive with the rapid growth of wireless communication technologies and services. Moreover, in such heterogeneous environment, there is a common belief that, recently, we are running out of the radio resources, especially the usable spectrum resource. Although some work have been done to integrate heterogeneous wireless networks, the application of quality of service (QoS) with dynamic spectrum resource allocation has been largely ignored. In this chapter, we present a novel distributed scheme based on the stochastic optimization formulation of the network selection problem for heterogeneous wireless networks. In addition, an optimal internetwork spectrum sharing scheme with spectrum pooling is proposed for heterogeneous wireless networks. The problems are formulated as restless bandits systems, respectively, of which the indexability property dramatically reduces the computational complexity. Extensive simulation results demonstrate the significant performance improvement compared with the existing schemes.

10.1 Introduction

In recent years, there has been significant growth in the research area of wireless communication networks. Different radio access technologies (RATs), including wireless wide area networks (WWANs) such as cellular networks [13], wireless metropolitan area networks (WMANs) such as WiMAX networks [14], and wireless local area networks (WLANs) such as IEEE 802.11e-based networks [2], operate simultaneously in the world with different characteristics. This makes it attractive to integrate the wide range of radio access technology standards, that forms the heterogeneous wireless network. Moreover, today's wireless spectrum is assigned to the holders on a long-term basis for large regions, that is, the wireless networks are characterized by the fixed spectrum assignment policy. However, there is a common belief that we are running out of the usable spectrum resource recently (i.e., the unlicensed spectrum bands suitable for radio communication are rare [32]). On the contrast, a large portion of the assigned spectrum is used sporadically, with the utilization ranging from 15% to 85% [1]. The limitation and the under utilization of the radio resource stimulate the research on better spectrum management policies and techniques. In future wireless networks, it becomes very promising to integrate different types of cognitive radio (CR) networks, that is, heterogeneous cognitive radio networks on which we focus in this chapter.

Although some work has been done to integrate heterogeneous wireless networks, most of the previous work considers network layer QoS, such as blocking probability and utilization, as the design criteria. However, multimedia applications, such as video telephony and surveillance, are very promising services that require more radio resource compared to other types of services in heterogeneous wireless networks. From a user's point of view, QoS at the application layer is more important than that at other layers and should be taken into account in the design of network selection schemes. Besides, recent work on spectrum pooling and internetwork spectrum sharing mostly concentrates on the system architecture and the design of flexible access algorithms and schemes. However, how to optimally allocate spectrum among secondary CR networks with spectrum pooling merits further investigation.

In this chapter, we present a novel distributed scheme based on the stochastic optimization formulation of the network selection problem of heterogeneous wireless networks. In addition, an optimal internetwork spectrum sharing scheme with spectrum pooling is proposed for CR networks. The rest of the chapter is organized as follows. In Section 10.2, the related work is studied. The system model is presented in Section 10.3. In Section 10.4, the network selection and spectrum sharing in heterogeneous cognitive radio networks are formulated as restless bandits problems. We also have a discussion on the process and implementation of the proposed schemes in Section 10.5. In Section 10.6, extensive simulation results are shown and discussed, and we draw a conclusion in Section 10.7.

10.2 Related Work

To improve the performance of heterogeneous networks and keep users always best connected (ABC) [10], a number of schemes are proposed to deal with the network integration problems. In [25, 24], several resource management and admission control schemes are proposed in cellular/WLAN integrated networks. Authors of Reference [3] propose an architectural framework for network selection and a comprehensive decision making process to rank candidate networks for the users. An optimal joint session admission control scheme is proposed in [30] for integrated cellular/WLAN systems with vertical handoff. Game theory is introduced to heterogeneous networks in [18] for radio resource management, including bandwidth allocation and admission control. Authors of Reference [8] propose a Markovian framework for the allocation of multiple services in multiple RATs (radio access technologies) and a model to embed the evaluation of several RAT selection policies considering different allocation criteria.

Recent work on spectrum pooling and internetwork spectrum sharing mostly concentrates on the system architecture and the design of flexible access algorithms and schemes. Authors in [26] describe a spectrum pooling scenario, where orthogonal frequency division multiplexing (OFDM) is considered due to the flexibility requirement in the design of secondary networks. OFDM-based WLANs, such as IEEE 802.11a and HIPERLAN/2, are suitable for spectrum pooling as they allow a very flexible frequency management on a carrier-by-carrier basis [11]. Physical layer issues, such as mutual interference and synchronization, and MAC layer issues, such as handoff and scheduling, are also well discussed in [26]. In [22], for a wireless multihop relay system, an optimal OFDM subchannels reassignment scheme employing OFDM-based spectrum pooling is proposed, based on the signal-to-noise ratio (SNR) information of each subchannel. For multiple secondary networks accessing the spectrum pool, a game theory-based analysis is presented in [7]. In addition, in Reference [5], authors propose a distributed spectrum management architecture where nodes share spectrum resource following spectrum rules, while ignoring the spectrum sharing among secondary networks. Authors in [21] present a strategy for the extraction of the channel allocation information. Other work on dynamic spectrum sharing investigates the fairness [28] and quality of service [29] in wireless networks.

10.3 System Model

In heterogeneous cognitive radio networks, multiple types of networks cooperate to provide seamless coverage for universal wireless access, and cooperate to share the spectrum bands in the spectrum pool. We assume that the operator may own more than one network (e.g., an operator may provide both cellular and WLAN hot spot access at the same time with overlapped coverage). Each new session is to be associated with one of the networks, with the aim of maximizing the reward, which is defined to be the combination of distortion and the cost to access the networks for multimedia services. This optimal network selection is based on the admissible sets of the networks with different RATs, which are closely related to the spectrum band that could be utilized. Consequently, the cooperation among the networks for sharing the spectrum in the spectrum pool is another key issue in the heterogeneous cognitive radio networks.

To enable these heterogeneous networks, which are also the secondary networks, to access the licensed frequency bands, the concept of spectrum pooling that represents the idea of merging spectrum from different owners (primary networks) into a common pool is introduced. From this common spectrum pool hosted by the licensed system public rental, secondary networks may temporarily rent spectral resources during idle periods of licensed users. The basic proposition is that the licensed system does not need to be changed, thus secondary networks need to be highly flexible in order to efficiently fill the spectrum gaps. In the internetwork spectrum pooling problem considered in this chapter, all secondary networks employ OFDM as the underlying physical layer transmission technique [26]. This is because the secondary network needs to be highly flexible with respect to the spectral gaps the licensed users leave during their idle periods. OFDM modulation is a candidate for such a flexible rental system as it is possible to leave a set of unused subcarriers to fill the spectral gaps. In this chapter, the spectrum access decision is made by measuring the channel state and cooperation among the OFDM-based secondary networks. We assume the available spectrum in the pool is divided into subbands with a fixed bandwidth w. The changing size of the spectrum pool makes the number of available spectrum subbands c variable. In the proposed optimal scheme, the spectrum is allocated with the unit of one subband.

10.3.1 Network Model

In this chapter, the total number of networks is assumed to be N. In network n, $1 \leq n \leq N$, we denote by $u^l(n, t_k)$ the number of class l users at epoch t_k, where $1 \leq l \leq L$, L is the number of available service types. Epoch is defined as the time point when the number of associated users in the networks or the number of available spectrum subbands changes. This concept will be discussed in more detail in Section 10.4.1.1 and Section 10.4.2.1. In addition, $c(n, t_k)$ is used to represent the number of subbands rented by the secondary network n at epoch t_k. Let $c(t_k)$ represent the number of available subband in the spectrum pool at epoch t_k.

10.3.2 Multimedia Distortion Model

In future wireless networks, multimedia is one of the typical services that require high data rate and low transmission latency. Recent advanced coding algorithms, such as H.264 and MPEG-4, use rate control mechanism to control the video encoder output bit rate and error resilience mechanism for error protection [12]. *Intra-refreshing*, also called intra-update, of macroblocks (MBs) is an important approach for rate control and error protection.

In heterogeneous wireless networks, different networks provide different data rates and different link quality to the mobile users. Given a data rate in a network, authors in [12] provide a closed form distortion model taking into account varying characteristics of the input video, coding algorithm, and the intra-refreshing rate. We will use this rate-distortion model in our study. The total distortion comprises the quantization distortion introduced by the lossy video encoder to meet a target bit rate and the distortion resulting from channel errors, which will be presented in the following subsections. The total distortion is

$$D(H_s, \psi, \xi) = D_s(H_s, \xi) + D_c(\psi, \xi), \tag{10.1}$$

where $D_s(H_s, \xi)$ is the source distortion depending, on source coding rate H_s and intra-refreshing rate ξ, and $D_c(\psi, \xi)$ is channel distortion depending on packet loss rate ψ and intra-refreshing rate ξ, respectively. Thus, the optimal ξ^* to minimize the total distortion is given by

$$\xi^* = \arg\min_{\xi} D(H_s, \psi, \xi). \tag{10.2}$$

To deal with the time-varying wireless connection states of the networks, we use adaptive intra-refreshing rate ξ to achieve the minimum distortion.

10.3.3   Admissible Sets of the Networks

In IEEE 802.11e-based wireless local area networks [2], throughput and delay are important QoS metrics. Authors of [15] derive the throughput of IEEE 802.11e, and an optimal operating point is determined in [33]. Assume that n is the network number of a WLAN, and $g(n)$ is a vector representing the numbers of different service types in network n. According to these results, we adopt the following admissible set for wireless LAN n.

$$\mathbf{S_n} = \left\{ \mathbf{g(n)} \in \mathbf{Z_+^J} : \mathbf{B^l(n)} \geq \mathbf{TB^l(n)}, \mathbf{E^l(n)} \leq \mathbf{TE^l(n)} \right\}, \tag{10.3}$$

where $B^l(n)$ and $E^l(n)$ are the throughput and delay of service type l in network n, respectively. $TB^l(n)$ is the throughput constraint and $TE^l(n)$ is the delay constraint for sessions of service type l in network n.

In this chapter we assume that the WiMAX networks use TDD (Time Division Duplexing) scheme based on OFDM/TDMA, although WiMAX supports both TDD and FDD (Frequency Division Duplexing) operations [14]. According to [16], the capacity for the WiMAX network n is

$$C_n = \sum_{\theta=0}^{\Theta+1} P(\theta) C(\theta), \tag{10.4}$$

where $C(\theta)$ is the capacity with mode θ, and Θ is the number of available modes. Thus, the admissible set of WiMAX networks n can be derived as [6]

$$\mathbf{S_n} = \left\{ \mathbf{g(n)} \in \mathbf{Z_+^J} : \sum_{l=1}^{\mathbf{L}} \mathbf{U^l(n, t)} \mathbf{W^l(n)} \leq \mathbf{C_n(t)} \right\}, \tag{10.5}$$

where $C_n(t)$ is the capacity of WiMAX network n at time t, L is the total number of service types, $U^l(n, t)$ is the number of sessions of service type l in network n at time t, and $W^l(n)$ is the bandwidth required by type l service in WiMAX network n.

10.3.4 Optimization Objectives

The optimization process in this chapter is divided into two steps, the optimization of network selection and of spectrum sharing.

We define the optimization goal of the network selection problem to be maximizing the total discounted reward that is defined in Section 10.4.1.3 as the combination of minimizing the multimedia distortion and the cost to access the network. The total reward along the time line is discounted by time with a discount factor β.

To quantify the reward gained by optimally sharing the spectrum bands in the spectrum pool, we use utility, which is a function of the spectrum efficiency and access price, as our optimization objective. In this chapter, among the various definitions of utility function in previous work, we adopt the following most commonly used quadratic utility function for secondary network n [17]

$$
\mathbf{U_n(c_n')} = \sum_{p=1}^{N^P} \mathbf{b_n c_n'(p) \eta_n(p)} - \frac{1}{2} \left(\sum_{p=1}^{N^P} \mathbf{c_n'^2(i)} + 2\theta \sum_{p \neq q} \mathbf{c_n'(p) c_n'(q)} \right)
$$
$$
- \sum_{p=1}^{N^P} b_n'(p) c_n'(p), \tag{10.6}
$$

where b_n denotes the price to be paid by a mobile user for accessing secondary network n, $b_n'(p)$ denotes the price for secondary network n to rent the spectrum band in the spectrum pool licensed to primary network p, N^P is the number of primary networks, $c_n'(p)$ is the number of subbands provided by primary network p and currently used by secondary network n, θ is the substitutability parameter, $-1 \leq \theta \leq 1$, and $\eta_n(p)$ denotes the spectrum efficiency of the transmission of secondary network n using the spectrum licensed to primary network p, which can be obtained from [9]

$$
\eta = \log_2(1 + K\gamma), \text{where } K = \frac{1.5}{\ln(0.2/\text{BER}^{\text{tar}})}, \tag{10.7}
$$

where γ is the receiving signal-to-noise-ratio (SNR) and BER^{tar} is the target bit-error-rate (BER).

In this chapter, the proposed optimal scheme of spectrum sharing is to maximize the time discounted summation utilities along the whole time line for the secondary networks.

10.4 System Formulation

The restless bandits problem is an extension to the classical multiarmed bandit problem [23], in which each lever provides a reward when it is pulled. It is a model of a project that is trying to achieve the balance of acquiring new knowledge and optimizing its decisions based on existing knowledge. The multiarmed bandit approach is to maximize the overall discounted reward based on the balancing.

According to the indexable rule of the multiarmed bandit problem, the optimal policy can be found by simply choosing the project with the largest index. Although it is a relatively simple solution to the multiarmed bandit problem, in our network selection problem, it is not realistic to allow only the active network to change state, which is one of the basic assumptions in multiarmed bandit problem. The restless bandits problem is proposed to deal with this problem [27, 4, 19, 20]. At each epoch $t_1, t_2, \ldots$, one or more projects out of N can be active, and the states of all N networks may change. We denote by M the number of active projects at each epoch. Reward is earned at each time slot by each project. There is also an indexable rule of the restless bandits problem. Projects are selected to be active according to their indices that are calculated by linear programming (LP) relaxation [4] based on the states, transition probabilities, and rewards.

10.4.1 Optimal Network Selection Formulation

The network selection problem in heterogeneous networks can be formulated as a restless bandits system, of which the indexability property can significantly reduce the optimization computational complexity.

10.4.1.1 Decision Epochs and Actions

We set the decision epochs to be the set of session arrival and departure time points, because the states change when a session arrives and departs. Denote by t_k the decision epochs, $k = 0, 1, 2, \ldots$, t'_k the arrival epochs, and t^*_k the departure epochs. The time intervals between two adjacent arrival epochs and two adjacent departure epochs are $(t'_k, t'_{k+1}]$, and $(t^*_k, t^*_{k+1}]$ respectively, and the durations of which are both exponentially distributed random variables with the expected number of epochs in each time unit, or traffic rate, $\nu = \sum_{l=1}^{L} \nu^l$ and $\sum_{l=1}^{L} \sum_{n=1}^{N} U^l(n, t_k) \mu^l$, respectively, where $U^l(n, t_k)$ is the number of type l sessions in network n at epoch t_k, ν^l and μ^l are the type l session arrival and departure rate, respectively. Consequently, the time intervals between epochs $(t_k, t_{k+1}]$ are exponentially distributed with the expected number of epochs in each time unit $\nu + \sum_{l=1}^{L} \sum_{n=1}^{N} U^l(t_k) \mu^l$. This is called the total traffic rate.

The action is the network selection decision at the current epoch. At each epoch t_k, one of the networks is selected to be active, meaning that it is ready to admit a new arrival session at the next epoch t_{k+1} if a new session arrives at t_{k+1}. At each arrival epoch t'_k, only the state of the network selected at the former epoch changes; At each departure epoch t^*_k, only the state of the network from which the session departs changes. For each network n at epoch t_k,

$$a_n(t_k) = \begin{cases} 1, & \text{if network } n \text{ is active at epoch } t_k, \\ 0, & \text{if network } n \text{ is passive at epoch } t_k. \end{cases} \tag{10.8}$$

The actions satisfy $\sum_{n=1}^{N} a_n(t_k) = 1$.

10.4.1.2 State Space and Transition Probabilities

The state of network n at epoch t_k is defined as $s(n, t_k) = [U^l(n, t_k)]_{l \in \{1,2,...,L\}}$, where L is the number of service types. Thus the state space of network n is the admissible set $\mathbf{S_n}$. The state of network n under action a evolves according to a Markov chain with the transition probability $p^a_{i,j}(n)$ from state $s_i(n) = [u^l_i(n)]_{l \in \{1,2,...,L\}}$ to $s_j(n) = [u^l_j(n)]_{l \in \{1,2,...,L\}}$. Define the expected interval duration between two epochs for the state s_i to be

$$\tau_i = \mathbf{E}\left(t_{k+1} - t_k | s_i(n, t_k)\right), \tag{10.9}$$

which is the inverse of the total traffic rate

$$\tau_i = \left(\nu + \sum_{l=1}^{L} U^l_i(n)\mu^l\right)^{-1} \tag{10.10}$$

Let us define the transition probability matrix of network n with action a to be $P^a(n) = [p^a_{i,j}(n)]_{S(n) \times S(n)}$, where $S(n)$ is the number of available states $s(n)$ of network n. Denote by $\chi(l)$, $1 \le l \le L$, the L-element row vector of which the lth element is one and the other elements are zero, thus the transition probabilities can be represented as follows:

$$p^a_{i,j}(n) = \begin{cases} \nu_l \zeta(s_j(n))a\tau_i, & \text{if } s_j(n) = s_i(n) + \chi(l), \\ U^l_i(n)\mu^l\tau_i, & \text{if } s_j(n) = s_i(n) - \chi(l), \\ 1 - \nu_l \zeta(s_j(n))a\tau_i - U^l_i(n)\mu^l\tau_i, & \text{if } s_j(n) = s_i(n), \\ 0, & \text{otherwise,} \end{cases} \tag{10.11}$$

where $\zeta(x)$ is defined by

$$\zeta(x) = \begin{cases} 1, & \text{if } x \in \mathbf{S_n}, \\ 0, & \text{otherwise.} \end{cases} \tag{10.12}$$

10.4.1.3 System Reward

The optimization goal is to maximize the total discounted reward which is defined as

$$Z = \sum_{k=0}^{T-1} \sum_{u=1}^{U(t_k)} \beta^{T-k-1} R_u(t_k), \tag{10.13}$$

where T is the number of epochs considered, β is a discount factor, and $R_u(t_k)$ is the reward of session u at epoch t_k

$$R_u(D(u), B(u)) = [-c_1 \lg(D(u)) - c_2 B(u) + c_3] \tau_i, \tag{10.14}$$

where $D(u)$ is session u's distortion, $B(u)$ is the price paid by session u, which is related to the current serving network. $c_1 \geq 0$, $c_2 \geq 0$, and c_3 are constant coefficients. By adjusting the coefficients, the balance of distortion and price can be achieved.

Since sessions of the same service type in the same network have the consistent properties, the distortion minimization for them will choose the same intra-refreshing rate, and achieve the same minimized distortion. Besides, the costs of these sessions are also the same. Consequently, (10.13) can be also written as

$$Z = \sum_{k=0}^{T-1} \sum_{n=1}^{N} \sum_{l=1}^{L} \beta^{T-k-1} U^l(n, t_k) R^l(n), \tag{10.15}$$

where $R^l(n)$ is the reward by session of type l in network n. The objective of our problem is to maximize the total reward to achieve

$$Z^* = \max_{A \in \mathbf{A}} Z(A). \tag{10.16}$$

10.4.1.4 Indices and Policies

The restless bandits approach has an indexable rule that reduces the computational complexity dramatically. For network n in state i_n, we denote by the index $\delta_n(i_n)$. According to the restless bandits approach, the optimal policy A^* is a set of optimal actions. Let the element of A^* in row n and column k be $a_n^*(t_k)$, which represents the optimal action for network n at epoch t_k, thus

$$a_n^*(t_k) = \begin{cases} 1, & \text{if } \delta_n \text{ is the smallest among } \{\delta_1, \delta_2, \ldots, \delta_N\}, \\ 0, & \text{else.} \end{cases} \tag{10.17}$$

Define the set of all available policies to be $\mathbf{A} = \{A\}$. Thus, $A^* = \arg\max_{A \in \mathbf{A}} Z(A)$. In our network selection problem, at each epoch, the network with the smallest index δ_n is set to be active, while other networks are passive. At the next epoch, if a session arrives, the active network will admit the new session; if a session departs, only the corresponding network needs to do the deassociation action.

10.4.2 Optimal Spectrum Sharing Formulation

In this section, the spectrum sharing problem of the different cognitive radio networks is also formulated as a restless bandits system.

10.4.2.1 Epochs and Actions

The epochs t_k, $0 \le k \le T - 1$, are the time points at which the state of any of the networks changes. Thus, there are four types of epochs:

1. Epoch type I, t_k^1: A user arrives at any secondary network.

2. Epoch type II, t_k^2: A user departs from any secondary network.

3. Epoch type III, t_k^3: A new spectrum subband is available for the secondary networks.

4. Epoch type IV, t_k^4: One or more spectrum subband that were previous used by the secondary networks are to be occupied by the primary networks.

The time duration between two epochs is decided by the time points when the user/band arrives/departs. Since the user/band arrival/departure is a stochastic process, the value of the duration is not a constant but a variable, which is related to the user/band arrival/departure rate.

For network n, define the action $a(n, t_k)$ at epochs t_k

$$a(n, t_k) = \begin{cases} 1, & \text{if network } n \text{ is active at epoch } t_k, \\ 0, & \text{if network } n \text{ is passive at epoch } t_k. \end{cases} \tag{10.18}$$

The active network at epoch t_k will make use of the new available spectrum subband at the next epoch if the next epoch is of type III. At each epoch, the reward R_{i_n} is earned by network n in state i, and the state changes according to the transition probabilities that will be discussed later. One subband is allocated to only one network at the same time, thus the actions satisfy

$$\sum_{n=1}^{N} a(n, t_k) = 1, \tag{10.19}$$

where N is the number of secondary networks. Define this action space to be **A**.

10.4.2.2 State Space and Transition Probabilities

We consider the state space of secondary network n as $[u^1(n, t_k), u^2(n, t_k), \ldots, u^L(n, t_k), c(n, t_k)]$, which is the combination of the numbers of users of different classes in network n, $u^l(n, t_k)$, $l = 1, 2, \ldots, L$, and the number of subbands used by secondary network n, $c(n, t_k)$, while the numbers of users and the number of subbands are not independent. The relationship between $u^l(n, t_k)$

and $c(n, t_k)$ is based on the admissible set of network n, and the maximum number of users in network n and the maximum number of subcarriers that can be used in network n is finite, thus the state space is finite. Denote by this state space $\mathbf{S_n}$ for network n.

The transition probability of network n from state i to j under action $a(n, t_k)$ is

$$p_{ij}^a(n, t_k) = P(j|i, a(n, t_k)). \tag{10.20}$$

Assume the class l user arriving rate in network n to be $\nu_u^l(n)$, the departure rate of class l user in network n to be $\mu_u^l(n)$, the new spectrum subband arriving rate to be ν_c, and the spectrum subband departure rate to be μ_c. Let $\bar{u}^l(n)$ represent the expected number of class l user in network n, and $\bar{c}$ present the total available subcarrier number. Thus, the total epoch rate is

$$r = \sum_{n=1}^{N} \sum_{l=1}^{L} \nu_u^l(n) + \sum_{n=1}^{N} \sum_{l=1}^{L} \bar{u}^l(n)\mu_u^l(n) + \nu_c + \bar{c}\mu_c, \tag{10.21}$$

Define the vector of the user numbers in network n under state i to be $\mathbf{U}_i(n) = [u_i^1(n), \ldots, u_i^l(n), \ldots, u_i^L(n)]$. By introducing the vector χ^l, $1 \leq l \leq L$, which is an L-element row vector of which the lth element is one and the other $L - 1$ elements are zero, the transition probabilities from state i to j under action a can be represented as

$$p_{ij}^a(n) = \begin{cases} u_i^l(n)\mu_u^l(n)/r, & \\ \quad \text{if } u_j = u_i - \chi^l, u_j \geq 0 \text{ and } c_j = c_i, & \\[2mm] c_i(n)\mu_c/r, & \\ \quad \text{if } u_j = \min\{u_i, \max_{u \in \mathbf{S_n(c_j)}} u\} \text{ and } c_j = c_i - 1, & \\[2mm] 1 - [u_i^l(n)\mu_u^l(n) - c_i(n)\mu_c - \zeta_i^l\nu_u^l(n) - a\nu_c]/r, & \\ \quad \text{if } u_j = u_i \text{ and } c_j = c_i, & \\[2mm] \zeta_i^l\nu_u^l(n)/r, & \\ \quad \text{if } u_j = u_i + \chi^l, u_j \leq u * (n) \text{ and } c_j = c_i, & \\[2mm] a\nu_c/r, & \\ \quad \text{if } u_j = u_i, c_j = c_i + 1 \text{ and } c_j \leq c * (n), & \\[2mm] 0, & \\ \quad \text{otherwise}, & \end{cases} \tag{10.22}$$

where $u_i^l(n)$ is the number of class l users in network n under state i, $c_i(n)$ is the number of spectrum bands using by network n under state i, $u * (n)$ is the maximized number of users that can be accepted by network n, $c * (n)$ is the maximized number of subcarriers that can be used in network n, and ζ_i

is defined as

$$\zeta_i^l = \begin{cases} 1, & \text{if } [\mathbf{U}_i(n) + \chi^l, c_i(n)] \in \mathbf{S_n} \\ \\ 0, & \text{otherwise}, \end{cases} \tag{10.23}$$

where $\mathbf{S_n}$ denotes the state space of network n. Here, we omit t_k for simplicity.

10.4.2.3 System Reward and Policy

With the optimization objective discussed in Section 10.3.4, we define the system reward $R_{i_n}(t_k)$ for network n in state i at epoch t_k to be its utility, that is,

$$R_{i_n}(t_k) = \mathbf{U_n}(\mathbf{c'_n}). \tag{10.24}$$

We denote by $\mathbf{H}$ the set of all admissible policies. An admissible policy $h \in \mathbf{H}$ is a $T \times N$ matrix, whose element of the kth row and the nth column is $a(n, t_{k-1})$, representing the action taken by secondary network n at epoch t_{k-1}. Here T is the maximum number of epochs considered. h satisfies:

$$h \times \underbrace{(1, 1, \ldots, 1)'}_{T} = \underbrace{(1, 1, \ldots, 1)'}_{T}, \tag{10.25}$$

which means at each epoch, there is only one active secondary network. In this chapter, the optimization goal is to maximize the discounted total reward under policy h, that is, the optimized reward

$$R^* = \max_{h \in \mathbf{H}} R(h) = \sum_{k=0}^{T-1} \sum_{n=1}^{N} \beta^{T-t_k} R_{i_n}(t_k) \tag{10.26}$$

where β is the discount factor, which indicates how much the reward obtained before contributes to the total reward. $\beta = 0$ means the total reward only includes the current reward, $\beta = 1$ means the reward obtained before is of equal importance to the current reward, and $0 < \beta < 1$ means the reward obtained before contributes to the total reward, with a less weight than the current reward. The earlier the reward obtained, the less the weight is, that is, the reward is discounted by time with the factor β.

10.4.3 Solving the Restless Bandits Problems

The standard restless bandits problem allows M out of N objects to be active at epoch t_k. The reward $R^a(n)$ is earned by each object, with its state changing according to the transition probability matrix $P^a(n)$. The total reward is time-discounted by the discount factor β. The aim is to find the optimal policy $A^* \in \mathbf{A}$ to maximize the expected reward $R(A)$. Due to the indexability property of the restless bandits problem, it is proved that the optimal policy is simply selecting the network with the lowest index to be the active one. Authors in [4] provide more detailed description of solving the restless bandits problems.

10.5   Schemes Implementation

In this section, we have a discussion on the implementation of the proposed schemes. Since the optimization of the problem is simply finding out the network with the lowest index, the key step is to calculate the indices. Note that all the possible indices for the networks in all the available states could be off-line computed and stored in a table, thus the networks only need to lookup the table according to the current state on-line [4].

10.5.1   Optimal Network Selection for Heterogeneous Systems

In this section, the optimization objective is to improve the application layer QoS. The system reward is defined based on the application layer QoS criterions. In the proposed optimal network selection scheme, at each epoch, a request from the session is sent to all the networks. If this is an arrival epoch, the new session is to be associated to the current active network, and an optimal intra-refreshing rate is selected for the transmission; if this is a departure epoch, a session leaves from its network. Then every network calculates its own index based on the current state, and shares with others in a distributed way. The network selection problem is formulated as a restless bandits system, which dramatically simplifies the optimal selection to be just select the network with the lowest index as the one to admit the arrival session. The index of each network is computed distributively according to the current state of the network and the state transition probabilities. By comparing the indices, the network with the lowest index is selected to be the active one for the new session association decision at the next epoch.

The network selection is in a distributed and cooperative way, which can be divided into the offline stage and the online stage. In the offline stage, indices are calculated for all states and actions, and are stored in a table. In the online stage, a network looks up its table to find out the index corresponding to the current state and action. In the proposed optimal scheme, at each epoch, a request from the session is sent to all the networks. If this is an arrival epoch, the new session is to be associated to the current active network, and an optimal intra-refreshing rate is selected for the transmission; if this is a departure epoch, a session leaves from its network. Then every network calculates its own index based on the current state, and shares with others in a distributed way. By comparing the indices, the network with the lowest index is selected to be the active one for the new session association decision at the next epoch.

10.5.1.1 Optimal Network Selection

The network selection is in a distributed and cooperative way, which can be divided into the offline stage and the on-line stage. In the offline stage, indices are calculated for all states and actions, and are stored in a table. In the online stage, a network looks up its table to find out the index corresponding to the current state and action. The offline computation is as follow.

1. According to the admissible sets of the networks and the session arrival/departure rate, the state space and transition probability matrices under different actions are determined.

2. For each network n and each possible state $i_n \in \mathbf{S_n}$, input the state transition probability $p_{i_n j_n}^a$, the reward $R_{i_n}^a$, the discount factor β and the initial state probability vector $\boldsymbol{\alpha}$, then off-line compute the finite set of the indices $\{\delta_{i_n}\}$. Store these indices and the corresponding $p_{i_n j_n}^a$, $R_{i_n}^a$ and $\boldsymbol{\alpha}$ in a table.

After the offline initialization, at epoch t_k, online computation is as follows.

1. Denote by n_a the current active network. If this is an arrival epoch and the active network n_a is capable to admit the new arrival session according to $\mathbf{S_{n_a}}$, n_a admit the session and update the its state s_{n_a}. If this is an arrival epoch but the active network n_a is not capable to admit the new session, the new session is to be rejected. If this is a departure epoch, the session leaves from the associated network, and the state of the network is updated.

2. Each network n shares its state s_n as the initial state probability vector $\boldsymbol{\alpha}$ with the others.

3. With $\boldsymbol{\alpha}$, each network looks up the index table to find out the corresponding index δ_{i_n}.

4. The networks share their indices δ_{i_n} in a distributed way.

5. Each network arranges the list of the indices from the lowest to the highest. A network is set to be active if its index is in the first place.

10.5.1.2 Optimal Intra-Refreshing Rate

Given the source-coding bit rate H_s and the packet loss rate ψ for session of type l, the intra-refreshing rate ξ is off-line optimized for different situations to minimize the total distortion. Thus, the minimized distortion $D^* = D(H_s, \psi, \xi^*)$ can be calculated as a part of the reward $R_{n,l}$. This reward is used for the policy optimization.

10.5.2 Cooperative Spectrum Sharing for Cognitive Radio Networks

We optimize the internetwork spectrum sharing with the objectives to maximize the spectrum efficiency and minimize the cost to use the spectrum bands. The internetwork spectrum sharing problem is formulated as a restless bandits system. We consider four types of epochs:

1. Epoch type I: A user arrives at any secondary network.

2. Epoch type II: A user departs from any secondary network.

3. Epoch type III: A new spectrum subband is available for the secondary networks.

4. Epoch type IV: One or more spectrum subband that were previous used by the secondary networks are to be occupied by the primary networks.

The active network at each epoch will make use of the new available spectrum subband at the next epoch if the next epoch is of type III, and a reward is earned by each network. One subband is allocated to only one network at the same time. The spectrum allocation is based on the indices of the secondary networks, which are computed offline for each available state of each secondary network, and stored in a table. In the online stage, it is only needed to lookup the table to decide the current index according to the state.

In this scenario, the spectrum allocation scheme is not to allocate subbands to specific users, but to the secondary networks. Since more spectrum enables higher throughput or data rate, if the access fee is not considered, the networks are all apt to claim more spectrum from the spectrum pool. However, available spectrum is a kind of limited radio source, thus the access fee is charged by the primary networks, and the optimal spectrum allocation takes into account these factors.

The new arrival spectrum in the spectrum pool is divided into subbands, each of which could incorporate one or more subcarriers. These subbands are allocated to the secondary networks one by one. Larger subband size enables faster processing, and smaller size improves the fineness of the allocation. The spectrum allocation is based on the indices of each secondary network, which are computed offline for each available state of each secondary network, and stored in a table. In the online stage, it is only needed to lookup the table to decide the current index according to the state.

10.5.2.1 Offline Computation

In the network initialization procedure, input the state transition probability $p_{ij}^a(n)$ of each secondary network $n \in \mathbf{N}$, the reward R_{i_n}, the discount factor β, and the initial state probability vector $\boldsymbol{\alpha}$, then offline compute the finite set of the indices $\{\delta_{i_n}\}$. Store these indices and the corresponding $p_{ij}^a(n)$, R_{i_n} and $\boldsymbol{\alpha}$ in a table.

10.5.2.2 Online Allocation

The online process is described as follows:

1. The secondary networks cooperate to sense the spectrum with their shared information.

2. According to the sensing results, the secondary networks cooperate to decide the spectrum bands that could be put into the spectrum pool.

3. At each epoch, the networks update their states and determines their indices according to the states.

4. The indices are shared among the networks.

5. Each secondary network arranges the list of the indices from the lowest to the highest. A network will be active in the following epoch if its index is in the lowest item of the list.

6. The secondary networks may have their state changed. The action and state changes differently for different types of epochs:

   - If this is an epoch of type I, according to the user's request, the corresponding network admit the user and updates its network state.

   - If this is an epoch of type II, the corresponding network de-associates the user and updates its network state.

   - If this is an epoch of type III, the active network is to occupy this new available subband and update its state.

   - If this is an epoch of type IV, the corresponding networks stop transmission on the spectrum and update their states.

7. Repeat the above steps.

10.6 Simulation Results and Discussions

In this section, extensive simulation results are presented to illustrate the performance of the proposed optimal network selection and spectrum sharing schemes. The video and VoIP (voice-over IP) session arrival time is Poisson distributed with the expected rates ν_1 and ν_2, respectively. The session departure time is also Poisson distributed, with the expected rate μ. The area considered is covered by three networks, a WLAN, a WiMAX network, and a cellular network. We adopt the parameters of the networks as in [24, 31]. The data rate of the VoIP service is 64 Kbps, while video service data rate varies in different networks. In WiMAX and WLAN networks, it is 1.17 Mbps. In

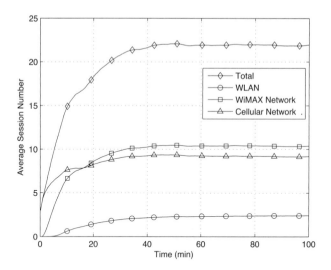

FIGURE 10.1

Average number of sessions associated to each network.

the cellular network, it is 240 Kbps because of relatively low bandwidth in the cellular network. The spectrum subband width is 2.5 MHz. The available spectrum subband arrives with the arrival rate of 24 subbands per hour and departure rate of 12 subbands per hour.

10.6.1 Optimal Network Selection

During the network selection process, at each epoch, a network selection decision is made, and the number of associated sessions is updated. To illustrate the dynamics of the system, we plot the session number, which is an average value of 2000 trials, in Figure 10.1. In the initial state, there are two VoIP and one video sessions in the networks. Assume $\mu = 0.2$, $\nu_1 = 1.6$ and $\nu_2 = 3.2$. From Figure 10.1, we can see that the number of sessions goes up first and becomes converged after about 60 minutes, when the balance between the expected numbers of sessions departs and arrives is achieved, and the total session number does not change dramatically any more. We can also observe that the WLAN, which provides the highest reward, is more likely to be selected when it's not saturated. After WLAN's saturation, the WiMAX network, whose reward is higher than the cellular network but lower than the WLAN, becomes the first choice.

As shown in Figure 10.2, with the increase of session departure rate μ, the session number decreases in the networks, and consequently, the reward

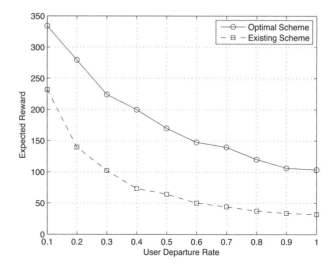

FIGURE 10.2
Expected reward comparison for different session departure rates.

decreases. We set $\nu_1 = 1.6$ and $\nu_2 = 3.2$ in this simulation. We can see that the reward of the optimal scheme is always better than the existing scheme.

10.6.2 Optimal Spectrum Sharing

The reward defined by the utility functions is presented to compare the performance of the schemes in Figure 10.3. In this simulation, we use 2.5 MHz subband width, and assume that each secondary network has one subband available and one user associated at first. With the increase of user number and subband number, the average reward goes up and then converges after about 20 minutes, when the arrival rate is equal to the departure rate approximately, and the user/subband number balance is achieved. In the following simulation, we use the time average value of the reward after 30 minutes. It is shown in this figure that the proposed scheme always performs much better than the existing scheme. This is because our scheme can optimally select the secondary network with the strongest requirement of new available spectrum bands.

The number of secondary networks could affect the performance dramatically. In Figure 10.4, for $n > 3$, we assume b_n and b'_n to be 14 Mbps^{-1} and 14 MHz^{-1}, respectively. With the increase of the number of secondary networks, the average reward drops because of the decrease of the expected number of subbands that can be allocate to one network. It is shown from the figure that the proposed restless bandits problem-based scheme improves the perfor-

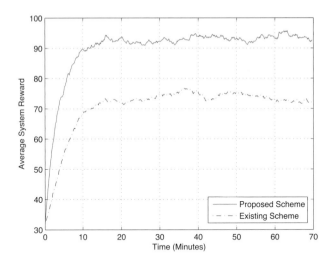

FIGURE 10.3
Average system reward comparison along the time line.

mance significantly compared with the existing scheme. We can also observe from the figure that smaller subband width improves the average utility performance even more, because of the relatively accurate spectrum allocation. However, on arriving the similar amount of available spectrum, more times of online spectrum allocation may be needed for smaller subband width, thus the computational load may be heavier.

10.7   Conclusion

In this chapter, we have proposed the optimal network selection and spectrum sharing schemes in heterogeneous cognitive radio networks considering application layer QoS, including multimedia distortion, access price, and the cost to rent the spectrum resource from the primary networks. The network selection and the spectrum sharing problems were formulated as restless bandits systems, respectively, of which the indexability property dramatically reduces the computational complexity. The process of the proposed schemes were discussed with offline and online steps. Simulation results demonstrated the significant performance improvement compared with the existing schemes.

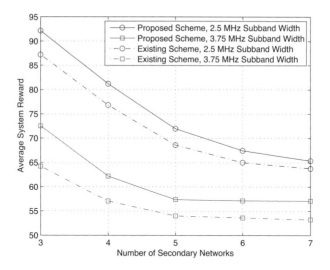

FIGURE 10.4
Average system reward comparison for different numbers of secondary networks.

Bibliography

[1] I.F. Akyildiz, W.-Y. Lee, M.C. Vuran, and S. Mohanty. NeXt generation/dynamic spectrum access/cognitive radio wireless networks: A survey. *Comput. Netw.*, 13(50):2127–2159, September 2006.

[2] ANSI/IEEE Std. 802.11e, Draft 5.0. Wireless medium access control (MAC) and physical layer (PHY) specification: Medium access control (MAC) enhancement for quality of service (QoS). July 2003.

[3] F. Bari and V.C.M. Leung. Automated network selection in a heterogeneous wireless network environment. *IEEE Netw.*, 21(1):34–40, Feb. 2007.

[4] D. Berstimas and J. Niño-Mora. Restless bandits, linear programming relaxations, and a primal dual index heuristic. *Operations Research*, 48(1):80–90, 2000.

[5] L. Cao and H. Zheng. Distributed rule-regulated spectrum sharing. *IEEE J. Sel. Areas Commun.*, 26(1):130–145, 2007.

[6] A. I. Elwalid and D. Mitra. Effective bandwidth of general Markovian traffic sources and admission control of high speed networks. *IEEE/ACM Trans. Netw.*, 1(3):329–343, Jun. 1993.

[7] R. Etkin, A. Parekh, and D. Tse. Spectrum sharing for unlicensed bands. *IEEE J. Sel. Areas Commun.*, 25(3):517–528, Apr. 2007.

[8] X. Gelabert, J. Peréz-Romero, O. Sallent, and R. Agustí. A markovian approach to radio access technology selection in heterogeneous multiaccess/multiservice wireless networks. *IEEE Trans. Mobile Comput.*, 7(10):1257–1270, Oct. 2008.

[9] A.J. Goldsmith and S.-G. Chua. Variable rate variable power MQAM for fading channels. *IEEE Trans. Commun.*, 45(10):1218–1230, Oct. 1997.

[10] E. Gustafsson and A. Jonsson. Always best connected. *IEEE Wireless Commun.*, 10(1):49–55, Feb. 2003.

[11] M. Haddad, A. Hayar, and M. Debbah. Spectral efficiency of spectrum-pooling systems. *IET Commun.*, 2(6):733–741, July 2008.

[12] Z. He, J. Cai, and C. Chen. Joint source channel rate-distortion analysis for adaptive mode selection and rate control in wireless video coding. *IEEE Trans. Circ. Sys. Video Tech.*, 12(6):511–523, June 2002.

[13] H. Holma and A. Toskala. *WCDMA for UMTS: Radio Access for Third Generation Mobile Communications*. John Wiley, New York, 2004.

[14] IEEE Std. 802.16-2004. IEEE standard for local and metropolitan area networks, part 16: Air interface for fixed broadband wireless access systems. Oct. 2004.

[15] Y.L. Kuo, C.H. Lu, E. Wu, and G.H. Chen. An admission control strategy for differentiated services in IEEE 802.11. In *Proc. IEEE Globecom'03*, pages 707–712, San Francisco, CA, Dec. 2003.

[16] Q. Liu, S. Zhou, and G. B. Giannakis. Queuing with adaptive modulation and coding over wireless links: Cross-layer analysis and design. *IEEE Trans. Wireless Commun.*, 4(3):1142–1153, May 2005.

[17] D. Niyato and E. Hossain. Competitive pricing for spectrum sharing in cognitive radio networks: Dynamic game, inefficiency of nash equilibrium, and collusion. *IEEE J. Sel. Areas Commun.*, 26(1):192–202, Jan. 2008.

[18] D. Niyato and E. Hossain. A noncooperative game-theoretic framework for radio resource management in 4G heterogeneous wireless access networks. *IEEE Trans. Mobile Comput.*, 7(3):332–345, Mar. 2008.

[19] J. L. Ny, M. Dahleh, and E. Feron. Multi-agent task assignment in the bandit framework. In *Proc. 45th IEEE Conf. Decision and Control*, pages 5281–5286, San Diego, CA, Dec. 2006.

[20] J. L. Ny and E. Feron. Restless bandits with switching costs: Linear programming relaxations, performance bounds and limited lookahead policies. In *Proc. 2006 American Control Conf.*, pages 1587–1592, Minneapolis, Minnesota, June 2006.

[21] M. Öner and F. Jondral. On the extraction of the channel allocation information in spectrum pooling systems. *IEEE J. Sel. Areas Commun.*, 25(3):558–565, Apr. 2007.

[22] A. Pandharipande and C. Keong Ho. Spectrum pool reassignment for wireless multi-hop relay systems. In *Proc. Int. Conf. Cognitive Radio Oriented Wireless Netw. and Commun.*, pages 1–5, Singapore, May 2008.

[23] H. Robbins. Some aspects of the sequential design of experiments. *Bulletin of the American Mathematical Society*, 55:527–535, 1952.

[24] W. Song, Y. Cheng, and W. Zhuang. Improving voice and data services in cellular/WLAN integrated networks by admission control. *IEEE Trans. Wireless Commun.*, 6(11):4025–4037, Nov. 2007.

[25] W. Song, H. Jiang, W. Zhuang, and X. Shen. Resource management for QoS support in cellular/WLAN interworking. *IEEE Netw.*, 19(5):12–18, Sep. 2005.

[26] T.A. Weiss and F.K. Jondral. Spectrum pooling: An innovative strategy for the enhancement of spectrum efficiency. *IEEE Commun. Mag.*, 42(3):S8–S14, Mar. 2004.

[27] P. Whittle. Restless bandits: Activity allocation in a changing world. In J. Gani, editor, *A Celebration of Applied Probability*, volume 25 of *J. Appl. Probab.*, pages 287–298. Applied Probability Trust, 1988.

[28] Y. Xing, R. Chandramouli, and S. Mangold. Dynamic spectrum access in open spectrum wireless networks. *IEEE J. Sel. Areas Commun.*, 24(3):626–637, Mar. 2006.

[29] Y. Xing, C.N. Mathur, M.A. Haleem, R. Chandramouli, and K.P. Subbalakshmi. Dynamic spectrum access with QoS and interference temperature constraints. *IEEE Trans. Mobile Comput.*, 6(4):423–433, Apr. 2007.

[30] F. Yu and V. Krishnamurthy. Optimal joint session admission control in integrated WLAN and CDMA cellular networks with vertical handoff. *IEEE Trans. Mobile Computing*, 6(1):126–139, Jan. 2007.

[31] S. Zhang, F.R. Yu, and V.C.M. Leung. Joint connection admission control and routing in IEEE 802.16-based mesh networks. In *Proc. IEEE Int. Conf. Commun. (ICC'08)*, pages 4938–4942, Beijing, P.R. China, May 2008.

[32] Q. Zhao and B. M. Sadler. A survey of dynamic spectrum access. *IEEE Signal Proc. Mag.*, 24(3):79–89, May 2007.

[33] H. Zhu and I. Chlamtac. A call admission and rate control scheme for multimedia support over IEEE 802.11 wireless LANs. *Wireless. Netw.*, 12(4):451–463, July 2006.

11

Network Selection and Congestion Avoidance Control in Multi-Access Networks

Dimitris E. Charilas

Department of Electrical and Computer Engineering,National Technical University of Athens, Greece
Email: dcharilas@mobile.ntua.gr

Ourania I. Markaki

Department of Electrical and Computer Engineering,National Technical University of Athens, Greece

Athanasios D. Panagopoulos

Department of Electrical and Computer Engineering, National Technical University of Athens, Greece

CONTENTS

Emerging 4G networks will be characterized by a heterogeneous environment where several access networks will be available. While one way to enhance the Quality of Service (QoS) offered to users in this context is through innovative protocols and new technologies, future trends should as well take into account the efficiency of resource allocation schemes. As a result, the goal of this chapter is to summarize techniques that may enable optimal distribution of resources in multiaccess network environments.

In this context, the chapter defines initially the network selection problem, focuses on specific approaches for network selection that may involve either Fuzzy Logic-based schemes or Multiattribute Decision Making (MADM) methods and presents the guidelines for the exploitation of the afore-mentioned tools. However, since network selection is only one aspect of the QoS provisioning problem, multiaccess network scenarios are also considered and the ways in which admission and load control mechanisms may be used to further ensure optimal QoS for all users are investigated. Finally, the authors investigate the effectiveness of proposed schemes in terms of pricing, apart from the provision of high levels of QoS.

11.1 Introduction

Nowadays, there is a plethora of independent radio access technologies (RATs), each supporting distinct coverage, mobility, data rates, and QoS. Future wireless networks have been envisaged as a convergence platform, where heterogeneous RATs will leverage on a converged all-IP core network to create an adaptive self-resilient network, such that services may be provisioned optimally through the most efficient access network. ITU's (International Telecommunication Union) vision of *Optimally Connected, Anywhere, Anytime* published in Recommendation [19] states that future wireless networks could be realized by a coalition of different RATs, each connected to a common IP-based core network. Upon such a scenario, the heterogeneity of access networks, terminals, and services should be exploited to enable better utilization of radio resources in order to improve the overall system capacity and QoS of users.

The corner stone of all visions on future wireless networks is that the latter will constitute a multiaccess network environment, where multiple heterogeneous access networks will be available. This assumption brings up the issue of selecting the most appropriate access network to cover a specific application's requirements and suggests the investigation of the relevant network selection approaches that have been adopted by widely accepted telecommunication standards, which in the frame of this chapter is performed briefly in the following paragraphs.

Currently, the 3rd Generation Partnership Project (3GPP) is defining an

Access Network Discovery and Selection Function (ANDSF) [3] to assist mobile devices in discovering and deciding which network to access. In fact, ANDSF is used for the Interworking of 3GPP and non-3GPP networks; however, the same principles may be applied to any type of network. ANDSF is based on a simple client–server architecture with Access Network Info Request and Response messages: mobile devices may contact the ANDSF server by sending an Access Network Info Request message to it. The ANDSF server responds back to the mobile device by sending the Access Network Info Response message. In this message, ANDSF provides two types of information, that is, network discovery and network selection information. Network discovery information is intended to help the mobile device to discover networks in its neighborhood, and thus it may contain information on the network type, network ID, used radio frequency, and channel. The 3GPP TS 24.312 technical specification [3] defines additionally management objects (MO), consisting of relevant parameters for intersystem mobility policy and access network discovery information that can be used and managed by the ANDSF. With network selection information on the other hand, an operator can affect which networks the mobile devices are using. Network selection information may remain static for a considerably long time.

IEEE 802.21 [5] is developing standards to enable handover between heterogeneous link layers, including both IEEE 802 and non-IEEE 802 networks. Furthermore, the 3GPP stage 2 technical specification [1] covers the architecture of 3GPP Interworking WLAN (I-WLAN) with 2G and 3G networks. This specification also discusses network selection issues and borrows ideas from the cellular Public Land-based Mobile Network (PLMN) selection principles [2]. In 3GPP PLMN selection, a mobile node may automatically select cells that belong to its Home PLMN, Registered PLMN, or an allowed set of Visited PLMNs.

Finally, addressing the fact that network selection behavior of the mobile device is defined only for 3GPP access types (i.e., I-WLAN), the 3GPP TR 22.912 specification [4] specifies the network selection procedures requirements for non-3GPP access types (e.g., Bluetooth, WLAN, and wired connections), covering both automatic and manual selection, as well as operator and end-user management. The aim is to ensure predictable behavior and also allow the user or application to select the most appropriate type of access for the service required.

As already indicated in the chapter's overview, network selection approaches and mechanisms are however only one aspect in the frame of enhancing the QoS experienced by the end users: A remarkable potential for further enhancing QoS provisioning is further related to resource allocation mechanisms and, more specifically congestion, avoidance control schemes. Radio Resource Management (RRM) strategies are responsible for the efficient utilization of the air interface resources in the Radio Access Network (RAN), aiming at developing schemes to guarantee a certain prior agreed QoS, to maintain the planned coverage area, to offer high capacity, etc. RRM plays a major role in QoS provisioning for wireless communication systems, since the performance of RRM techniques has a direct impact on each user's individual performance and on the overall network performance as well [41].

A major track of research lies in the allocation of bandwidth as a resource and the implementation of adaptive admission control algorithms based on available bandwidth and QoS requirements. Resource management issues in wireless networks based on cellular architecture involve specifically dynamic channel assignment, dynamic transmit power control, load balancing, mobility management, and so on. RRM techniques control the amount of the assigned resources to each user with the objective of maximizing some function such as the total network throughput, total resource utilization, or total network revenue, subject to some constraints such as the maximum call blocking/dropping rate, and/or the minimum signal to interference ratio [16]. In the light of the heterogeneous future wireless networks, the significance of RRM strategies increases significantly [35].

The scope of this chapter is to address the issue of enhanced QoS provisioning in modern wireless networks through the deployment of network selection and resource allocation algorithms and schemes. The chapter consists of two main parts. In the first one, the key factors of network selection are examined and some guidelines for exploiting tools such as Fuzzy Logic and MADM methods are exposed. MADM methods, which involve the selection of a series of criteria that impact the decision process and the comparison of the alternatives taking into account the relative importance of these criteria, are deployed in various approaches for network selection. In these approaches, a weight is assigned to each criterion to reflect its importance in the final decision, and then alternatives are evaluated based on their overall performance. MADM therefore provides a solid framework for network selection, since the latter constitutes a multicriteria decision problem. On the other hand, the use of Fuzzy Logic addresses the issue of fuzziness (i.e., imprecision), and may allow for the inclusion of subjective criteria in the decision process.

In the second section of the chapter the authors provide an overview of congestion avoidance control schemes and propose a framework for congestion avoidance control to which they embed the idea on federation agreements between service providers. A set of simulations is also carried out to test the efficiency of this framework. The framework is finally associated with network selection, aiming at setting the grounds for future research and highlighting possible applications of the network selection process, since the latter currently appears only during access admission control or handovers.

11.2 Network Selection in Heterogeneous Environments

11.2.1 Network Selection Problem

The selection of the most efficient and suitable access network to meet a specific application's QoS requirements has become a significant topic, the actual focus of which is maximizing the QoS experienced by the user. The end-users can potentially take wise decisions on which access network to connect to on the basis of several merit functions including the current load of the network and the cost-for-connectivity. The main concept of the Network Selection problem [9] is shown in Figure 11.1. The goal is to aid the user/mobile terminal in connecting to the radio access network that will best serve the requested service, according to a series of metrics that may refer to technical characteristics, economic aspects, etc. However, since multiple factors have to be taken into account, it is no longer easy to rank the candidate networks according to preference on a single criterion. In fact, multiple criteria have to be combined and scaled in a meaningful way. In addition, various criteria in the decision process may oppose to each other (e.g., a desirable increase in QoS may be accompanied by an undesirable increase of price). Thus, trade-offs are sometimes required.

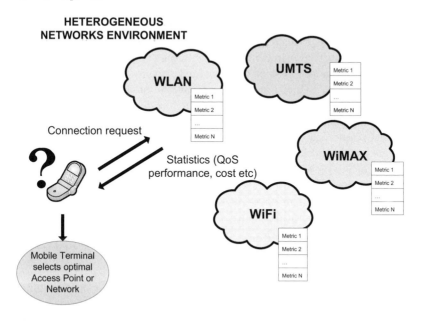

FIGURE 11.1
Network selection problem.

11.2.2 Multiattribute Decision Making for Network Selection

Several researchers have considered in the frame of the network selection problem the use of MADM algorithms to evaluate and rank candidate networks in a preference order [12, 38, 40, 42]. Numerous types of MADM algorithms exist, and several of them may be suitable for solving a decision problem so that the decision maker may encounter the task of selecting among a number of feasible methods the most appropriate one. As already mentioned, MADM methods involve the selection of a series of criteria that impact the decision process, in this context the network selection process, the calculation of the weights of the aforementioned criteria, and the final ranking of the alternatives. These steps are described in detail in the following paragraphs.

11.2.2.1 Selection Criteria

Evaluating the performance of a wireless network presupposes the existence of adequate metrics that reflect the network's actual capability to satisfy its users. *Key Performance Indicators* (KPIs), a set of measurements used to keep track of the network status over the time, may be used as such metrics. KPIs can be split in two types as to whether they describe the network's resources (e.g. bandwidth availability, coverage, etc.) or the QoS provisioned. The main KPIs related to QoS can be measured in any type of packet-switched network. Some of them that may as well serve as evaluation criteria are indicators of Delay, Jitter, BER, Utilization, Throughput, Coverage, Reliability, etc.

Another important criterion that affects the evaluation of the wireless networks' efficiency is the cost of the provided services. Note that most parameters (e.g., Delay, Jitter) are considered as the smaller-the-better, meaning that lower values are desirable. On the contrary, parameters such as throughput, coverage, and reliability are considered as the larger-the-better, since the higher the value, the more satisfied the user will be. Of course, MADM requires also an initial indication on each parameter's relative importance with regard to other parameters. As far as network selection is concerned, such an indication may be acquired from questionnaires or measurements [11, 27]. This issue is explained in detail in the next paragraph, which deals with the extraction of the criteria weights.

11.2.2.2 Extraction of Weights

Once the selection criteria have been determined, the next step is to define the importance (aka weight), of each one of them in the final outcome. In most cases encountered in the literature, the weights of the selection criteria are defined through the creation and analysis of questionnaires, which capture the user's overall perception of a service. However, such approaches depend only on user feedback to determine the relative weights and thus cannot be considered precise, since user's perception and opinion is subjective. Alternative approaches consider the system's characteristics and adapt the values of weights according to each cell's performance [11, 27], in order to provide more objective and thus meaningful weights.

To address the former insufficiencies, in this chapter the authors rely for weight specification on mathematical tools, such as fuzzy logic and multiattribute decision making, which have the capability to incorporate the imprecision that is innate in the decision maker's subjective judgments. As a result, in this section, the authors propose a framework for extracting weights through the combination of two mathematical methods, called Principal Component Analysis (PCA) and Analytic Hierarchy Process (AHP).

PCA reveals the internal structure of data in a way that best explains their variance, and thus can be used in the frame of network selection as a means of defining the relative importance of the selected criteria, exploiting actual data from the networks [37]. More specifically, PCA constitutes a method for identifying patterns and expressing data, in such a way so as to highlight their similarities and differences. This enables use of the variance of data as a means of comparison. By reconstructing data based on selected eigenvectors, a pattern is revealed that directly indicates how scarce a data set is compared to another. A great advantage of this approach is that it removes the impact of diverse measurement units.

Still, while the weights of all QoS parameters may be defined through PCA, the same does not apply for the weight of cost. The reason for this exception is, on the one hand, that cost values do not differ during a service session and therefore no diversity applies, and on the other hand, the fact that the importance of this factor depends on the preferences of the user and not on the mathematical relations extracted from data sets. Each network employs its own fixed price per byte pricing scheme and is subjected to different background traffic patterns. On the other hand, every user wants his data to be delivered in timely way at the lowest price. The more delay the user experiences, the less he is willing to pay. The user may employ different possible tactics to select the network that will maximize his satisfaction. For example, a user may choose to always select one designated network regardless of its current characteristics or to always minimize his expenses by choosing the cheapest network, or decide to continually follow the strategy of random network selection. Based on the above, users may be classified according to their preferences as *risk neutral, risk seeking, or risk adverse.* The significance of this differentiation will become clear in the next section of the chapter, where the authors will indicate how user preferences affect the user's behavior.

The second part of the proposed framework involves the extraction of the criteria weights through the application of AHP. The latter is used in various decision-making problems due to its ability to produce more objective weights as compared to merely relying on vague comparisons [34]; for this reason, it has also been selected for the purposes of this work. One of the main advantages of this method is the relative ease with which it handles multiple criteria. In addition to this, AHP is easy to understand, and it can effectively handle both qualitative and quantitative data. Finally, its use does not involve complex mathematics.

The AHP decision problem is structured hierarchically at different levels. The top level of the hierarchy represents the overall goal, while the lowest level is composed of all possible alternatives. One or more intermediate levels embody the decision criteria and subcriteria. The relative importance of the decision elements (weights of criteria) is assessed indirectly through a series of comparative judgments: the decision-maker provides his preferences by comparing all criteria, subcriteria, and alternatives with respect to upper level decision elements. Several application examples of AHP can be found in the literature [11, 12, 25, 27].

Since AHP presupposes certain preknown relations between the selection criteria, the outcome of the application of PCA can be directly inserted to AHP through the use of the Direct Weighting scale. The two methods can thus be combined; of course, one could choose to use AHP directly. In any case, the weights of QoS parameters deriving through AHP have to be normalized, taking into account cost, an additional selection parameter that bears its own prefixed weight.

11.2.2.3 Ranking of Alternatives

The final step in the process of network selection lies in the evaluation of the alternatives. Thus, in this section the authors present a series of MADM methods (e.g., SAW, ELECTRE, and TOPSIS) for ranking and evaluating alternatives. These methods estimate the performance of alternatives based on the selected evaluation criteria, and then rank the latter according to certain indices. The outcome of this process is, of course, the selection of the most suitable network for the specific service requirements.

Before any of these methods can be used, a *normalized decision matrix* has to be constructed. The problem of network selection can be, in fact, modeled as $P = (A, C, w)$, where

- $A = \{1, ..., N\}$ denotes the set of alternatives, in this case candidate networks

- $C = \{1, ..., M\}$ denotes the set of criteria impacting the decision process of network selection,

- $w = \{1, ..., M\}$ denotes the set of weights assigned to the selected criteria depending on the information about the specific service requested or the user profile, so that $\sum_{i=1}^{M} w_i = 1$

Consider the simple example where the decision on network selection has to be based on the values of the delay (D), jitter (J), utilization (U), throughput (T), and cost (C) parameters. Based on the scores achieved for each one of the aforementioned criteria, the ith candidate network can be represented by a vector as

$$DM_i = [\ D_i \quad J_i \quad B_i \quad U_i \quad T_i \quad C_i\ ]$$

For N alternative networks to be considered in the selection process, the decision matrix DM is formulated as follows:

$$DM = \begin{pmatrix} D_1 & J_1 & B_1 & U_1 & T_1 & C_1 \\ D_2 & J_2 & B_2 & U_2 & T_2 & C_2 \\ \vdots & \vdots & \vdots & \vdots & \ddots & \vdots \\ D_N & J_N & B_N & U_N & T_N & C_N \end{pmatrix}$$

At this point attention must be drawn to the fact that the utility associated with each alternative may be a monotonically decreasing function of parameters that are viewed as smaller-the-better (delay, jitter, BER, and cost) and a monotonically increasing function of parameters that are considered as larger-the-better (throughput). However, since a monotonically increasing or decreasing level of importance is desired for all criteria considered, the authors adopt the modification proposed by [10]. The main idea of this modification lies in the concept of a reference network, which is considered as an access network that demonstrates the desired performance with regard to all criteria in question as this is perceived by the user.

The reference access network is used to calculate the absolute difference between the value of each one of the attributes of matrix DM and the corresponding reference attribute value, allowing to assume that all criteria have a monotonically decreasing utility: the larger the attribute value, the farther it is from the desired or reference value. The application of most of the aforementioned MADM methods requires furthermore that the values assigned to all criteria are measured on a common scale. In order, thus, to remove the impact of the diverse measurement units, each one of the adjusted attribute values in row i of a specific column j is normalized using Equation (11.1):

$$\overline{v_i} = value_{i,norm} = \frac{\max\limits_{j=1,\dots,N}\{V_{j,adj}\} - V_{i,adj}}{\max\limits_{j=1,\dots,N}\{V_{j,adj}\} - \min\limits_{j=1,\dots,N}\{V_{j,adj}\}}, \tag{11.1}$$

where $V_{i,adj} = V_i - V_{i,ref}$ and $\max\limits_{j=1,\dots,N}\{V_{j,adj}\}$, $\min\limits_{j=1,\dots,N}\{V_{j,adj}\}$, indicate, respectively, the maximum and minimum measurement achieved for the specific criterion. As a result, every attribute is measured in dimensionless units, making it possible to perform interattribute comparisons. Note that the normalization procedure is either optional or necessary depending on the MADM method to be used afterwards (TOPSIS has its own normalization process). The normalized decision matrix is finally represented as follows:

$$DM_{norm} = \begin{pmatrix} \overline{D_1} & \overline{J_1} & \overline{B_1} & \overline{U_1} & \overline{T_1} & \overline{C_1} \\ \overline{D_2} & \overline{J_2} & \overline{B_2} & \overline{U_2} & \overline{T_2} & \overline{C_2} \\ \vdots & \vdots & \vdots & \vdots & \ddots & \vdots \\ \overline{D_N} & \overline{J_N} & \overline{B_N} & \overline{U_N} & \overline{T_N} & \overline{C_N} \end{pmatrix}$$

Afterwards, a ranking MADM method may be applied. Explanations of these methods as well as examples of their application may be found in numerous works in the literature [12, 25, 38, 40, 42].

11.2.3 Fuzzy Inference Systems (FIS) for Network Selection

11.2.3.1 Expressing Imprecision with Fuzzy Numbers

There are some situations in which information may be hard or even impossible to quantify due to its nature, and thus, it can only be expressed in linguistic terms (e.g., when evaluating the sound quality of a telephony session, terms like "good," "fair," "poor" can be used). In other cases, precise quantitative information cannot be provided because it is either unavailable or the cost for its computation is too high and an approximate value can be tolerated. In this case, linguistic terms can be used as well instead of numeric values.

Such qualitative information can be mathematically modeled through the use of Fuzzy Sets Theory. The latter handles fuzziness and represents qualitative aspects as linguistic variables, that is, variables whose values are not numbers but words or sentences according to a natural or artificial language. More specifically, qualitative information expressed through linguistic terms (e.g., "low," "medium," and "high") can be converted to fuzzy numbers using a suitable conversion scale. Processing of the relevant information takes place using these fuzzy numbers, which are finally converted to crisp numbers through the process of defuzzification.

Regarding network selection, criteria such as user satisfaction or cost are rather subjective and in certain cases it is best to be expressed in linguistic terms that correspond to fuzzy numbers. [42] addresses the issue of imprecise criteria in the frame of handover selection, whose values cannot be easily obtained. A combination of fuzzy logic and genetic algorithms is also proposed in a similar context by [7].

11.2.3.2 Modeling Network Selection as an FIS

This part of the chapter brings to the reader's attention an alternative approach for establishing a highly dynamic comparison framework. This time, instead of using MADM methods, the authors clarify how a decision system can be established with the help of fuzzy logic and a set of rules, the essence of which will be explained in the following paragraphs. As shown in Figure 11.2, a fuzzy decision system is made up of:

- A *fuzzifier* that converts input values into linguistic variables matched to membership degrees based on membership functions,

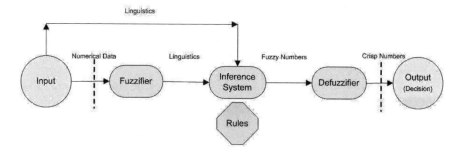

FIGURE 11.2
Fuzzy system components.

- An *inference system* that applies the fuzzy rules, and

- A *defuzzifier* that deduces the final decision from the intermediate ones.

Fuzzification makes possible the comparison of the decision parameters. Each input is characterized by a membership function, which measures the degree of membership of a given value to the fuzzy subsets defined. If a variable value changes slightly, the degrees of membership to the fuzzy subsets also vary slightly. In fuzzy logic, membership degrees' variations are continuous contrary to conventional logic in which membership degrees can be either 0 or 1.

The fuzzy controller uses a list of "IF" – "THEN" rules, that may be based on prior field experience, questionnaires, network measurements, etc., to control the system. The following example of such rules may be considered. According to a survey regarding the provision of a service, end users have provided feedback that may be translated into the following set of rules:

- If the QoS offered is "High" and the cost is "Very Low," then the user's satisfaction is considered as "Very High".

- If the QoS offered is "Low" and the cost is "Medium," then the user's satisfaction is considered as "Low."

- If the QoS offered is "High" and the cost is "Very High," then the user's satisfaction is considered as "High."

This simple example constitutes of 2 criteria and 5 linguistic variables, providing a total of 5^2 possible combinations. Of course, having all 25 different rules is not obligatory, since for a more complex example the overall complexity would reach unacceptable levels. In general, the maximum number of rules required is N^M, where N is the number of linguistic variables and M the number of criteria. Using a suitable tool, different combinations can be explored. More specifically, changing the value of one criterion while maintaining the others, may invoke a certain change in the final decision, which is quantified by the system. An efficient tool for performing such an analysis is the *Matlab Fuzzy Logic Toolbox*, as it provides computational and visual aid. A tutorial on this tool is available by [29], while more research approaches towards this direction may be found in the literature [21].

The advantage of such an approach is that it solely requires some initial estimations. Afterwards, rules can be easily formulated or modified. Also, one may easily preview the anticipated impact of a modification in the value of one or more criteria. The behavior of a fuzzy system can be changed by modifying the appropriate rules. Learning techniques can be also applied to improve the decision rules by optimizing the membership functions of the latter.

In the following paragraphs, we are going to explain in detail how an appropriate FIS for network selection can be built. The network ranking in this case can be achieved through fuzzy logic. More specifically, the output of a rule is computed using the Mamdani method [29]. The idea behind using a Mamdani rule base is that the rules for many systems can be easily described by humans in terms of fuzzy variables (linguistics). Thus, one can effectively model a complex nonlinear system with common sense rules on fuzzy variables. As an example, the reader may consider a system with seven inputs (packet delay, packet jitter, packet loss, available bandwidth, coverage, reliability, and cost), and one output. Assuming that each input data is divided into five partitions ("Very Low," "Low," "Medium," "High," "Very High") represented by membership functions and the output data into seven partitions ("Very Low," "Low," "Medium Low," "Medium," "Medium High," "High," and "Very High"), this system requires $5^7 = 78.125$ rules to express all possible combinations, becoming extremely complex and thus non scalable.

The FIS may be simplified, and the number of rules may be reduced by choosing a hierarchical approach instead, that is, by dividing the original FIS in three smaller ones, as shown in Figure 11.3. The three output variables are in this case QoS, Performance, and Rank, as the value of each one is composed by lesser parameters in the lower level of the hierarchy. In this way the number of required rules is reduced to a total of $5^4 + 5^2 + 5^3 = 775$. Through this simplification rules can now be defined for all three systems. A simplified example of such rules may be:

IF **QoS** is High and **Cost** is High and **Performance** is Medium THEN **Rank** is High.

The last step is to provide some data for all seven criteria and for all available networks to be used as input to the FIS. A fair question that could be raised here is how this data can be acquired. A simple approach would be to use as input the mean values of measurements recorded during past sessions for all the networks. On the other hand, a more sophisticated approach would be

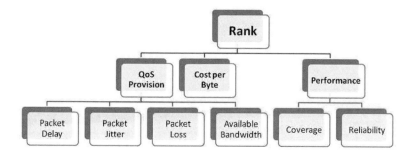

FIGURE 11.3
Selection criteria hierarchy.

to deploy regression or more complex forecasting techniques. The objective of this process would be to estimate the networks' expected performance based on past observations and measurements. For a given set of input data, the desired output, aka, the ranking of the network, is finally calculated by the FIS using the series of rules described previously. Of course, the network with the highest ranking value is the optimal one.

11.2.3.3 Adaptive Fuzzy Model (ANFIS)

In this section, the authors examine the application of the Adaptive Network based Fuzzy Inference System (ANFIS) to the problem of network selection as an alternative approach to the previous one. ANFIS is an FIS whose membership function parameters are adjusted using either a back propagation algorithm alone or in combination with a least-squares type of method [28]. A fuzzy system of this kind is able to learn from the data it is modeling, since it has a structure similar to that of a neural network. The parameters associated with the membership functions change through the learning process. A typical fuzzy rule in a Sugeno fuzzy system [28] has the format:

> *If x is A and y is B then $z = f(x,y)$*, where A and B are fuzzy sets.

The firing strengths (weights) are usually obtained as the product of membership grades of the premise part, and the output is the weighted average of each rule's output. Following the previous example, we may implement ANFIS as seven inputs and one output system, using the fuzzy logic tool box of MATLAB. Having created the data set, the next step is to "train" the system. Using the subtractive clustering technique, the ANFIS automatically selects the membership function and also generates the new FIS. Training here is a way to express the set of rules that can be applied over the generated FIS. The rules and the membership functions are updated in each iteration to reduce the error in optimization. An application of this process can be found in the work of Kher [21].

11.3 Congestion Avoidance in Heterogeneous Networks

11.3.1 Congestion Avoidance Control Mechanisms

Congestion avoidance, which involves call admission control and load control, constitutes part of system's RRM. The aim of congestion control is to regulate the operation of the total network, so as to ensure the benefit of service in the existing connections and simultaneously some compromise for the admittance of new connections. The system, therefore, based on the congestion avoidance control, accepts or rejects a connection according to some strategy.

Call Admission Control (CAC) has been extensively studied in wireless networks as an essential tool for congestion control and QoS provisioning. Different aspects of CAC design and performance analysis have been investigated in the literature; However, the problem of CAC in wireless networks is quite sophisticated due to the unique features of the latter such as channel multiple access interference, channel impairments, handoff requirements, and limited bandwidth. [6] provides a full categorization of CAC schemes.

CAC plays an important role in QoS provisioning in terms of the signal quality, call blocking and dropping probabilities, packet delay, loss rate, and transmission rate. In the first and second generation of wireless systems, CAC was developed for a single-service environment. In third-generation (3G) wireless systems, due to the heterogeneity of offered services in terms of user requirements, various QoS profiles have been adopted, leading unavoidably to more complex CAC schemes [6].

Load Control (LC), on the other hand, ensures the effective performance of a wireless network by keeping the load of the network within normal boundaries. It performs traffic balancing between nodes or cells of the same mode preventing congestion situations. Reactive load control is employed to cope with overload situations of the network, when the users' QoS is at high risk. In such cases, the load control performs several actions to decrease the amount of traffic in the congested cell. The unwanted congested states may be prevented by the load control mechanism that continuously monitors the system. In the following sections, the authors present selectively a series of common CAC and LC approaches.

11.3.2   Approaches and Algorithms

11.3.2.1   Common CAC Approaches

Some CAC schemes rely on signal quality to decide whether a new call can be admitted or not; a new call is admitted only if the value of Signal Interference Ratio (SIR) is higher than a minimum value. Various SIR-based CAC schemes have been proposed in the literature [22, 24, 31, 36]. The residual capacity (R_k) of a cell is calculated periodically, and when a new user arrives at cell k, it is checked whether R_k is greater than zero; if it is, the new call is accepted; otherwise it is rejected.

$$R_k = \begin{cases} \left| \dfrac{1}{SIR_{TH}} - \dfrac{1}{SIR_k} \right|, \text{if } \left| \dfrac{1}{SIR_{TH}} - \dfrac{1}{SIR_k} \right| > 0 \\ \\ 0, \text{otherwise.} \end{cases} \tag{11.2}$$

In the previous equation, SIR_k is the measured SIR in cell k and SIR_{TH} is the threshold. An extension of this approach adopts the use of the interference coupling between adjacent cells b, as

$$R_k = \begin{cases} \min\{R_k^{(j)}, j\epsilon K(k)\}, \text{if } \min\{R_k^{(j)}, j\epsilon K(k)\} > 0, \\ \\ 0, \text{otherwise} \end{cases} \tag{11.3}$$

where

$$R_k^{(j)} = \begin{cases} \left| \dfrac{1}{SIR_{TH}} - \dfrac{1}{SIR_k} \right|, \text{if } j = k \\ \\ \left| \dfrac{1}{b} \left(\dfrac{1}{SIR_{TH}} - \dfrac{1}{SIR_k} \right) \right|, \text{if } j \neq k \end{cases} \tag{11.4}$$

whereas $K(k)$ is a subset that contains cell k and its neighboring cells. Certain approaches accept calls with a SIR value that ensures blocking probability below a specified threshold, where $P_{BLK}(k) = Pr\{R_k = 0\}$. Bandwidth-based CAC schemes, on the other hand, express the SIR requirement as

$$P(I > \frac{W}{R}) < L, \tag{11.5}$$

where W is the system bandwidth, R is the transmission rate, and L is a system parameter that determines the signal quality reliability.

11.3.2.2   Common LC Approaches

In Wireless Local Area Networks (WLANs), the most widely used criterion for load control is the signal power as well, since the received power must overcome a certain threshold so that users may successfully connect to Access

Points (APs). The number of users that may be served by an AP, as well as the traffic type, plays an important role in determining the load that can be supported by a system. Generally two criteria may be applied in deciding whether a new user request should be accepted or not [15, 33]. The first one involves the Relative Occupied Bandwidth (ROB) [17], which is calculated as

$$B_{\text{occu}} = \frac{T_{\text{busy}}}{T} \cdot 100, \quad (11.6)$$

where T is a time window and T_{busy} is the time during which the wireless media is occupied. In other words, ROB expresses the utilization percentage of the wireless media in a given duration.

The second criterion involves the Average Collision Time (ACT) [17], which measures the average rate of collisions in a given period T as

$$R_c = \frac{N_c}{N_t}, \quad (11.7)$$

where N_c is the number of collisions and N_t the number of all transmissions. Other LC schemes rely on the total throughput to decide whether it is possible to accept new connections or not. According to these schemes, in WCDMA systems, a new connection is rejected if the following condition applies in the uplink [8, 20, 30]:

$$n_{\text{ul}} + \Delta L > n_{\text{ul,threshold}} \quad (11.8)$$

where

$$n_{\text{ul}} = \frac{E_b/N_0}{W/R} \cdot N \cdot u \cdot (1 + i), \quad (11.9)$$

being the load factor, E_b/N_0 the energy per bit, W the chip rate, N the number of all users in the same cell, $i = \dfrac{\text{other cell interference}}{\text{own cell interference}}$, R the bitrate, ΔL is a load increment, and u the activity factor. Similarly, the condition for the downlink is

$$n_{\text{dl}} + \Delta L > n_{\text{dl,threshold}}, \quad (11.10)$$

where

$$n_{\text{dl}} = \sum_{j=1}^{N} \cdot u_j \cdot \frac{(E_b/N_0)_j}{W/R_j} \cdot [(1 - a_j) + i_j], \quad (11.11)$$

being the load factor, and a the orthogonality factor.

11.3.3 Modeling Federation Agreements

One of the key elements that may be introduced to enhance the QoS offered to users is the concept of federation agreements between different providers, in a way similar to roaming, which as a service allows a subscriber of one network operator to use the services of another operator when inside the latter's coverage area [32]. The legal roaming business aspects negotiated between the roaming partners for billing of the services obtained are usually stipulated in so-called roaming agreements. In a similar way, a *federation agreement* can be seen as a contract between two providers that specifies the conditions, that is, the economic and legal terms, as well as the authentication and authorization procedures, under which the latter can collaborate. In the rest of the chapter we will assume that such a contract between collaborating providers preexists, and that the actual agreement is established upon availability of resources.

The goal of such an agreement is to let another provider handle a session request that cannot be adequately served by the "home provider." At this point, it is clarified that under the term "home provider" we refer to the provider with whom the customer maintains a signed contract. According to the proposed approach, however, we consider that in case of insufficient resources, the customer is free to pursue a higher QoS with another provider, given that there is some kind of federation agreement between the visited and the home provider (possibly under a small monetary penalty). In order for this scheme to succeed, of course, financial incentives need to be in place, so that all parties have something to gain. This approach is clarified in the following paragraphs by examining two possible scenarios, in which the idea of a federation agreement is applied.

Scenario 1: A customer, having a contract with provider A, wishes to make a phone call; however, the default provider is not able to accept the customer's request as this would congest the network. Normally, Provider A would reject the session request, thus obtaining no income for serving the call. However, in case there is a federation agreement with Provider B, Provider A may forward the customer details to the latter. Upon availability of network resources and existence of financial motives, Provider B may accept the session request in question and collect, as well, the generated income, submitting only a small fee to Provider A, as the latter is actually the mediator among the customer and visited provider. In this way both providers acquire income that they would not have acquired otherwise, and the customer is also served. This scenario is illustrated in Figure 11.4.

Scenario 2: In this scenario, the same customer is now served by Provider A; however, the customer is dissatisfied with the QoS he experiences, and therefore decides to pursue higher QoS at another provider (even at a greater cost). To this end, on the basis of a federation agreement, Provider A forwards the customer details to Provider B, inquiring whether the latter is able to serve the customer more efficiently. Upon a positive response, a handover is performed, and the customer is now served by Provider B. Compared to the previous scenario, the difference in this case is that, since the customer leaves on his own free will, he is obliged to pay a small monetary fee (penalty) to Provider A.

It may seem that Provider A loses income by "giving up" his customer, but that is not actually the case; within the frame of such a federation agreement, Provider A receives both the customer's penalty and the fee submitted by Provider B, without allocating any resources. At the same time, he manages

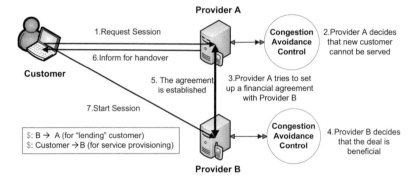

FIGURE 11.4
Provider change due to lack of resources.

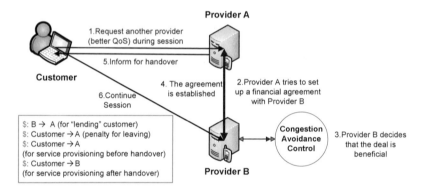

FIGURE 11.5
Provider change due to customer dissatisfaction.

to avoid possible congestions, because of which customers could be dissatisfied in the first place. On the other hand, through serving the call, Provider B receives, as previously, an income not to be acquired in any other way. Finally, the customer experiences a higher quality of service albeit even at a greater cost, since this was his initial objective. The scenario is illustrated in Figure 11.5.

So far the authors have clarified how the concept of federations can be theoretically modeled. An important issue that rises, however, in the scenarios described is how the new provider can be correctly identified. Obviously, if only economic metrics are considered, each provider will redirect unwanted customers to the collaborating provider that will deposit the highest fee; this unfortunately will only randomly act in the customer's best interest. Therefore, new mechanisms need to be developed that encompass a series of criteria to point out the most suitable alternative choice.

This is where the network selection concept and the techniques presented in the first part of this chapter may be introduced, lending to the former a far greater role than simply indicating the optimal network during CAC: in a heterogeneous environment scenario that supports federations, network selection needs to be a repetitive process that takes place each time a decision has to be reached. This need highlights the advantages of the tools (MADM and FIS), described earlier, as they allow for complex decisions to be reached dynamically and spontaneously in dynamic and scalable systems.

11.3.4 Pricing Models

Since the adoption of congestion avoidance schemes unavoidably impacts both the providers' income and the customers' expenses, it is considered crucial to perform a brief analysis on the pricing models that can be applied so as to ensure optimal distribution of resources [18, 14]. In order to develop highly dynamic and flexible systems, instead of adopting fixed prices, pricing mechanisms that tailor the user's perception of a service, that is, the satisfaction the user receives, to economic metrics, need to be sought.

In this direction, the authors suggest two possible pricing models; the results regarding the efficiency of any congestion control scheme highly depend on the choice of pricing model that has been made, since it affects all parties. More specifically, service provisioning may be charged:

- *Based on the levels of QoS reached*: The fewer resources a network provider offers to customers, the less the latter will be obliged to pay. This acts as an incentive for the providers to offer a high level of services to their customers, since an excessive increase in the number of customers will not generate more income [13, 39], where an equilibrium between money and QoS is sought. This model can present numerous variations, depending on the ways the level of received QoS is estimated. In some cases this is achieved with the use of the sigmoid function [23], involves the number (or rate) of lost packets.

- *Per transmitted packet*: In this case, the amount of money a customer pays depends on the number of packets that are exchanged during the session. This approach is adopted in [26], where the authors also clarify that if several transmissions are needed, the user has to pay several times.

Apart from the pricing model adopted, certain parameters of economic nature need to be specified as well. These are the *penalty* paid by the customer for leaving his default provider and the *customer "lending" fee* among collaborating providers. The latter can be either fixed or a percentage of the income received by the new provider. Optimality of the relevant choices can be achieved by adjusting these values to the specific characteristics of the heterogeneous environment under consideration.

For the purposes of this study (i.e., for the simulation to be carried out), the authors utilize the second pricing model described above, assuming, however, that charging takes into account only the received packets instead of all the transmitted ones. This assumption discourages providers from accepting all service requests, since congestion may lead to packet loss and thus income reduction. The following pseudocode explains how the cost for customer i and the income for provider j are calculated.

```
for all providers 1..j..m {
    cost(i) = cost(i) + #rcvPackets(i ⟶ j) * rate(j) + penalty(i ⟶ j)
}
for all customers 1..i..n {
    income(j) = income(j) + #rcvPackets(i ⟶ j) * rate(j) + penalty(i ⟶ j)
    for all providers 1..k..m except j {
        income(j) = income(j) + lendingFee(i)(j ⟶ k) − lendingFee(i)(k ⟶ j)
    }
}
```

11.3.5   Simulation Scenarios

In this section of the chapter, the authors set up an environment of heterogeneous networks and exploit the ideas presented before to test the efficiency of certain congestion avoidance schemes, while adopting at the same time the concept of federation agreements between service providers. The performance of these schemes will be investigated in terms of QoS provisioning and pricing.

The simulations are carried out in NS2, while the scenario under consideration involves three different service providers (802.11b networks) and 100 customers randomly placed in a $500x500$ topology, one third of which request TCP services and two thirds of which request CBR services. We assume by default that the customer has a contract, and is therefore associated with the provider that is closest to him/her. However, in case a provider is forced to reject a session request in order to avoid possible congestion or a customer is dissatisfied, the provider may redirect the customer to another provider, according to the scenarios described before.

The first scheme to be implemented is a quite simple CAC scheme that is based on the assumption that each provider accepts a new session request only if the bandwidth utilization percentage is found below a certain threshold, to which we will refer from now on as the *acceptance threshold*. On the side of the customer a similar threshold is also defined, so as to reflect the latter's requirements with regard to the bandwidth utilization the provider bears. Based on these assumptions, four possible combinations can be identified:

1. Both the customer and provider are satisfied: in this case the session begins.

2. Only the customer is satisfied: the customer is redirected to another provider and no penalty is imposed on him.

3. Only the provider is satisfied: the customer turns to a different provider and pays the penalty as well.

4. Both the customer and provider are dissatisfied: as in case 2, the customer changes provider without paying any penalty.

The second scheme under consideration is a LC scheme, according to which each provider performs periodically a congestion check, and in case this is found positive, a set of calls is terminated. We will refer to this period as the *time window*. Once again, a customer may choose to leave at his own will or be dismissed by the provider. There are several ways to select the calls that should be terminated. For the purposes of this study we have chosen to terminate the most consuming sessions in terms of bandwidth so as to avoid congestions.

So far, the analysis has illustrated how basic CAC and LC can be integrated in a wireless scenario. What remains to be clarified is the way in which a new provider is chosen. Given that, in our example, each provider has always two choices (since there are three providers in total), the optimal choice has to be defined through a set of decision criteria. Ideally, a complex decision making algorithm has to be formulated. For simplicity reasons, we assume that in the specific scenario the economic incentives are identical for both choices, so only QoS parameters are to be taken into account.

Nevertheless, as a provider may guarantee high levels of QoS for nearby nodes, but not for distant ones as well, an additional aspect to be considered is the proximity factor (pf), which may be used to express the percentage (%) of degradation in QoS, as a result of the distance. Therefore, the metrics estimated by each provider have to be decreased by a certain amount that increments according to the distance of the customer and the potential new provider as follows:

$$\widetilde{Metric\%} = Metric\% - pf \cdot \text{distance}(m) \qquad (11.12)$$

In this work we intend to examine the efficiency of some basic selection modes, by investigating the following three alternative handover scenarios:

- First scenario: redirect to closest provider.

- Second scenario: redirect to provider with least bandwidth utilization.

- Third scenario: redirect to provider with least packet loss possibility.

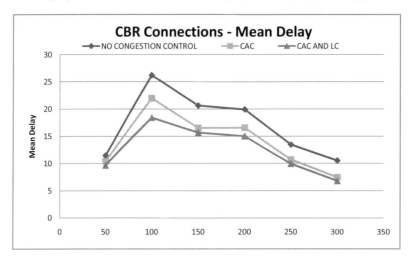

FIGURE 11.6
Mean delay (second) for CBR connections.

11.3.6   Simulation Results and Discussion

In this part of the chapter we expose the results of the simulation carried out using the aforementioned simulation schemes. The figures presented hereafter aggregate indicative results for all three providers with the aim of illustrating how efficiently congestion is dealt with in each case. Note that for both CBR and TCP connections, the mean values of key QoS metrics are illustrated over time. The duration of the simulation is in all cases equal to 300 seconds. Besides the mechanisms' performance with regard to congestion, we also examine the latter's effectiveness in terms of cost (for customers) and income (for providers). We assume that each provider charges different rates which lie between 10^{-4} and 2×10^{-4} monetary units per received packet, while the penalty is set at 0.3 monetary units and the customer lending fee at 5% of received income.

Mean values are initially gathered for all three control schemes to be simulated, that is, (1) the application of no congestion control at all, (2) the sole application of CAC and finally, (3) the application of both control schemes. As it can be observed in Figure 11.6 to 11.9, congestion occurs at around 100 seconds for CBR connections and 200–250 seconds for TCP connections. Note that for CBR connections these mean values concern the parameters of Delay and Packet Loss, while for TCP connections they refer to the metrics of Packet Loss and Throughput.

As indicated by the fluctuation of the measured values for Delay and Packet Loss in Figures 11.6 and 11.7, the application of the proposed CAC scheme significantly enhances the overall performance of CBR connections (19%). In fact the restriction imposed by the acceptance threshold used in the CAC scheme

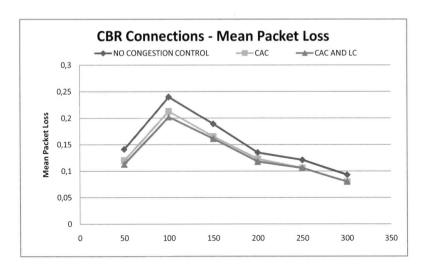

FIGURE 11.7
Mean packet loss (%) for CBR connections.

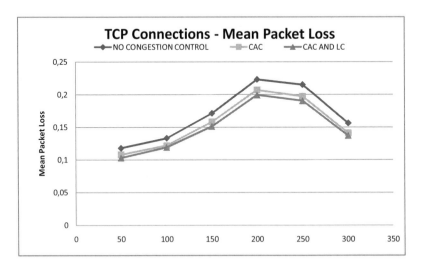

FIGURE 11.8
Mean packet loss (%) for TCP connections.

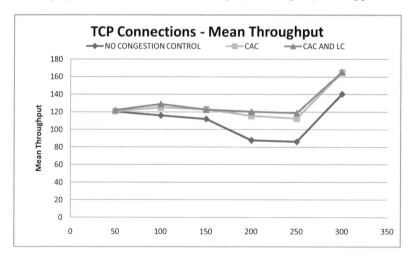

FIGURE 11.9
Mean throughput (Kbps) for CBR connections.

leads to a more balanced customer distribution among the three providers that results in lower Delay and Packet Loss rates. A slight improvement is further achieved by applying the LC scheme (overall improvement 21%). On the other hand, due to penalties, the overall cost for customers is also increased in these two cases by 10% and 11%, respectively, thus by a percentage significantly lower than that of the QoS. As far as TCP connections are concerned, similar conclusions can be drawn: the increase in QoS reaches 21 to 22%, while the corresponding increase in cost is 7–8% respectively. At the same time, the mean income increase for all providers is around 9%.

As the CAC and LC schemes implemented incorporate the concept of federation agreements, one may state that the simulation results point out that by exploiting the latter it is possible to serve efficiently all customers while at the same time providing a more balanced pricing equilibrium. Even though there is an increase in cost, both more satisfactory services are offered and higher income is guaranteed for the providers since all customers are finally served. As indicated by the simulation, in case no federation agreement is applied in the former scenario, a percentage of users of around 10% is denied admission and is therefore not served.

Moving on with the analysis, we examine how the three handover scenarios described earlier perform for the combination of both the CAC and LC schemes, as the latter provides better results in the former experiment. Figure 11.10 to Figure 11.13 illustrate the mean values of the same series of metrics over time for all three scenarios.

As seen from Figure 11.10 to Figure 11.13, the first scenario, which corresponds to the selection of the nearest available provider, stands out as the

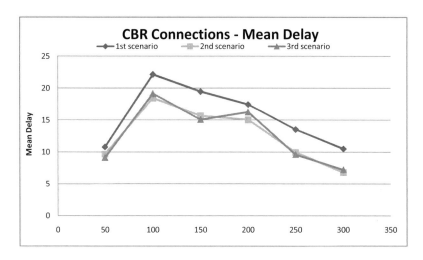

FIGURE 11.10
Handover scenarios mean delay (second) for CBR connections.

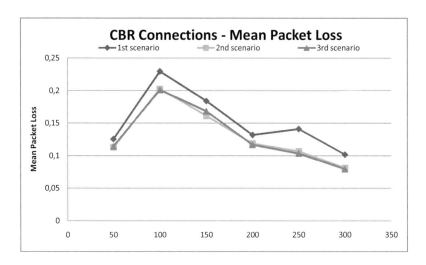

FIGURE 11.11
Handover scenarios mean packet loss for (%) CBR connections.

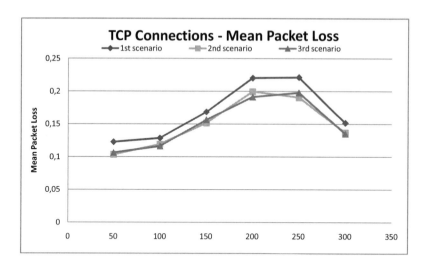

FIGURE 11.12
Handover scenarios mean packet loss for (%) TCP connections.

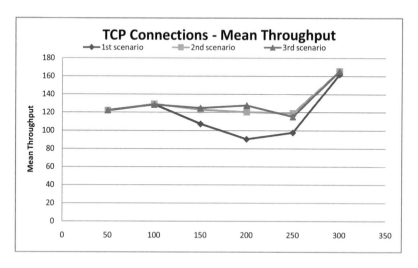

FIGURE 11.13
Handover scenarios mean throughput (Kbps) for TCP connections.

one with the worst performance for both CBR and TCP service types. This was anticipated since the decision relied solely on the network topology, neglecting the networks' current status. The two remaining scenarios, indicating handover to the provider with the lowest throughput utilization and packet loss possibility, respectively, present a similar but still better behavior. More specifically, the first scenario does not offer a satisfactory level of QoS to customers, even though it does guarantee the lowest cost for them. On the other hand, scenarios 2 and 3 offer a QoS improvement of around 15% for CBR and 20% for TCP, and cause at the same time an increase of cost that reaches 11% and 4%, respectively, making them more efficient.

As a concluding remark, the authors highlight that the decision mechanism employed may have a strong impact on the system's performance, since appropriate handovers can drastically increase the customer's satisfaction. Note also that the adoption or not of a specific mechanism depends on the end user's profile; if the majority of users are characterized as risk seeking and thus prefer higher QoS over cost, then handovers should occur more frequently. On the other hand, for risk adverse users who prefer to save some money, handovers should be realized only if the expected QoS improvement is anticipated to be significantly greater that the extra cost.

11.4   Future Research Directions

In this chapter we exposed some basic ideas on network selection and congestion avoidance in heterogeneous environments. The goal was to provide an adequate background, so as to propose mechanisms that can be enriched and applied in complex decision making scenarios. Since this chapter proposed a series of MADM methods for network evaluation and ranking, future research on the issue of network selection should further investigate the performance of these methods, so as to point out the most suitable ones with regard to the problem at hand. Particular emphasis should also be placed on the way in which the weights of network selection criteria are specified. Moreover, the alternative approach of FIS for the problem of network selection should be investigated in terms of its performance and the conversion scales to be utilized.

As far as congestion avoidance control is concerned, more complex congestion avoidance control schemes should be formulated to guarantee the most efficient distribution of resources in dynamic heterogeneous environments. Pricing issues, discussed briefly in the frame of this chapter, should also be further studied and tailored to the specific characteristics of each individual system. Finally, the idea of federation agreements should be further examined as on the one hand it offers unlimited possibilities for algorithmic research and on the other it involves system parameters that may significantly enhance service provisioning and thus the satisfaction the end users receive.

11.5　Conclusion

This chapter addressed the issues of network selection and congestion avoidance control in multiaccess network environments. Due the heterogeneity of different RATs, complex decision schemes were discussed, emphasizing the use of fuzzy logic and MADM methods as tools that may allow for more efficient and objective decisions. The authors pointed out the main challenges involved in the network selection process and exposed some ideas on how the afore-mentioned tools can be successfully utilized. In the second part of the chapter, a framework that considers the possibility of customer exchange and agreements between providers was presented, and possible applications of the process of network selection in this context were highlighted. Finally, the framework was tested through a series of simulations, and results were discussed.

Bibliography

[1] 3GPP. 3GPP System to Wireless Local Area Network (WLAN) InterworkingSystem Description; Release 6; Stage 2. 3GPP TS 23.234, 2005.

[2] 3GPP. Non-Access-Stratum (NAS) functions related to Mobile Station (MS) in idle mode. 3GPP TS 23.122 V6.5.0, 2005.

[3] 3GPP. Access Network Discovery and Selection Function (ANDSF) Management Object (MO). 3GPP TS 24.312 V8.3.0, 2009.

[4] 3GPP. Study into network selection requirements for non-3GPP access. 3GPP TR 22.912 V8.0.0, 2009.

[5] IEEE 802.21. IEEE Standard for Local and Metropolitan Area Networks - Part 21: Media Independent Handover Services. IEEE Std 802.21-2008, 2007.

[6] M. Ahmed. Call admission control in wireless networks: a comprehensive survey. *IEEE Communications Surveys and Tutorials*, 7(1):51–69, 2005.

[7] M. Alkhawlani and A. Ayesh. Access network selection based on fuzzy logic and genetic algorithms. *Advances in Artificial Intelligence*, 8(1), 2008.

[8] S. AlQahtani and A. Mahmoud. Performance analysis of two throughput-based call admission control schemes for 3g wcdma wireless networks supporting multiservices. *Computer Communications*, 31(1):49–57, 2008.

[9] J. Arkko, B. Aboba, J. Korhonen, and E. Bari. Network Discovery and Selection Problem. IETF RFC 5113, 2008.

[10] F. Bari and V. Leung. Application of ELECTRE to network selection in a hetereogeneous wireless network environment. In *Proc. of IEEE Wireless Communications and Networking Conference*, 2007.

[11] D. Charilas, O. Markaki, D. Nikitopoulos, and M. Theologou. Packet-switched network selection with the highest QoS in 4G networks. *Computer Networks*, 52(1):248–258, 2008.

[12] D. Charilas, O. Markaki, J. Psarras, and P. Constantinou. Application of fuzzy AHP and ELECTRE to network selection. In *Proc. of the 1st International Conference on Mobile Lightweight Wireless Systems*, 2009.

[13] D. Charilas, A. Panagopoulos, P. Vlacheas, O. Markaki, and P. Constantinou. Congestion avoidance control through non-cooperative games between customers and service providers. In *Proc. of the 1st International Conference on Mobile Lightweight Wireless Systems*, 2009.

[14] S. Das, H. Lin, and M. Chatterjee. An econometric model for resource management in competitive wireless data networks. *IEEE Network*, 18(6):20–26, 2004.

[15] Y. Dong, D. Makrakis, and T. Sullivan. Effective admission control in multihop mobile ad hoc networks. In *Proc. of the International Conference of Communications Technology*, 2003.

[16] M. Ghaderi and R. Boutaba. Call admission control in mobile cellular networks: A comprehensive survey. *Wireless Communications and Mobile Computing*, 6(1):69–93, 2006.

[17] D. Gu and J. Zhang. A new measurement-based admission control method for IEEE802.11 wireless local area networks. In *Proc. of the 14th IEEE Personal, Indoor and Mobile Radio Communications (PIMRC)*, 2003.

[18] J. Hou, J. Yang, and S. Papavassiliou. Integration of pricing with call admission control to meet QoS requirements in cellular networks. *IEEE Transactions on Parallel and Distributed Systems*, 13(9):898–910, 2002.

[19] ITU. Framework and overall objectives of the future development of IMT-2000 and systems beyond IMT-2000. ITU-R M.1645, 2003.

[20] L. Jing, F. Chen, Y. Dacheng, and Jian. UMTS soft handover algorithm with adaptive thresholds for load balancing. In *Proc. of the 62nd IEEE Vehicular Technology Conference (VTC)*, 2005.

[21] S. Kher, A. Somani, and R. Gupta. Network selection using fuzzy logic. In *Proc. of the 2nd International Conference on Broadband Networks*, 2005.

[22] M. Kim, B. Shin, and D. Lee. SIR-Based call admission control by intercell interference prediction for DS-CDMA systems. *IEEE Communications Letters*, 4(1):29–31, 2000.

[23] H. Lin, M. Chattergee, and S. Das. ARC: An integrated admission and rate control framework for competitive wireless CDMA data networks using noncooperative games. *IEEE Transactions on Mobile Computing*, 4(3):243–258, 2006.

[24] Z. Liu and M. Zarki. SIR-based call admission control for DS-CDMA cellular systems. *IEEE Journal on Selected Areas in Communications*, 12(4):638–644, 1994.

[25] S. Mahmoodzadeh, J. Shahrabi, M. Pariazar, and M. Zaeri. Project selection by using fuzzy AHP and TOPSIS technique. In *Proc. of World Academy of Science, Engineering and Technology*, 2007.

[26] P. Maille and B. Tuffin. Price War in Heterogeneous Wireless Networks. COST IS0605 (Econ@tel) Meeting, 2009.

[27] O. Markaki, D. Charilas, and D. Nikitopoulos. Enhancing quality of experience in next generation networks through network selection mechanisms. In *Proc. of the Mobile Terminal Assisted Enhanced Services Provisioning in a B3G Environment Workshop, IEEE PIMRC*, 2007.

[28] Mathworks. ANFIS and the ANFIS Editor GUI: Tutorial (Fuzzy Logic Toolbox), http://www.mathworks.com/access/helpdesk/help/toolbox/fuzzy/fp715dup12.html .

[29] Mathworks. Fuzzy Inference Systems: Tutorial (Fuzzy Logic Toolbox), http://www.mathworks.com/access/helpdesk/help/toolbox/fuzzy/fp351dup8.html .

[30] Z. Meng, C. Tao, and H. Jiancun. Throughput-based and power-based load control. In *Proc. of the International Conference on Wireless Communications, Networking and Mobile Computing*, 2005.

[31] H. Perros and K. Elsayed. Call admission control schemes: A review. *IEEE Communications Magazine*, 34(11):82–91, 1996.

[32] O. Pohjola. *Game analysis: Roaming agreements*. Nokia Research Center, 2003.

[33] G. Razzano and A. Curcio. Performance comparison of three call admission control algorithms in a wireless ad hoc network. In *Proc. of the International Conference Communications Technology*, 2003.

[34] T. Saaty. Relative measurement and its generalization in decision making: why pairwise comparisons are central in mathematics for the measurement of intangible factors - The analytic hierarchy/network process. *Review of the Royal Spanish Academy of Sciences Series A Mathematics*, 102(2):251–318, 2008.

[35] W. Shen and Q. Zeng. Resource allocation schemes in integrated heterogeneous wireless and mobile networks. *ACM Journal of Networks*, 2(5):78–86, 2007.

[36] S. Singh, V. Krisnamurthy, and H. Poor. Integrated voice/data call admission control for wireless DS-CDMA systems. *IEEE Transactions on Signal Processing*, 50(6):1483–1495, 2002.

[37] Smith, L. A Tutorial on Principal Components Analysis, `http://www.cs.otago.ac.nz/cosc453/studenttutorials/principalcomponents.pdf`.

[38] P. Tran and N. Boukhatem. Comparison of MADM decision algorithms for interface selection in heterogeneous wireless networks. In *Proc. of the 16th International Conference on Software, Telecommunications and Computer Networks*, 2008.

[39] P. Vlacheas, D. Charilas, E. Tragos, and O. Markaki. Maximizing quality of service for customers and revenue for service providers through a noncooperative admission control game. In *Proc. of ICT Mobile Summit*, 2008.

[40] Y. Yu, B. Yong, and C. Lan. Utility-dependent network selection using MADM in heterogeneous wireless networks. In *Proc. of the 18th Annual IEEE International Symposium on Personal, Indoor and Mobile Radio Communications*, 2007.

[41] J. Zander. Radio resource management in future wireless networks: Requirements and limitations. *IEEE Communications Magazine*, 35(8):30–36, 1997.

[42] W. Zhang. Handover decision using fuzzy MADM in heterogeneous networks. In *IEEE Wireless Communications and Networking Conference*, 2004.

12

Efficient Spectrum Band Packing for Hosting Multiple Wireless Systems

Lichun Bao

Computer Science Department, University of California, Irvine, US
Email: lbao@ics.uci.edu

Shenghui Liao

Department of Electrical Engineering and Computer Science, University of California, Irvine, US

Qiang Fu

Victoria University of Wellington, New Zealand

CONTENTS

The radio frequency (RF) spectrum is one of the most expensive and tightly regulated resources in wireless communication domain. Recently, the spectrum scarcity problem has become more pronounced due to the crowding number of wireless devices and emerging wireless technologies. It is critical that the current wireless systems incorporate organic self-adaptive mechanisms to allow the coexistence of themselves and future wireless systems.

In this chapter, we follow the traditional wisdom of cognitive radios by categorizing wireless systems as primary and secondary users, and propose a deterministic spatial spectrum reuse scheme by exploiting the fact that some well-established wireless systems have asymmetric characteristics in spectrum band utilization and wireless station deployment, suitable for the coexistence of secondary systems. Specifically, we examine the widely accepted and deployed GSM and WiFi systems as primary and secondary users, respectively, and design the coexistence system in such a way that they cause negligible interferences to each other. Our scheme eliminates the spectrum sensing or monitoring complexities. Experimental results show that the proposed scheme is a feasible approach to the spectrum scarcity problem.

12.1 Introduction

Governments consider the electromagnetic or the RF spectrum to be a public resource, and delegate special organizations to manage the RF spectrum, such as the Federal Communications Commission (FCC) [21], the Canadian Radio-Television and Telecommunications Commission (CRTC) [13], the European Telecommunications Standards Institute (ETSI) [19], and the Association of Radio Industries and Businesses (ARIB) [6], etc.

The radio spectrum, especially the ideal gigahertz bands for personal communication systems in the urban and metropolitan environments, has gradually been allocated completely. Therefore, new emerging wireless technologies face increasing difficulties in finding wireless bands with sufficient capacity to operate. Fortunately, according to a recent spectrum usage investigation conducted by the FCC [22], the RF spectrum is far from fully utilized, and the typical channel occupancy was less than 15%, while the peak usage was only close to 85%. Such findings open up new venues to accommodate the system and traffic demands using adaptive and efficient spectrum reuse mechanisms so as to fully utilize the wireless bands.

Femtocells, cognitive radios, and spectrum access scheduling are three prominent solutions to improve spectrum utilization. Femtocells provide smaller coverage to supplement those of public terrestrial base stations under the same wireless network system architecture, whereas cognitive radios opportunistically share the RF spectrum that was originally allocated to the primary spectrum users. In spectrum access scheduling, individual wireless systems are aware of the existence of other wireless carriers in the same RF band, and modify their channel access behavior such that they time-share the bandwidth.

In this chapter, we introduce a spectrum reuse scheme that allows the coexistence of heterogeneous wireless systems in the licensed spectrum bands without complicated spectrum coordination mechanisms. In such a scheme, the wireless systems are categorized into two types – the primary users of the spectrum and the secondary users. Because cellular wireless systems are the primary shareholders of the wireless spectrum, we treat them as the primary users herein.

Cellular systems normally comprise a WWAN (wireless wide area network) and WWANs that typically adopt the FDD (Frequency Division Duplexing) spectrum allocation and utilization scheme. In FDD, the communication channels consist of one uplink and one downlink band, each containing multiple narrower channels for individual mobile stations. FDD-based WWANs have the following characteristics that we could exploit for our spectrum sharing strategies:

- *Asymmetric user population:* It is worth noticing that cellular systems consist of base stations and mobile clients. Base stations are few and far between, whereas mobile clients are potentially unlimited. Because wireless interference only concerns the receivers, and there are different numbers of receivers in the uplink and downlink bands, the uplink band is apparently a convenient choice for spectrum reuse by the secondary users.

- *Asymmetric user mobility:* Usually, especially in commercial cellular networks, base stations are stationary in contrast to mobile stations.

- *Interference resistant technologies:* Newer wireless technologies allow much higher interference resistant mechanisms so as to reduce the potential interference level to other systems. When considering the coexistence of multiple systems, the choice of secondary wireless systems also plays a critical role in the overall system feasibility.

Exploiting these characteristics, we examine the primary systems in much details for spectrum reuse purposes, instead of treating the primary system as a black box, and enable the coexistence of secondary wireless systems by utilizing the uplink band of the primary system with spread spectrum technologies and careful location planning.

The rest of the chapter is organized as follows: Section 12.2 presents a detailed survey about three other spectrum reuse solutions, namely Femtocell, cognitive radio, and spectrum access scheduling. Section 12.3 reviews the current and emerging wireless technologies, namely GPRS, WiFi, WiMAX, and LTE-Advanced. In Section 12.4, we describe the wireless system design perspectives and the spatial spectrum reuse approach in detail. The coexistence mechanisms of GSM and WiFi systems is presented according to our spatial spectrum reuse approach. Section 12.5 evaluates the communication performance of the coexisting systems using simulations. Section 12.6 summarizes the chapter.

12.2 Related Work and Background

12.2.1 Femtocells

Studies on wireless usage show that more than 50% of voice calls and more than 70% of data traffic originates indoors [12]. However, many cellular users experience little or no service in indoor areas, resulting in failed or interrupted wireless communication, or wireless communication of less than desirable quality. Therefore, the Femtocell technology fills in the gap by installing short-range, low-cost and low-power base stations for better signal coverage, especially in indoor environments [55]. The small base stations communicate with the cellular network over a broadband connection such as DSL (Digital Subscriber Line), cable modem, or a separate RF backhaul channel.

The value propositions of Femtocell are the low upfront cost to the service provider, increased system capacity due to smaller cell footprint at reduced interference, and the prolonged handset battery life with lower transmission power. When the traffic originating indoors can be absorbed into the Femtocell networks over the IP backbone, cellular operators can provide traffic load balancing from the traditional heavily congested macro-cells towards Femtocells, allowing better reception for mobile users.

The Third Generation Partnership Project (3GPP) published the world's first Femtocell standard in 2009, covering aspects of Femtocell network architecture, radio interference, Femtocell management, provisioning, and security. Several cellular operators provide Femtocells, for instances, Sprint's Airave Femtocell [50] and Verizon's Wireless Network Extender using CDMA, both in the United States [52]. More adaptive reconfigurable Femtocells allow execution of multiple wireless systems on them [24, 16, 51].

Femtocells only improve the spectrum reuse efficiency by reducing the cost and power of cellular base stations, and do not modify the spectrum sharing schemes for multiple wireless systems to access the same RF bands. Hence, there is still room to improve RF channel utilization efficiency for Femtocells.

12.2.2 Cognitive Radio

In recent years, cognitive radio has been extensively studied in order to address the spectrum reuse issue [26, 4, 5], which was first introduced by Mitola [40, 41, 42, 43]. In the cognitive radio approach, wireless users are catego-

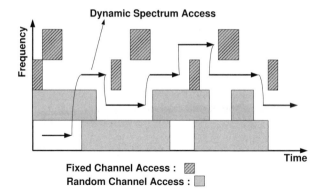

FIGURE 12.1
Cognitive radio networks concepts.

rized into two groups of radio spectrum users – ones that have the legitimate primary right of access, called "primary users," and others that do not, called "cognitive users." Whereas the primary spectrum users access the RF channels in their normal ways, secondary users use their spectrum cognitive and agile capabilities to discover and use the underutilized RF bands, originally allocated to the primary users, therefore achieving spectrum reuse for efficiency purposes.

Dynamic spectrum access techniques using cognitive radios face a couple of difficult challenges. First, cognitive radios require spectrum sensing, learning, decision, and monitoring capabilities. Second, cognitive radios require the cognitive channel access coordination functions to avoid channel access conflicts between the primary and secondary spectrum users.

Figure 12.1 presents the cognitive radio concepts in both frequency and time domains. The gray or shadow areas indicate the RF bands in use by the primary users, while cognitive radios were to discover such spectrum usage patterns and reuse the remaining RF resources, called "spectrum holes," adaptively. By monitoring and learning about the current radio spectrum utilization patterns, the decision logic in cognitive radios can take advantage of the vacant "spectrum holes" [17] in different locations and during time periods, and opportunistically tune their transceivers into these spectrum holes to communicate with each other [48]. Therefore, the channel access mechanisms are opportunistic in nature, and pose significant system requirements to the cognitive radios due to their radio spectrum agility.

Several organizations including the Defense Advanced Research Projects Agency (DARPA) [14], SDR Forum [49], the Institute of Electrical and Electronics Engineers (IEEE) [30], the National Institute of Information and Communications Technology (NICT) [44], the Electronics and Telecommunications Research Institute (ETRI) [18], and the Federal Communications Commission (FCC) [21] have started to investigate cognitive radios. IEEE P1900 is a standard series focusing on the areas of dynamic spectrum access (DSA), cognitive radio (CR), interference management, coordination of wireless systems, and advanced spectrum management. Since March 2007, the 1900 series have been placed under the newly formed IEEE Standards Coordinating Committee 41 (SCC 41), "Dynamic Spectrum Access Networks" [34]. IEEE 1900.4 published in 2009. It aims to standardize the overall system architecture and information exchange between the network and mobile terminals, so that the elements can be optimally chosen from available radio resources [34]. Approaches, concepts, and architectures serving as a basis for the 1900.4 WG have been extensively studied in the European projects End-to-End Reconfigurability (E^2R II) and End-to-End Efficiency (E^3) [15].

A few network architectures based on cognitive radios have been proposed [5]. The spectrum pooling architecture is based on orthogonal frequency division multiplexing (OFDM) [53, 54]. The Cognitive Radio approach for usage of the Virtual Unlicensed Spectrum (CORVUS) system exploits unoccupied licensed bands in a coordinated manner by local spectrum sensing, primary user detection, and spectrum allocation to share the radio bandwidth [8, 11]. IEEE 802.22 is a new working group of the IEEE 802 LAN/MAN standards committee, which aims at constructing a Wireless Regional Area Network (WRAN) utilizing white spaces (channels that are not already used) in the allocated TV frequency spectrum [33].

In order to coordinate between cognitive radios, a control channel, called a rendezvous, is mandatory to exchange channel quality and utilization information [35]. Because the spectrum holes are dynamically changing, the assigning of a rendezvous channel is a challenging issue [23, 35]. In [11, 28, 37], the rendezvous was achieved by dedicating a certain radio band, whereas in [38], a DOSS (Dynamic Open Spectrum Sharing) was proposed using tri-band spectrum allocation, namely the control band, the data band, and the busy-tone band. In [9], a common Coordinated Access Band (CAB) is proposed to regulate authorities such as the FCC in order to utilize CAB to coordinate spectrum access. In [46], a similar channel called the Common Spectrum Coordination Channel (CSCC) is proposed for sharing unlicensed spectrum (e.g., 2.4GHz ISM and 5GHz U-NII). Spectrum users have to periodically broadcast spectrum usage information and service parameters to the CSCC, so that neighboring users can mutually observe via a common protocol. In addition, the duration of the spectrum availability is also essential in order to avoid conflicts with the primary users. The authors of [29] apply statistical analysis of spectrum utilization.

12.2.3 Spectrum Access Scheduling

Spectrum Access Scheduling (SAS) is an approach that interleaves the channel access among multiple wireless systems in a TDMA fashion to solve the spectrum scarcity problem [7]. In spectrum access scheduling, individual wireless systems are aware of the existence of other wireless carriers in the same RF

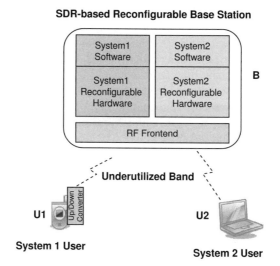

FIGURE 12.2
SDR-based reconfigurable base station for heterogeneous wireless systems.

band, and modify their channel access schedule such that they *time-share* the bandwidth. That is, all wireless systems are considered as first-class citizens of the spectrum domain, and they intentionally allow each other chances for channel access in a TDMA fashion, thus improving the spectrum utilization efficiency.

Different from the traditional TDMA scheme that works with homogeneous wireless stations, the spectrum access scheduling approach enables heterogeneous wireless systems to cooperate over the same bands. Because of inherent differences in system and protocol operational modes, spectrum access scheduling brings up new challenges as to the desired mechanisms for protocol coexistence, and leads to further questions about the changes needed on hardware platforms.

Different from cognitive radio approaches, which are opportunistic and non-collaborative in general, the spectrum access scheduling approach proactively structures and interleaves the channel access moments of heterogeneous wireless systems, using collaborative designs. In the spectrum access scheduling approach, a crucial architectural component – the base stations of wireless systems, is modified to implement spectrum access scheduling, using Software Defined Radios (SDRs). Specifically, the base station supports and executes heterogeneous wireless systems simultaneously, and alternates their channel access in fine-tuned temporal granularity so that the mobile stations of all heterogeneous wireless systems may communicate with the base station.

Figure 12.2 illustrates the hardware and software elements of a base station using the SDR platform for coexistence of heterogeneous wireless systems over a common carrier. In Figure 12.2, the base station B operates two wireless systems which both use the underutilized band. The antenna of the unit $U1$ is extended with a up/down converter for switching the native frequency band of the system 1 to and from the underutilized band, so that both the unit $U1$ and the unit $U2$ work over the underutilized band simultaneously with the base station B. The use case mostly affects the base station of the overall system architecture, and utilizes only one common spectrum band for operations in a micro-time scale. Note that although Vanu nodes also use SDR platform and support multiple concurrently active wireless standards [51], they do not modify the characteristics of the wireless systems, nor have any interactions between the heterogeneous wireless systems [36].

The use case in Figure 12.2 could be more complicated if we consider user mobility, which also leads to wireless system mobility between different SDR platforms. In these cases spectrum access scheduling would have to address issues related with quality of service provisioning, SDR hardware reconfiguration, etc.

12.3 Current and Emerging Wireless Technologies

12.3.1 General Packet Radio Service

General Packet Radio Service (GPRS) is an enhancement over the existing GSM systems by using the same air interface and channel access control procedure. Specifically, we discuss the GPRS systems based on the GSM-900 bands. In GSM-900, the downlink (DL) and uplink (UL) frequency bands are 25 MHz wide each and each band is divided into 200 KHz channels. The frequency separation between the corresponding downlink and uplink channels is 45 MHz. Figure 12.3 shows a GSM-900 TDMA frame and its slots. The duration of a frame is 4.615 ms with 8 time slots (bursts), each of which lasts for 0.557 ms [39].

The GPRS downlink and uplink channels are centrally controlled and managed by the base stations (BSs). GPRS uses the same physical channels as in GSM, but organizes them differently from GSM. In GPRS, the Data Link Layer (DLL) data frame is mapped to a radio block, which is defined as an information block transmitted over a physical channel of four consecutive frames [39]. With regard to the GPRS channel access mechanisms, we first look at the normal GPRS downlink channel operations, as shown in Figure 12.4.

In Figure 12.4, "TN" means the time slot number in each 8-slot time frame. Each time slot could be dedicated for

- A single purpose. For example, "TN0" marked with "PB" are used as the PBCCH (Packet Broadcast Control CHannel) logical channel to beacon

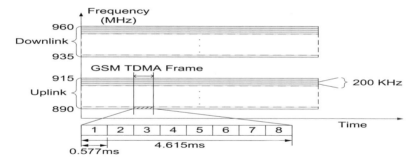

FIGURE 12.3

Downlink and uplink bands for GSM system.

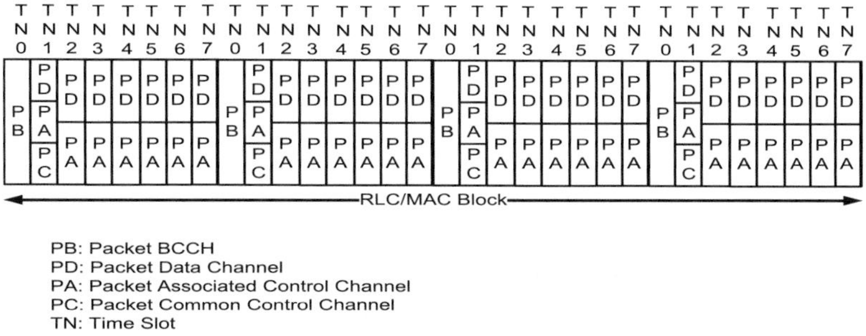

PB: Packet BCCH
PD: Packet Data Channel
PA: Packet Associated Control Channel
PC: Packet Common Control Channel
TN: Time Slot

FIGURE 12.4

Possible configuration of a GPRS downlink radio channel.

the GPRS packet system information at the position of block 0 of a 52-frame multiframe.

- Multiple purposes. For example, "TN1" marked with "PD/PA/PC" is used as PDTCH (Packet Data Traffic CHannel) to transfer data traffic, or as PACCH (Packet Associated Control CHannel) to be associated with a GPRS traffic channel to allocate bandwidth, or as PCCCH (Packet Common Control CHannel) for request/reply messages to access the GPRS services. The PCCCH includes three subchannels, Packet Random Access CHannel (PRACH), Packet Access Grant CHannel (PAGCH), and Packet Paging Channel (PPCH) [39].

GPRS stations use the PRACH to initiate a packet transfer by sending their requests for access to the GPRS network service, and listen to the PAGCH for a packet uplink assignment. The uplink assignment message includes the list of PDCH (Packet Data CHannels) and the corresponding Uplink Status Flag (USF) values per PDCH. GPRS stations keep listening to the USFs of the allocated PDCHs. If the corresponding USF is set, it means that the GPRS station has now been granted access to the next PDCH block. Not

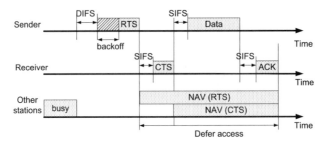

FIGURE 12.5
IEEE 802.11 DCF channel access coordination based on RTS/CTS and NAV.

all the GPRS stations have the capability to simultaneously transmit and receive. Half-duplex mobile stations can communicate in only one direction at a time. In a full-duplex system, stations allow communication in both directions simultaneously.

12.3.2 IEEE 802.11 or WiFi

The channel access method in IEEE 802.11 Distributed Coordination Function (DCF) is based on Carrier Sensing Multiple Access (CSMA) for sharing a common channel [31]. It is essentially a time-division multiplexing method, only that the time slots are virtual and flexible to the transmission time of each data frame.

DCF use five basic mechanisms to inform and resolve channel access conflicts:

1. Carrier sensing (CS) before each transmission.

2. Collision avoidance using RTS/CTS control messages.

3. Interframe spacings (IFSs) to prioritize different types of messages.

4. Binary exponential backoff (BEB) mechanism to randomize among multiple channel access attempts.

5. Network Allocation Vector (NAV) for channel reservation purposes.

Figure 12.5 illustrates the CSMA/CA access method with NAV. Using the RTS and CTS frames, which carry the NAV information, the sender and the receiver can reserve the shared channel for the duration of the data transmissions, thus avoiding possible collisions from other overhearing stations in the network. The NAV-based channel reservation mechanism will be utilized in spectrum access scheduling for allocating time periods for heterogeneous wireless system operations.

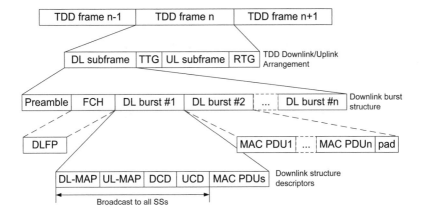

FIGURE 12.6
Fixed WiMAX/TDD frame structure and burst information.

12.3.3 IEEE 802.16 or WiMAX

12.3.3.1 IEEE 802.16-2004

We discuss the following IEEE 802.16-2004 (fixed WiMAX) system for spectrum access scheduling: (a) a single cell operating in the Point-to-MultiPoint (PMP) mode, (b) no mobility, (c) use of 2.4 GHz unlicensed bands, and (d) use of Time Division Duplex (TDD) as the channel duplexing scheme.

Figure 12.6 illustrates a WiMAX frame structure using the TDD scheme [3]. A frame consists of a downlink (DL) subframe and an uplink (UL) subframe, interleaved by two transition gaps, the RTG (receive/transmit transition gap) and TTG (transmit/receive transition gap). Both gap durations are adjustable according to user's needs. A downlink subframe starts with a long preamble for synchronization purposes. A Frame Control Header (FCH) burst follows the preamble, and contains the Downlink Frame Prefix (DLFP), which specifies the downlink burst profile. In the first downlink burst, optional DL-MAP and UL-MAP indicate the starting time slot of each following MAC PDU data burst in downlink and uplink transmissions, respectively. The additional information contained in Downlink Channel Descriptor (DCD) and Uplink Channel Descriptor (UCD) tells the physical layer characteristics of the downlink and uplink channels, such as the modulation algorithm, forward error-correction type, the preamble length.

12.3.3.2 IEEE 802.16m or WiMAX 2.0

IEEE 802.16m uses scalable OFDMA for both uplink and downlink multiple access. It further supports SDMA (Multiuser MIMO) in the uplink and downlink. IEEE 802.16m supports both TDD and FDD duplexing schemes as well as half-duplex FDD terminal operation in FDD networks.

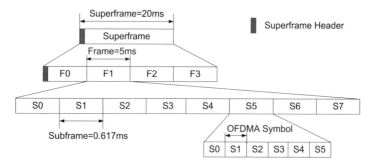

FIGURE 12.7
Basic frame structure for IEEE 802.16m.

The basic frame structure of IEEE 802.16m consists of superframe, frame, and subframe as shown in Figure 12.7. The basic frame structure is applied to both FDD and TDD duplexing schemes, including half-duplex FDD. Each superframe is 20 ms long and consists of four consecutive radio frames, which are 5 ms long. The beginning of a superframe is marked with a superframe header. The superframe header carries short-term and long-term system configuration information. Each radio frame is further divided into subframes where each subframe is comprised of an integer number of OFDMA symbols. The number of subframes per radio frame depends on the channel bandwidth and the cyclic prefix length. There are four types of subframes: (1) type-1 subframe, which consists of 6 OFDMA symbols, (2) type-2 subframe, which consists of 7 OFDMA symbols, (3) type-3 subframe, which consists of 5 OFDMA symbols, and (4) type-4 subframe, which consists of 9 OFDMA symbols. Type-4 is only applied to the uplink subframe for the 8.75 MHz channel bandwidth.

A data burst occupies either one subframe or multiple contiguous subframes. The one subframe case is the default Transmission Time Interval (TTI) transmission and the multiple contiguous subframes case is the long TTI transmission. The long TTI in FDD is equal to 4 subframes for both downlink and uplink. The long TTI in TDD is equal to the all the downlink/uplink subframes in the downlink/uplink in a frame. Figure 12.8 is an example of the TDD and FDD frame structure for 5, 10, and 20 MHz channel bandwidth with a CP of $\frac{1}{8}$Tu (Tu is useful symbol time). With OFDM symbol duration of 102.857 μs and a CP length of $\frac{1}{8}$Tu, the length of Type-1 and Type-3 subframes are 0.617 ms and 0.514 ms, respectively. TTG and RTG are 105.714 μs and 60 μs, respectively. Other numerology may result in a different number of subframes per frame and symbols within the subframes. In FDD, the structure of a frame, such as the number of subframes, has to be identical for the downlink and uplink for each frame [32].

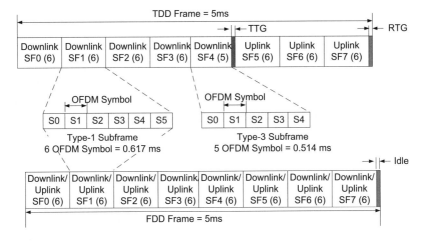

FIGURE 12.8
TDD (DL to UL ratio of 5:3) and FDD frame structure with a CP of $\frac{1}{8}$Tu.

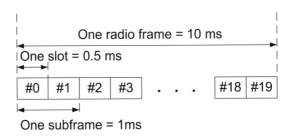

FIGURE 12.9
Type 1 frame structure for FDD mode.

12.3.4 LTE Release 10 and Beyond (LTE-Advanced)

12.3.4.1 Frequency Division Duplexing

The LTE-Advanced supports both half-duplex and full-duplex Frequency Division Duplexing (FDD). For the base station, a duplexer is needed for both half-duplex and full-duplex FDD. For the handset, a duplexer is needed for full-duplex FDD, but there is no specified transmit/receive isolation for the half-duplex handset.

Frame structure Type 1 (as shown in Figure 12.9) is defined for the FDD mode. One radio frame has an overall length of 10 ms. One radio frame consists of 10 subframes, each of length 1 ms. Each subframe consists of two slots, each of length 0.5 ms. Each slot consists of 7 OFDM symbols (6 OFDM symbols in case of extended cyclic prefix). Downlink control signaling is time-multiplexed with data in a subframe basis with control signaling transmitted in the first up to three OFDM symbols of each subframe and data transmitted in the remaining part of the subframe (up to 13 OFDM symbol). Uplink control signaling is frequency-multiplexed with data for other handsets (no time separation) when the handset has no data to be transmitted. Uplink control signaling is typically transmitted at the edges of the overall system bandwidth [2].

12.3.4.2 Time Division Duplexing (TDD)

In the LTE-Advanced TDD (Time Division Duplexing) mode, a duplexer is not needed for both the base station and the handset. The TDD guard time is configurable in the range from one to ten OFDM symbols (approximately 70 to 700 μs) to meet different deployment scenarios. A variable DL/UL ratio is supported, with the supported ratio ranging from 9/1 (nine DL subframes and one UL subframe) to 4/6 (four DL subframes and six UL subframes). In addition, adjacent cells using the same carrier frequency typically use the same DL/UL configuration.

One radio frame has an overall length of 10 ms. One radio frame consists of 10 subframes, each of length 1 ms. Each subframe consists of two slots, each of length 0.5 ms. Each slot consists of 7 OFDM symbols (6 OFDM symbols in case of extended cyclic prefix). For the frame structure of TDD, it is possible to have two DL/UL switching points (one downlink part and one uplink part) or four DL/UL switching points (two downlink parts and two uplink parts) per frame. Switching downlink to uplink takes place in a special subframe that consists of the three fields: the Downlink Pilot Time Slot (DwPTS), the Guard Period (GP), and the Uplink Pilot Time Slot (UpPTS) [1]. Downlink control signaling is time-multiplexed with data on a subframe basis with control signaling transmitted in the first up to three OFDM symbols of each subframe and data transmitted in the remaining part of the subframe (up to 13 OFDM symbols). Uplink control signaling is frequency-multiplexed with data for other handsets (no time separation) when the handset has no data to be transmitted. Uplink control signaling is typically transmitted at the edges of the overall system bandwidth [2].

12.3.4.3 Resource Element and Resource Block

A resource element is the smallest unit in the physical layer and occupies one OFDMA or SC-FDMA symbol in the time domain and one sub-carrier in the frequency domain as shown in Figure 12.10. A Resource Block (RB) is the smallest unit that can be scheduled for transmission. An RB physically occupies 0.5 ms (1 slot) in the time domain and 180 kHz in the frequency domain. The number of subcarriers per RB and the number of symbols per RB vary as a function of the cyclic prefix length and subcarrier spacing [1].

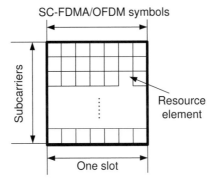

FIGURE 12.10
Uplink and downlink resource grid.

TABLE 12.1
Design perspectives to achieve coexistence of heterogeneous systems.

Point of View	Approaches	
Architectural	Change parts of the wireless systems for coexistence, such as modifying base stations or mobile handsets alone to enable spectrum agility.	Design the whole system to be spectrum agile.
Structural	Leverage existing protocol mechanisms, such as protocol messages, conditions or signals to coordinate channel access schedules.	Build-in interoperability mechanisms at the beginning of the protocol design phase, so that the new wireless system lives with other systems in constant dialog and harmony.
Temporal	Share at micro-scale, which requires protocols to multiplex the spectrum resource at fine-grained millisecond levels, close to the hardware clock speed.	Share at macro-scale, which requires setting up advance timetable at hour or day level for different wireless system to operate without running into each other's ways.
Spectral	Monopoly, which allows a wireless system to occupy the spectrum completely for the protocol operations.	Commonwealth, which allows multiple systems to fragment the channel in frequency domain.

12.4   System Design Specifications

12.4.1   Design Perspectives

In Femtocell, cognitive radio, spectrum access scheduling, and spatial spectrum reuse research, there are many design perspectives from the architectural, temporal, radio spectral, and protocol design points of view. The multiple design choices are shown in Table 12.1. Essentially, we categorize them in terms of:

- *Architectural choices:* We can either change parts of the existing wireless systems or the whole system to be spectrum agile. For example, the spectrum access scheduling approach changes the base stations in order to allow the coexistence of heterogeneous systems on the same spectrum bands. In spatial spectrum reuse, we can add a spectrum up/down converter on the stations in order to shift the radio carriers from the stations' native operating bands to other bands.

- *Protocol design:* We can allow the coexistence of heterogeneous wireless systems either by leveraging their protocol features so that they accommodate each other, or by considering the coexistence issues at the beginning of the protocol designs. Apparently, the former approach allows backward compatibility, and we adopt this approach for spatial spectrum reuse.

- *Temporal arrangement:* The time scale at which heterogeneous wireless systems share the spectrum can either be large, or be small in terms of milliseconds at the packet transmission level. The spectrum access scheduling approach allows system coexistence at the millisecond level, and the spatial spectrum reuse approach makes systems coexist after the systems are turned on.

- *Spectral multiplexing:* The spectrum bands available for heterogeneous wireless systems can either be shared by one system at a time, or be shared by several systems at a time using finer granularity of spectrum separations.

We can see that a popular cognitive radio system design tends to have all units to be spectrum agile, and operate at macrotime scales (minutes or hours), whereas Femtocells exploit the spectral multiplexing approach by deploying Femtocells at remote or indoor environments which the main wireless infrastructure cannot reach. SDR has the particular communication characteristics that can be realized through specialized software running on flexible signal processing hardware. This is very different from the traditional approach of using specialized hardware and has the benefit of sharing of a single platform for multiple communication purposes.

An SDR architecture consists of a digital subsystem and an analog subsystem as shown in Figure 12.11 and pushes the analog stage as close as possible to the antenna [47, 10]. The analog functions are restricted to those that cannot be performed digitally such as antenna, RF filtering, RF combination, etc. The digital subsystem includes application software, middleware, and hardware. The combination of hardware and middleware is often termed a framework. In the ideal SDR architecture, the hardware is completely abstracted away from the application software by the middleware layer. It particularly means that the application part should operate independent of the specifications of hardware resource. The reconfigurable hardware can be a digital signal processor (DSP), a field programmable gate array (FPGA), or an embedded central processing unit (CPU).

Middleware includes the operating system, hardware drivers, and resource management. The middleware layer can wrap up the hardware elements into

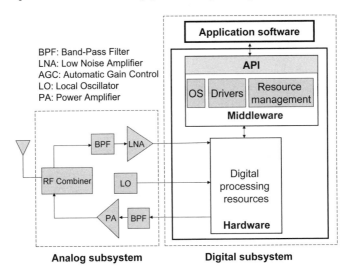

FIGURE 12.11
SDR with layered hardware and software.

objects and provides services to allow the objects to communicate with each other through a standard interface, such as Common Object Request Broker Architecture (CORBA). The resource manager provides a control function to every hardware resource in the SDR-based architecture. The control function of the resource manager includes system control management, configuration management, fault management, performance management, virtual channel management, network management, and security management [47].

12.4.2 Spatial Spectrum Reuse

Spatial Spectrum Reuse (SSR) is an approach that allows the coexistence of heterogeneous wireless systems in the licensed spectrum bands without complicated spectrum coordination mechanisms. In such a scheme, we consider two general categories of wireless systems – the primary users of the spectrum and the secondary users and implement the following strategies to enable their coexistence:

- To exploit the spectrum allocation and utilization characteristics of the primary wireless systems. A widely used allocation scheme for communication channels is FDD, in which the communication channels consist of one uplink band and one downlink band, each containing multiple narrower channels for individual mobile stations. Because wireless interference only concerns the receivers, and there are different numbers of receivers in the uplink and downlink bands, respectively, the uplink band

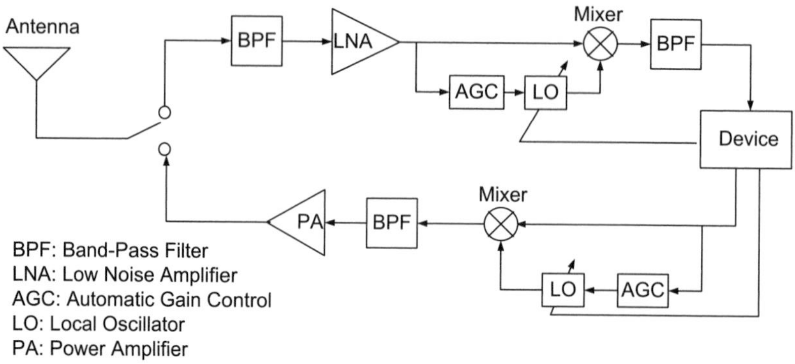

FIGURE 12.12
Frequency converter for secondary wireless system channel alignment.

is apparently a convenient choice for spectrum reuse by the secondary users.

- To purposefully deploy the secondary wireless systems far away from the primary users such that the wireless interference is reduced dramatically.

- To utilize interference resistant communication technologies in the secondary wireless systems so that both interferences from and to the primary users are reduced.

Of the aforementioned strategies, the first choice highlights our novel exploitation of the spectrum allocation and utilization characteristics of the primary wireless system for spectrum reuse purposes, instead of treating the primary system as a blackbox. The other two strategies reduce the interferences between the two categories of spectrum users.

In order to operate on the spectrum bands of primary wireless systems, the secondary wireless systems require a frequency converter to shift the operational channels onto the spectrum bands of primary wireless systems [27]. We adopt the SDR platform as our implementation hardware.

Figure 12.12 shows the schema of the up/down frequency converters on the mobile station to shift the secondary wireless system carriers to the primary wireless system band. In the signal reception direction, the BPF (Band Pass Filter) selects the desired signal, then the LNA (Low Noise Amplifier) amplifies the desired signal while simultaneously minimizing the noise component. Because the input signal could be at different amplitudes, the AGC (Automatic Gain Control) tunes the amplitude of the output of the LO (Local Oscillator) to generate the compensating frequencies to mix with the output signal of the LNA. Afterward, the mixer converts the received signal to the desired frequency band, and the desired signal is extracted by the BPF and sent into the device. The signal transmitting process is similar to the receiving process in the reverse direction.

TABLE 12.2

Comparison between Femtocell, Cognitive Radio (CR), Spectrum Access Scheduling (SAS), and Spatial Spectrum Reuse (SSR).

	Femto-cell	CR	SAS	SSR
Design Objectives	Spectral reuse in *homogeneous* systems.	Spectral reuse in *heterogeneous* systems.	Spectral reuse in *heterogeneous* systems.	Spectral reuse in *heterogeneous* systems.
User Types	Primary users (PUs).	PUs and Cognitive users (CUs).	PUs.	PUs and Secondary users (SUs).
Among Heterogeneous System Users	N/A	Non-collaborative: CUs have to avoid interference with PUs.	Collaborative: heterogeneous systems are aware of each other at the base station.	Purposefully deploying the secondary systems far away from the primary system.
Spectrum Access	Deterministic access.	Opportunistic access.	Deterministic time sharing at the millisecond level.	Deterministic access.
Technique	Overlay with macro-cell.	CRs underlay, overlay and interweave with Pus.	Users interweave with each other.	Secondary systems overlay with primary systems.
RF Interface Structure	Leverage existing protocol mechanisms.	Build-in interoperability mechanisms at the beginning of the protocol design phase.	Leverage existing protocol mechanisms for coexistence of heterogeneous systems.	Leverage existing protocol mechanisms for coexistence of heterogeneous systems.

Essentially, the formula of the converter in Figure 12.12 is

$$f_{\text{chann}} = f_{\text{oper}} + f_{\text{LO}}$$

where f_{chann} is the channel frequency that goes into and from the antenna, f_{oper} is the operating frequency of the device, and f_{LO} is the add-on frequency generated by the local oscillators. Table 12.2 is the comparison between spectrum reuse schemes mentioned in this chapter so far.

12.4.3 Case Study

To analyze our spatial spectrum reuse scheme, we consider two specific wireless systems – the GSM wireless system as the licensed primary spectrum user, and WiFi based on DSSS (direct sequence spread spectrum) as the secondary users, and provide the system implementation and analysis for their coexistence. The spatial spectrum reuse approach is suitable for other wireless technologies such as the technologies mentioned in Section 12.3. However, the detail of coexistence for other wireless technologies is omitted in this section because of its similarity with the case study mentioned in this section.

GSM is by far the most successful wireless communication standard, and its systems have been widely deployed in the world. There are several RF bands for GSM operations, namely the 900 MHz, 1800 MHz, and 1900 MHz bands. Without loss of generality, we only consider GSM systems at the 900 MHz band, denoted as GSM-900. GSM-900 allocates two separate spectrum slots for scheduling uplink (UL) and downlink (DL) transmissions, the bandwidths of which are about 25 MHz, respectively.

WiFi systems are the dominant wireless networks deployed for data communication purposes. There are multiple physical layer standards for WiFi systems that use different modulation and coding mechanisms, namely DSSS, FHSS (frequency hopping spread spectrum), and OFDM (orthogonal frequency division multiplexing), mostly operating at the 2.4 GHz RF bands. In order to enable the coexistence between the WiFi and GSM systems, we specifically require WiFi systems to operate with DSSS, as standardized in IEEE 802.11b, offering data rates at 1, 2, 5.5 and 11 Mbps. The bandwidth of IEEE 802.11b is around 22 MHz, which can fit perfectly in the 25 MHz uplink or downlink bands of GSM-900 systems.

Under the GSM and WiFi system settings, the following content of this section describes the coexistence between the secondary WiFi system and the primary GSM-900 systems in the 900 MHz bands, so as to augment the network operation bandwidth of WiFi systems in addition to the crowded ISM band at 2.4 GHz with negligible impact on the GSM systems. Specifically, we choose GSM uplink for the WiFi system operations.

12.4.3.1 Opportunities for Coexistence

In 3GPP TS 45.005, there are fourteen bands defined, of which GSM-900 and GSM-1800 are used in most parts of the world. Without loss of generality, we use the standard GSM-900 bands for the coexistence study of two wireless systems, GSM-900 and WiFi. GSM-900 uses 890-915 MHz to send information from the mobile station to the base station (uplink), and 935-960 MHz for the other direction (downlink), providing 124 RF channels (channel numbers 1 to 124) spaced at 200 KHz. Duplex spacing of 45 MHz is used [39]. On the other hand, an IEEE 802.11b system typically covers a small area using the 2.4 GHz ISM band. There are eleven channels identified for the Direct Sequence Spread Spectrum (DSSS) system, three of which are non overlapping channels, each channel occupying approximately 22 MHz [31].

Interestingly, the RF station mobility and populations in the uplink and downlink bands of GSM-900 systems present different characteristics. In the uplink band, the RF receivers are GSM base stations, which are few, geographically sparse and stationary in contrast to mobile stations. In the downlink, the receivers are GSM mobile handsets with potentially random and mobile locations and in large numbers. Therefore, in order to enable the coexistence of GSM-900 and WiFi systems, we will have to choose the uplink for the secondary wireless system deployments. This is because

- *Spectrum allocation and utilization characteristics of the primary wireless system:*

    ○ GSM base stations are sparsely and stationarily located, thus providing a static spectrum utilization environment for the secondary wireless systems.

    ○ The GSM mobile stations are the transmitters in the uplink, shared with the WiFi systems. Therefore, the transmissions of the WiFi

FIGURE 12.13
WiFi deployment scheme for coexistence with GSM.

systems do not impact GSM mobile stations because the reception
channels of the GSM mobile stations are intact, even though GSM
mobile stations may be next to the WiFi stations.

- *Locations of the secondary system deployments:* In our scheme, we place
  WiFi systems at sufficient distance from the GSM base stations, therefore limiting the interference to the GSM base stations by the WiFi
  stations.

- *Interference resistant technologies:*

  - In our scheme, the WiFi systems are based on DSSS, in which wire-
    less signal energy is spread over the whole 22 MHz band, whereas
    the GSM signals are transmitted over a narrower 200 KHz chan-
    nel, therefore further reducing the interference between the two
    systems.

  - A GSM-900 mobile station has a nominal maximum output power
    of 2W, although usually lower than 100 mW under power control
    functions. In WiFi systems, the maximum output power is limited
    under 200 mW. Therefore, the disparity between transmission pow-
    ers presents another opportunity for the WiFi systems to coexist
    with the GSM systems.

Figure 12.13 shows the WiFi system deployment strategy for coexistence with
the GSM systems, in which the WiFi systems (nodes G, H and K) are beyond a certain range of all GSM base stations (nodes A and B), and could

potentially colocated with multiple GSM mobile stations (nodes C, D, E and F). The region in which the WiFi systems can be deployed is denoted by the shaded area in Figure 12.13.

12.4.3.2 Locations of the WiFi System Deployments

The deployment locations of the WiFi systems have to be sufficiently distant from the GSM base stations so that the interference from the WiFi systems is negligible to the GSM systems. The closest distance from the WiFi systems to the primary GSM systems can be derived from the maximum tolerance of interference by the GSM base station receivers. The upper bound of total WiFi systems interference power must satisfy the quality described in Equation (12.1), based on the concept of signal to interference plus noise ratio (SINR).

$$10^{\frac{I_w}{10}} \leq 10^{\frac{S_r - CIR}{10}} - 10^{\frac{N_0 + NF + 10\log_{10} B_N}{10}} \tag{12.1}$$

where I_w is total WiFi systems interference power, S_r is the receiver sensitivity, CIR is the carrier to cochannel interference ratio, N_0 is the thermal noise spectral density in dBm/Hz, NF is the noise figure in dB, and B_N is the noise bandwidth of the receiver in Hz. The noise figure NF of a receiver is the allowed signal degradation when the signal passes through the receiver circuit [45]. In Equation (12.1), the thermal noise is characterized as AWGN (Additive White Gaussian Noise). The bandwidth of the receiver in GSM systems is $B_N = 200$ KHz, so the thermal noise power is $N_0 + 10\log_{10} B_N = -174$ dBm $+ 10\log_{10}(200,000) = -121$ dBm at room temperature.

With regard to the other parameters in Equation (12.1), Table 12.3 provides the reference values according to the ETSI standard [20]. According to Equation (12.1), the total interference power I_w needs to satisfy $I_w \leq -119.9$ dBm.

TABLE 12.3 Parameters for GSM-900 in the ETSI standard.

Parameters	Meaning	Values
S_r	Receiver sensitivity	-104 dBm
CIR	Carrier to cochannel interference ratio	9 dB
NF	Noise figure	7 dB

On the other hand, we can derive the total WiFi systems' interference power to a GSM base station using Equation (12.2), which assumes that WiFi stations are ambiently deployed at distance farther than d_{wg} from the GSM base station.

$$I_w = \int_0^{2\pi} \int_{d_{wg}}^{\infty} \rho_w P_r(x) x \, dx d\varphi \tag{12.2}$$

where P_r is WiFi systems' power received by the GSM base station, and ρ_w is density of WiFi stations. Specifically, the d_{wg} is closest distance that a WiFi system can get to the GSM base station. We assume that $\rho_w = 1/(\pi R_c^2)$, in which R_c is the carrier sensing range. This is because only one WiFi station can be active at any time moment using the CSMA MAC control protocol in a WLAN system. Therefore, $\rho_w = 1/(\pi R_c^2)$ is the upper bound of the WiFi station density in our calculations.

We assume that the WiFi signal follows the log-distance path loss model [45], thus the received WiFi system signal power $P_r(x)$ can be calculated by Equation (12.3).

$$P_r(d) = \frac{P_t G_t G_r \lambda^2}{(4\pi)^2 d_0^2 L} \left(\frac{d_0}{d}\right)^n \qquad (12.3)$$

where P_t is the transmit power by the WiFi system, G_t is the transmitter antenna gain, G_r is the receiver antenna gain, d is the transmit–receive separation distance in meters, d_0 is the close-in reference distance, L is the system loss factor not related to propagation, and λ is the wavelength in meters. Table 12.4 provide the value of link budget parameters.

TABLE 12.4 Link budget parameters.

Parameters	Meaning	Values
P_t	Transmitted power	100 mW
G_t	Transmitter antenna gain	1
G_r	Receiver antenna gain	1
L	System loss factor	1
d_0	Close-in reference distance	1 m
λ	Wavelength	0.33 m
n	Path loss exponent	$2 \le n \le 6$

TABLE 12.5

Relation between path loss exponent and distance when $P_t = 100$mW.

Path loss exponent	Closest distance d_{wg}
3	369 Km
3.5	3.92 Km
4	429 m

Using Equations (12.1), (12.2), and (12.3) and regular parameter settings given by Tables 12.3 and 12.4, we can derive the minimum distance between the GSM base station and the WiFi system, as shown in Table 12.5. For instance, when the path loss exponent is 4, WiFi system can be deployed at about 450 meters away from the GSM base stations.

In practice, the closest distance value d_{wg} can be much smaller because

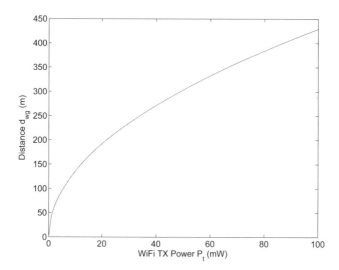

FIGURE 12.14
WiFi power level versus closest distance (d_{wg})

the upper bound of x in Equation (12.2) is limited by the size of the WiFi system deployments, as well as other GSM base stations. Furthermore, the closest distance value d_{wg} relates to the transmitted power of WiFi system. Using Equations (12.1), (12.2), (12.3) and the pass loss exponent $n = 4$, we can derive the minimum distance between the GSM base station and the WiFi system based on different WiFi power levels, as shown in Figure 12.14.

12.4.3.3   Interference Analysis

According to the WiFi system deployment strategy, which is to deploy WiFi systems at sufficient distance from GSM base stations (see Table 12.5 and Figure 12.14), we know that WiFi systems create negligible interference to the GSM systems. Therefore, our interference analysis in this section provides a basic understanding of the impact of GSM systems to WiFi systems, only. Specifically, we analyze the impact of GSM mobile stations because these are the GSM transmitters in the GSM uplink band, in which the WiFi systems operate.

We assume the channel is an AWGN channel, and that the interference from GSM mobile stations behaves as AWGN, then the SINR of the WiFi systems is given by Equation (12.4).

$$\text{SINR} = \frac{P_d}{P_n + P_i} G_p \tag{12.4}$$

where P_d is the power of WiFi signals, P_n is the noise power, P_i is the total

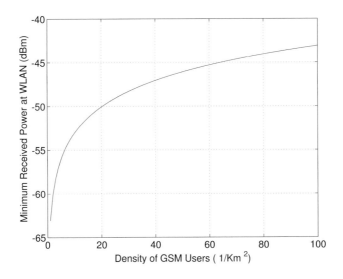

FIGURE 12.15
Density of GSM mobile stations versus WiFi minimum received power.

power of the GSM interference, and G_p is the processing gain of the WiFi receiver [56]. Because we use DSSS in our WiFi systems, the processing gain G_p depends on the spreading code to symbol rate ratio, which is an approximate measure of the interference rejection capability [45]. For instance, Barker code provides $G_p = 10.4$ dB gain at 1 Mbps in WiFi systems.

In addition, the total interference from GSM mobile stations P_i is calculated as follows, Equation (12.5),

$$P_i = \int_0^{2\pi} \int_{d_{gw}}^\infty \rho_g P_r(x) x \; dx d\varphi \tag{12.5}$$

where ρ_g is the density of GSM mobile stations, d_{gw} is the closest distance from GSM mobile stations to the WiFi receiver, and P_r is computed according to Eq. (12.3) with $P_t = 33$ dBm (i.e., 2 W) assuming the log-distance path loss model.

In WiFi systems, an SINR of at least 10 dB is usually needed to decode IEEE 802.11b signals correctly. A 802.11b signal can be -0.4 dB weaker than an interferer to have 1% frame error rate [25]. According to above discussions, Figure 12.15 illustrates the relation between the GSM user density and the minimum power required by WiFi receivers in order to decode the WiFi packet at 1% frame error rate in 1 Mbps DSSS based WiFi systems. Equivalently, the higher GSM mobile station density also drives down the size of WiFi cells. Figure 12.16 shows that the maximum distance between WiFi transmit/receive stations is decreased when the GSM user density increases.

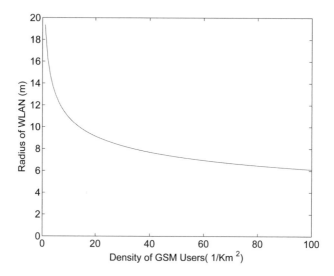

FIGURE 12.16
Density of GSM mobile stations versus radius of WiFi system.

12.5 Performance Evaluation

We evaluate the performance of our WiFi and GSM coexistence scheme by simulations using NCTUns 5.0. Since GPRS systems use the same air interface as GSM systems do, we chose GPRS uplink to coexist with WiFi in the GSM-900 band in the simulations. This is because we can easily control the data rate in GPRS applications.

We used a network setting with one WiFi system and one GPRS cell as shown in Figure 12.17. The antennas were omnidirectional with 0dB gain. We used the log-distance path loss model, in which the path loss exponent was $n = 4$, and the closest distance between a WiFi system and a GPRS base station was $d_{wg} = 430$ m. In the WLAN system, one stationary host (Host 2) attached to the WLAN AP communicates with the WiFi mobile station. In the GPRS system, we place one SGSN (Serving GPRS Support Node) and one GGSN (Gateway GPRS Support Node) on the infrastructure side of the base station to handle GPRS related data packets. GGSN is the Internet gateway router that is responsible for sending data packets to the Internet. In addition, a stationary host (Host 1) attached to the GGSN communicates with all GPRS mobile stations under traffic flows. In the simulations, there are more GPRS mobile stations than what is shown in Figure 12.17.

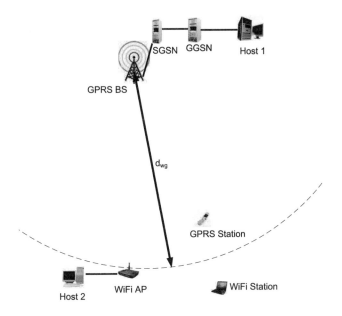

FIGURE 12.17
Network topology with WiFi and GPRS systems.

12.5.1 Impact of GPRS User Density

How the density of GPRS users affects the WiFi system is discussed first.
Figure 12.17 illustrates the network topology. We considered scenarios with
one WiFi access point, one WiFi mobile station, one GPRS base station, and
the variations in the number of GPRS mobile stations from 1 to 10. The
GPRS mobile stations were uniformly distributed in the GPRS cell. For the
WiFi traffic flow, the data payload size was set to 1000 bytes. For the GPRS
traffic flows, the time between consecutive packets was exponential distributed
with mean of 0.5 seconds, minimum of 0.2 seconds, and maximum of 5 seconds,
and the length of generated packets was exponentially distributed with mean
500 bytes, minimum 100 bytes, and maximum 1000 bytes. We did not use
mobility in the simulations.

The DSSS module in WiFi implements three CCA (clear channel assess-
ment) modes – (I) energy above threshold, (II) carrier sensing only, and (III)
carrier sensing with energy above threshold [31]. We chose the CCA mode (III)
in the simulations. The CCA mode (III) only detects strong enough DSSS signals. Usually IEEE 802.11b signals can be decoded correctly if the SINR is
at least 10 dB. Furthermore, GPRS mobile stations apply narrow-band modulation and WiFi systems apply relatively wide-band transmission, so WiFi
systems will have capabilities to reduce interference from the GPRS system.

We increased the number of GPRS users from 1 to 10 in the GPRS cell
to evaluate the system in the simulations. The WiFi load was CBR traffic

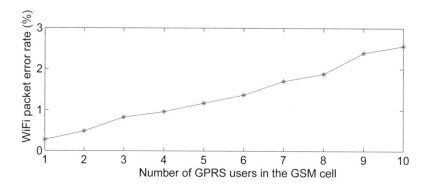

FIGURE 12.18
Packet error rate for WiFi network.

(11 Mbps). In Figure 12.18, we can observe that while the number of users increases, the packet error rate of WiFi also increases.

12.5.2   Impact of WLAN Load

In the first set of experiments, we deployed ten GPRS mobile stations randomly in the GPRS cell, and varied the CBR load of WiFi from 0.1 to 11 Mbps. The traffic flow setup is the same as in Section 12.5.1. However, in this section CCA mode (I) was used, in which the WiFi system only detects the energy level of the channel, and avoids sending packets when the channel already has GPRS packet transmissions. In addition, we simulated two scenarios, one that has the GPRS system coexisting with the WiFi system, and the other that has the WiFi system alone, so as to contrast the impact of GPRS systems on the WiFi system.

Figures 12.19(a) and 12.19(b) show the performance of the WiFi systems in terms of throughput and end-to-end delay, respectively. In both figures, we can observe that the performance difference of the two scenarios is negligible. This is because the received signals at the WiFi system were $SINR \geq 10$dB and the WiFi system can carrier sense the GPRS transmission with CCA mode (I) to avoid most of the packet collisions in the simulations.

In the second set of experiments, we set the GPRS mobile station close to the WiFi system to create higher interference for the WiFi system than the previous experiments. We collected the throughput performance of the WiFi system under several GPRS traffic loads, namely those of 1, 6, 7, 8, and 15 Kbps. Figure 12.20 shows that the throughput performance of the WiFi system is proportionally driven down by the increasing GPRS traffic loads.

More specifically, when the GPRS load is less than 6 Kbps, the variations of the GPRS loads have little impact on the throughput of the WiFi

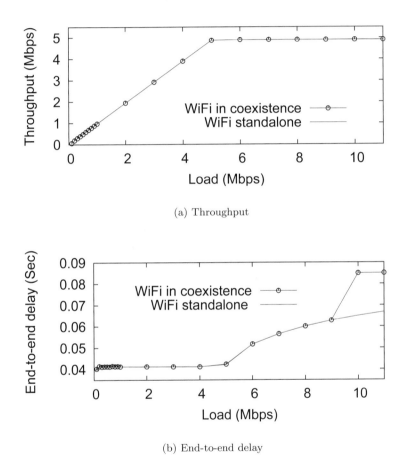

(a) Throughput

(b) End-to-end delay

FIGURE 12.19
Throughput and end-to-end delay for WiFi systems when GPRS base stations
are randomly distributed.

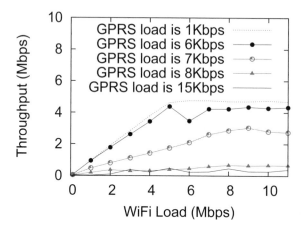

FIGURE 12.20
Throughput of WiFi systems when GPRS base stations are nearby.

system because the CCA mechanism can largely eliminate packet collisions in the WiFi system. However, when the GPRS load increases above 6 Kbps, the probability of GPRS packets corrupting WiFi system packet receptions increases. Furthermore, because the WiFi system has to spend more and more time on carrier sensing and backoff when the GPRS traffic load increases, the throughput of the WiFi systems also suffers. Even in some cases, the extended period of the interference from the GPRS systems caused the AP and the WiFi mobile station to disconnect and reconnect frequently during simulations, thus further driving down the throughput of the WiFi systems.

12.6   Conclusion

We have presented a spatial spectrum reuse approach to allow the coexistence of heterogeneous wireless systems, and address the conflict between the spectrum scarcity and under-utilization. The mutual impact to the system performance between the coexisting systems are analyzed and evaluated by simulating the coexistence of GSM with WiFi in the GSM-900 band specifically. The performance results of the simulations show that the proposed scheme is a feasible solution for spectrum reuse, and is worthy of practical applications.

Bibliography

[1] 3GPP. A Simple Network Management Protocol (SNMP). 3GPP TS 36.211 V8.8.0, Oct. 2009.

[2] 3GPP. Evolved Universal Terrestrial Radio Access (E-UTRA) - User Equipment (UE) radio transmission and reception. 3GPP TS 36.101 V9.1.0, Oct. 2009.

[3] IEEE Std 802.16-2004. IEEE Standard for Local and Metropolitan Area Networks–Part 16: Air Interface for Fixed Broadband Wireless Access Systems. IEEE Technical Report, Oct. 1, 2004.

[4] I.F. Akyildiz, W. Lee, M. C. Vuran, and S. Mohanty. A Survey on spectrum management in cognitive radio networks. *IEEE Communications Magazine*, 46:40–48, April 2008.

[5] I.F. Akyildiz, W.-Y. Lee, M.C. Vuran, and S. Mohanty. Next generation dynamic spectrum access cognitive radio wireless networks: A survey. *Computer Networks*, pages 2127–2159, 2006.

[6] ARIB. http://www.arib.or.jp/english/.

[7] L. Bao and S. Liao. Scheduling heterogeneous wireless systems for efficient spectrum access. *EURASIP Journal on Wireless Communications and Networking (JWCN), special issue on Wireless Network Algorithms, Systems, and Applications*, 2010.

[8] R.W. Brodersen, A. Wolisz, D. Cabric, S.M. Mishra, and D. Willkomm. Corvus: a cognitive radio approach for usage of virtual unlicensed spectrum. White paper, Berkeley Wireless Research Center (BWRC), 2004.

[9] M.M. Buddhikot, P. Kolody, S. Miller, K. Ryan, and J. Evans. DIMSUM-Net: New directions in wireless networking using coordinated dynamic spectrum access. In *Proc. IEEE WoWMoM*, pages 78–85, 2005.

[10] P. Burns. *Software Defined Radio for 3G*. Artech House, Norwood, MA, 2002.

[11] D. Cabric, S.M. Mishra, D. Willkomm, R. Brodersen, and A. Wolisz. A Cognitive radio approach for usage of virtual unlicensed spectrum. In *Proc. 14th IST Mobile and Wireless Communications Summit*, Jun. 2005.

[12] V. Chandrasekhar, J. G. Andrews, and A. Gatherer. Femtocell networks: A survey. *IEEE Comm. Magazine*, 46:59–67, Sept. 2008.

[13] CRTC. http://www.crtc.gc.ca/eng/home-accueil.htm.

[14] DARPA. http://www.darpa.mil/.

[15] E3. http://www.ict-e3.eu/.

[16] Ericsson's Femtocell Solution. http://www.ericsson.com.

[17] P. Kolodzy et al. Next Generation Communications: Kickoff Meeting. Proc. DARPA, Oct. 17, 2001.

[18] ETRI. http://www.etri.re.kr/eng/.

[19] ETSI. http://www.etsi.org/WebSite/homepage.aspx.

[20] ETSI. Digital Cellular Telecommunications System (Phase 2+); Radio Transmission and Reception. ETSI TS 100 910 V8.20.0, 2005.

[21] FCC. http://www.fcc.gov/.

[22] FCC. Spectrum Policy Task Force Report. ET Docket No. 02-155, 2002.

[23] FCC. Facilitating Opportunities for Flexible and Efficient and Reliable Spectrum Use Employing Cognitive Radio Technologies. ET Docket No. 03-108, 2005.

[24] Femtocell Forum. http://www.femtoforum.org/femto/Files/File/ Airvana_Femtocell_White_Paper_Oct_2007.pdf.

[25] R. Gummadi, D.Wetherall, B. Greenstein, and S. Seshan. Understanding and mitigating the impact of RF interference on 802.11 networks. *SIGCOMM*, pages 385–396, Aug. 2007.

[26] H. Harada. A Study on a new wireless communications system based on cognitive radio technology. IEICE Technical Report SR2005-17, 2005.

[27] S. Haykin. *Communication Systems*. John Wiley & Sons, New York, 4th edition, 2001.

[28] O. Holland, A. Attar, N. Olaziregi, N. Sattari, and A. H. Aghvami. A Universal Resource Awareness Channel For Cognitive Radio. In *The 17th Annual IEEE International Symposium on Personal, Indoor and Mobile Radio Communications (PIMRC)*, Helsinki, Finland, Sep. 11-14 2006.

[29] Alex C.-C. Hsu, David S. L. Wei, and C.-C. Jay Kuo. A cognitive MAC protocol using statistical channel allocation for wireless ad-hoc networks. In *Proc. of IEEE Wireless Communications and Networking Conference (WCNC)*, Mar. 11-15, 2007.

[30] IEEE. http://www.ieee.org/portal/site.

[31] IEEE 802.11 Working Group. Part 11: Wireless LAN Medium Access Control (MAC) and Physical Layer (PHY) Specifications. 2007.

[32] IEEE 802.16 Working Group. IEEE 802.16m System Description Document (SDD) `http://ieee802.org/16/tgm/core.html\#09\_0034`.

[33] IEEE 802.22 Working Group on Wireless Regional Area Networks (WRANs). `http://www.ieee802.org/22/`.

[34] IEEE SCC41. IEEE Standards Coordinating Committee 41 (Dynamic Spectrum Access Networks). `http://grouper.ieee.org/groups/scc41/`.

[35] P. J. Jeong and M. Yoo. Resource-aware Rendezvous Algorithm for Cognitive Radio Networks. In *The 9th International Conference on Advanced Communication Technology*, pages 1673–1678, Feb. 2007.

[36] F. K. Jondral. Software-defined radio – Basics and evolution to cognitive radio. *EURASIP Journal on Wireless Communications and Networking*, 2005 (3):275–283, 2005.

[37] S. Krishnamurthy, M. Thoppian, S. Venkatesan, and R. Prakash. Control Channel-based MAC-layer Configuration, Routing and Situation Awareness for Cognitive Radio Networks. In *Proc. of IEEE Military Communications Conference (MILCOM)*, pages 1–6, Oct. 2005.

[38] L. Ma, X. Han, and C.-C. Shen. Dynamic open spectrum sharing MAC protocol for wireless ad hoc networks. In *Proc. of IEEE Symposium on New Frontier in Dynamic Spectrum Access Networks*, Nov. 2005.

[39] P. McGuiggan. *GPRS In Practice: A Companion to the Specification*. John Willey & Sons, New York, 2004.

[40] J. Mitola. Cognitive radio for flexible mobile multimedia communications. In *Proc. of IEEE Mobile Multimedia Conference (MOMUC)*, pages 3–10, 1999.

[41] J. Mitola. *Software Radio: Wireless Architecture for the 21st Century*. ISBN: 0967123305. Mitola's Statisfaction, 1999.

[42] J. Mitola. *Cognitive Radio: An Integrated Agent Architecture for Software Defined Radio*. PhD thesis, Royal Institute of Technology (KTH), Sweden, 2000.

[43] J. Mitola and G.Q. Maguire. Cognitive radio: Making software radios more personal. *IEEE Personal Communications*, 6:13–18, 1999.

[44] NICT. `http://www.nict.go.jp/index.html`.

[45] T. S. Rappaport. *Wireless Communications: Principles and Practice*. Prentice Hall, Upper Saddle River, NJ, 2nd edition, 2002.

[46] D. Raychaudhuri and X. Jing. A spectrum etiquette protocol for efficient coordination of radio devices in unlicensed bands. In *Proc. of 14th IEEE PIMRC2003*, Sept. 2005.

[47] N. Ryu, Y. Yun, S. Choi, R.C. Palat, and J.H. Reed. Smart antenna base station open architecture for SDR networks. *IEEE Wireless Communications*, June 2006.

[48] A. Sahai, N. Hoven, and R. Tandra. Some Fundamental Limits on Cognitive Radio. In *Proc. of Allerton Conference*, Monticello, Oct. 2004.

[49] SDR Forum. http://www.wirelessinnovation.org.

[50] Sprint Inc. http://www.nextel.com/en/services/airave/index.shtml?id9=vanity:airave.

[51] Vanu Inc. http://www.vanu.com.

[52] Verizon Inc. http://www22.verizon.com.

[53] T. A. Weiss and F. K. Jondral. Spectrum Pooling: An innovative strategy for the enhancement of spectrum efficiency. *IEEE Communications Magazine, Radio Communications Supplement*, pages S8 – S14, Mar. 2004.

[54] T.A. Weiss, J. Hillenbrand, A. Krohn, and F.K. Jondral. Efficient signaling of spectral resources in spectrum pooling systems. In *Proc. 10th Symposium on Communications and Vehicular Technology (SCVT)*, Nov. 2003.

[55] S. Yeh, S. Talwar, S. Lee, and H. Kim. WiMAX Femtocells: A perspective on network architecture, capacity, and coverage. *IEEE Communications Magazine*, pages 58–65, Oct. 2008.

[56] D. G. Yoon, S. Y. Shin, W. H. Kwon, and H. S. Park. Packet error rate analysis of IEEE 802.11b under IEEE 802.15.4 interference. *IEEE Vehicular Technology Conference*, pages 1186–1190, May 2006.

Index